How to Implement
Lean Manufacturing

About the Author

LONNIE WILSON has been teaching and implementing lean techniques for more than 45 years. His experience spans 20 years with an international oil company where he held a number of management positions. In 1990 he founded Quality Consultants which teaches and applies lean techniques to small entrepreneurs and Fortune 500 firms, principally in North, Central, and South America. Mr. Wilson's focus is to assist firms who are now struggling plus those who have struggled and failed to transform their company into a lean enterprise and wish to try again.

How to Implement Lean Manufacturing

Lonnie Wilson

Second Edition

New York   Chicago   San Francisco   Athens
London   Madrid   Mexico City   Milan
New Delhi   Singapore   Sydney   Toronto

Cataloging-in-Publication Data is on file with the Library of Congress.

McGraw-Hill Education books are available at special quantity discounts to use as premiums and sales promotions, or for use in corporate training programs. To contact a representative, please visit the Contact Us page at www.mhprofessional.com.

How to Implement Lean Manufacturing, Second Edition

1 2 3 4 5 6 7 8 9 0   DOC/DOC   1 2 0 9 8 7 6 5 4

ISBN 9781265832414
MHID 1265832412

Sponsoring Editor
Robert Argentieri

Editorial Supervisor
Donna M. Martone

Acquisitions Coordinator
Amy Stonebraker

Project Manager
Tanya Punj,
Cenveo® Publisher Services

Copy Editor
Surendra Nath Shivam

Proofreader
Megha Saini,
Cenveo Publisher Services

Indexer
Cenveo Publisher Services

Production Supervisor
Pamela A. Pelton

Composition
Cenveo Publisher Services

Art Director, Cover
Jeff Weeks

Contents

Preface

Why I Am Writing This Book

I am writing it for a handful reasons. First, I have been asked to. In the case of this, the Second Edition, McGraw-Hill asked me to. And just like in the First Edition, even though I was asked to; there were larger, deeper, and more personal reasons that drove me to write the Second Edition as I will elaborate below.

There Is Lots to Share...

Second, there is a tremendous body of knowledge I wish to share. Some of its material that was not appropriate to put in the First Edition, mostly information on cultural change. Also since writing the First Edition, I have continued to work with clients and build on my body of knowledge and I have found subjects which I wished I would have included or explained more fully (more on that later). Finally, and amazingly, I found that the majority of lean transformations undertaken failed to succeed. Hence, the real driving force behind this work is designed to not only answer "What should I do to be successful in my lean implementation?" but also "What should I NOT do?".

It Requires Hard Work and Then Some...

Third, I continue to see jobs and work go overseas for all the wrong reasons. It usually is some manager, often with the best of intentions who is trying to save a few dollars on labor while he knowingly or unknowingly adds 6 to 8 weeks of lead time to the delivery of his product with all the seven wastes being brought to full maturity. Frequently, corporate policies force this action but nonetheless these managers are causing untold damage to their employees, their facilities' competitiveness, their customers, and to the U.S. economy as a whole. Fortunately I have seen of late some recent breakthroughs in thinking but for many workers who have lost jobs due to this myopic corporate policy, it is—too little—too late. Therefore, I hope to catalyze some thinking so that outsourcing is done with some thought to the **true economics** and not some back-of-the-envelope calculations done by some manager who is, without seeing the slightest contradictions, carefully assessing the impact of that outsourcing on his annual bonus. In addition, I still see managers everywhere looking for this "silver bullet" called Lean Manufacturing. They see it as a catchball for attacking all their business woes—including poor profitability and low levels of competitiveness—and transforming their business into the pinnacle of profitability. I want to stand up and yell at the top of my lungs and make it very clear that there is no silver bullet. It is not all that complicated but it does require a lot of awareness, courage, and hard work. In this regard, I now offer up three quotes here that experience has proven to be true.

And Thinking Is Also Required...

Fourth, I have often thought about this problem—that is, the one about the plant manager who wants to solve his problem quickly and easily without any type of detailed analyses. I have discussed it with others, including consulting colleagues, my wife, psychiatrists, psychologists, my pastor, as well as dozens of top-level managers. As might be expected, the root cause of this problem has myriad descriptions—some call it denial, others laziness, and still others say it results from a "quick-fix mentality." But the one I keep coming back to is simplistic thinking.

> "In the choice between changing one's mind and proving there's no reason to do so, most people get busy on the proof."
>
> John Kenneth Galbraith

> "Men stumble over the truth from time to time, but most pick themselves up and hurry off as if nothing happened."
>
> Sir Winston Churchill

> "Opportunity is missed by most people because it is dressed in overalls and looks like work."
>
> Thomas Edison

So just what is simplistic thinking? I like to say that there are two groups of people who can solve any problem—those who know nothing about the problem but its name and those who clearly know what is happening. For those who know nothing, everything is simple. If you have teenagers, all too often they fall into this category. Unfortunately, many managers seem to reside here as well. Fueled by success stories that are simplified by TV and literature in which—in a world of instant gratification—television detectives can solve any problem within an hour, and despite facing persistent problems in meeting ever-tougher business objectives, executives continually want to believe that life is simple. Well, it is not. Life is not only complicated; it is difficult. It's easy to focus on the success stories, simplify them, and ignore many of the bumps and bruises so inherent in any change transformation. Unfortunately, many believe lock, stock, and barrel in these fairy tale stories. Even worse, they believe that such stories detail all the efforts involved. This is never the case. Any journey always involves complications, confrontations, disagreements, and wrong-paths-taken. To ignore such things makes a story simpler and more publishable, but it does not make it any truer.

On the other hand, a select few have been blessed with both the clarity of thought and the ability to take seemingly complicated situations and reduce them to a simplicity that is not only amazing in its clarity but also revealing in its truth. Justice Oliver Wendell Holmes is credited with the saying, "I don't give a fig for the simplicity on this side of complexity, but I would die for the simplicity on the other side."

I find his statement extremely profound. And I am hopeful that this book will stimulate your thinking and maybe reduce some of these seemingly complicated things to a very actionable level.

Really, It's Not That Complicated, so Let's Get Started

Now for the fifth reason I'm writing this book. Well, really it is pretty complicated. However, the more complicated aspects of motivation, discipline, and cultural change management initially are quite straightforward and not too difficult. The very nature of your lean transformation is that it will initially address mostly technical topics. And these engineering aspects and basic concepts are really not new nor are they that complicated, at least not so complicated we cannot get started. Consequently, throughout this book two concepts shall appear:

- First, "Points of Clarity" will be scattered through the pages, where seemingly complicated concepts will be reduced to their simplest form—usually just one pithy sentence. After all, if things truly are simple, they can typically be expressed in only a few words.

- Second, the application of lean concepts can be reduced to a simple, basic prescription—a prescription that, once the concepts and limitations are understood, can be readily applied to a wide variety of situations.

Huge Gains Can Be Made

Sixth, I have found from experience that if lean manufacturing is implemented, it is highly possible to derive huge early gains from the effort. I will give several examples of this throughout the book. These large early gains, which I call "low-hanging fruit," are sometimes the fuel used to catalyze a truly deep and profound plunge into the heart of Lean, including the cultural change that is both so necessary and so beneficial. Only when the cultural change is completed will the benefits be realized and fully exploited. Unfortunately, harvesting this low-hanging fruit frequently feeds the bias of management, since they now have living proof that there really are "quick fixes," and so while the lean implementation is really in its infancy the focus is sometimes changed to something else causing the lean implementation, and its benefits, to predictably regress. The message is this: Often there are huge early gains to be made, but if these benefits are to be sustained, a cultural change must occur. This change does not come easily, nor quickly. All too often, in manufacturing as in life, things that come very easily often disappear just as quickly. This can also be true of gains made from the implementation of Lean.

New Stuff in the Second Edition

Lean Transformations Are Failing

In the First Edition, I frequently used the term "initiative" to describe the effort and the results of changing your manufacturing system to a lean system. When I was writing the Second Edition, Kelly Moore, a friend and a member of my Technical Review Team, requested I find a better word. Her concern was that most "initiatives" were forced upon people and not "initiated" for their benefit. True enough. Upon further thought, it dawned on me that the most basic of problems with "initiatives" was just that—they were initiated—and not much else. Motivated by Kelly's comments, I tried to find a better word. The most descriptive word I could find was "transformation." It most completely embodies what we are attempting to achieve: a complete changeover of our manufacturing system. I then used that word throughout this book; although at times, in certain sentences, it sounds a bit awkward.

In the fall of 2011, I wrote an article for *Industry Week Magazine* entitled, "How to Design a Lean Initiative so Failure Is Guaranteed." It was based on some recent studies conducted in the industry as well as my data gathered over more than 30 years. Although my data was not of research quality, it was far more than anecdotal. In addition, using my database, I statistically analyzed these data to further understand the cause–effect relationship to these lean failures. I found a number of factors that greatly contribute to the failings. First, I found 10 issues that were present in businesses, and they were so common and antagonistic to lean principles that they needed to be addressed before any significant progress toward Lean could be made; these are the Lean Killers. Second, I found that a number of very common aspects in the design of the lean roll out were so flawed that they virtually guaranteed failure. These are the Six Roll Out Design Errors.

These topics so resonated with the readership that I wrote additional follow-up articles. Finally, I made this topic the centerpiece of this, the Second Edition.

In each case, I have investigated and I have asked, "Is it the lean philosophy or it is the execution that is the root cause of the failing?". In every case, without exception, I found nothing wrong with the basic philosophy of Lean; rather I found that the execution was lacking.

And interestingly enough, when reviewing these factors that caused the lean implementation to be less-than-successful, I find a repeating list of common topics. Yet if these factors are managed well, I find a high incidence of success. Of these factors, very few are really earthshaking and none are beyond the reach of the typical businessman. Sounds like it should work with ease. The problem is that many of the issues that need to be solved and resolved to successfully implement a lean transformation are countercultural and many are counterintuitive.

The bad news is that these lean transformations are failing to be successful; the good news is that all the "issues" are correctable.

Excuses Abound...

In the last 30 years I have heard more excuses why Lean will not work in the United States, or why it won't work in this or that industry or in this or that plant. I am reminded by Benjamin Franklin who said:

"He that is good for making excuses is seldom good for anything else."

As to the excuses, they fail to explain how Japan, a little island nation, lying in ruins in 1945, with virtually none of the classic natural resources, has risen to become a world manufacturing power only 30 years later. The excuses fail to explain why a few Japanese firms began this manufacturing revolution and other Japanese firms learned and adopted the techniques, magnifying the Japanese successes. The excuses fail to explain why we have not been able to match their growth in worldwide manufacturing presence. The excuses fail to explain how some U.S. firms have adopted these techniques and prospered greatly. The excuses fail to explain why now there are superior Japanese automobile manufacturing plants on both U.S. and Japanese soil.

It's All about the Leadership and the Management

The time for excuses is over; it is time that U.S. managers and managers all over the world accept the real reason the Japanese have prospered.

They Are Taught to Be Better Managers and Leaders

There is no real secret. As better managers and leaders they have developed a superior manufacturing system and it is there for us to creatively steal from; there are no real secrets to success…just excuses that prevent it.

What I have found in my work is that there are a lot of transformations that are failing or have failed. With this I agree. However, regarding the excuses, I do not buy into any of them. What I find, without exception, is not a basic flaw in the lean philosophy rather it is a flaw in the design or execution of the change transformation itself. Consequently, I have classified the causes and separated them into three areas of concern, each a section in this book.

Part One, focusing on preparation prior to any roll out efforts, has two chapters. The first discusses key management behaviors and attitudinal issues that need to be addressed prior to starting a lean transformation. The second chapter has to do with the design of the initial roll out and explains exactly how you can design this so failure is assured!!

Part Two focuses on the cultural change that is needed to sustain any gains achieved. There are six chapters. One on cultural basics, including a new "prescription on how to change your culture." Additionally, there is one chapter on each of the five key cultural change leading indicators to success. The five leading cultural change indicators are leadership, motivation, problem solving, total facility engagement, and learning/ teaching/experimenting.

Part Three addresses the technical topics of Lean and the formula to the lean implementation. In it we address how to roll out Lean at the corporate level as well as how to roll it out for a single-value stream.

Furthermore, there are various topics from the First Edition that have been enhanced. Leadership is a key theme in this book. And we will distinguish the subject of leadership from management; both are crucial to success. The absence of strong leadership will guarantee a failure. The presence of leadership, especially a type we will learn to call *lean leadership*, will enhance the probability of success markedly. The chapter on leadership will also address the four "ah-ha" experiences that all lean leaders will go through on their way to making a true lean conversion.

There are two other topics I wished I had covered more deeply in the First Edition. These are the topics of Leader Standard Work and planning. Regarding planning we will discuss the power of planning as a key improvement strategy. We will also address the plan for the roll out of the lean transformation both at the corporate and at the value stream levels.

Finally I Hope to Further Enhance Your Understanding

I hope this book will enhance your understanding of Lean and its many applications, as well as its limitations. Lean, they say, is a journey without any destination. T. S. Eliot expressed this idea quite elegantly:

"We must not cease from exploration. And the end of all exploration will be to arrive where we began and to know the place for the first time."

My hope is that this book will enhance your own exploration.

Lonnie Wilson

Acknowledgments

I wish to thank the entire McGraw-Hill team for encouraging me to write, and assisting me in the production of this, the Second Edition of *How to Implement Lean Manufacturing*.

However, the genesis of that thankfulness lies in the many readers and clients who purchased and used the First Edition. For without strong sales, there would have never been a Second Edition. And to them I am eternally thankful for two very personal reasons. First, as an author, it was extremely rewarding. I was totally surprised when the book not only sold very well, but when it got into the top 10 in Amazon sales I was amazed. Then when it actually reached number 1 for numerous weeks, I was totally beyond myself. In addition, I was contacted by five different colleges and universities that used this for either their business or engineering curricula; that too was very energizing. Second, as a lean practitioner, it was professionally rewarding. Since its publication in 2009, the book received a number of very positive reviews and a great deal of feedback as the First Edition was used as it was designed, as an implementation guide: a "how to" book as I had hoped. Finally the reviews and feedback from readers and clients alike, coupled with my own "refining in the fires of practical applications," were invaluable in making the Second Edition a very good "next step" to those wishing to benefit from the lean transformation. For all this I wish to thank my readers and my clients for assisting me in not only enjoying the First Edition but also in making the Second Edition even better.

As I embarked upon this challenge, I chose a select group of lean practitioners covering a broad range of industries and lean experiences to assist me; they were my Technical Review Committee. My thanks go to Brandon Hughes, Jason Farley, Phil Coy, Michael Wiseman, Kelly Moore, Thom Longcore, Fred Kaschak, and Robert Simonis. I thank each of you for your time and contribution as I know it drew a piece of your time that was already in scarce supply.

Finally, above all else, I thank the love of my life, my wife Roxana for supporting and assisting me with both her patience and her love, through this whole journey of becoming an author.

CHAPTER 1

What Is the Perspective of This Book?

Aim of this chapter ... to explain the perspective of this book, as viewed by its author. The perspective of this book is fivefold. It is written by a consultant who is focused on engineering principles and very practical applications so progress can be made in short order. It addresses the leading indicators of cultural change so changes can not only be made but sustained. It is filled with theory but has a "how to" approach for easy application. Finally, it is definitely geared to the manufacturing world and focuses on applications of lean principles so that large early financial gains can be made and sustained.

From a Practical Perspective

First, this book is written through the lens of a seasoned consultant. That means it offers a great deal of practical advice, is filled with examples, and is geared toward making progress—in some cases very rapid progress. Every effort is made to keep the principles and applications as simple as they can be and, where applicable, these principles are reduced to *Points of Clarity*.

From an Engineering Viewpoint

Second, this book is written primarily from an engineering viewpoint. Lean Manufacturing, also known as the Toyota Production System (TPS), is part of a *business* system, so there clearly are business aspects to it. However, whereas an engineer should easily understand this book, a production supervisor and a human resource (HR) manager should also identify well with what is written here. In fact, almost anyone involved in processes will resonate with these materials.

With a Lot of Cultural Advice

The moment we change how we produce or how we inspect a product, we have changed the behavior of those people doing the work. Because our culture can be described as "just how we do things around here," we have effectively initiated a change in the culture. It may be a small way or a large way, but nonetheless we have changed it. For these reasons, we always need to keep our eye on the culture (see Chaps. 4–9).

Because, at the outset of any production improvements, a cultural change of some magnitude will naturally occur, it seems logical to address that issue up front. Unfortunately addressing cultural issues is a difficult task. It takes special skills to address the type of culture you have, the type you need, and the means you will use to change to what you need. Therefore, whereas there is as great deal of cultural information in this book, what is written is far from exhaustive. Take, for instance, Chap. 4 on cultures. It would be easy enough to fill an entire book with the aspects of measuring your existing culture, determining the absolutely correct culture for your business, and applying the skills and methods needed to change it, hence converting it to a strong, flexible, and appropriate culture. However, experience has shown us that despite this complexity, there are some very critical cultural aspects that must be addressed for a lean transformation to be successful. These include, but are not limited to, the "Five Cultural Change Leading Indicators." Leadership, motivation, problem solving, learning/teaching/experimenting environment, and total facility engagement are the five cultural change leading indicators. Each is addressed in its own chapter.

However, managing a culture is not typical Western thinking. In contrast, review the history of Toyota. They have worked on their culture for as long as they have existed as a company. It was part of the genius of the Toyoda family (founders of the Toyota Industries Corporation) that they understood the importance of the culture and sought its well-being from the beginning.

Point of Clarity The things that make the TPS unique among lean companies are not technical in nature but rooted in culture.

All these cultural skills (which are apparent within the TPS) are, in the final analysis, the aspects that separate the TPS from other manufacturing systems. All these skills require a strong, flexible, and appropriate culture. This development takes a great deal of time, effort, and management skill; it does not come quickly nor does it come easily. Yet it is these cultural aspects that make the TPS almost unique.

This Book Has a "How to" Perspective

This topic of "how to" is something that many will advise you to avoid. They say flowery things like "The TPS is a roadmap, not a prescription," or "Lean is a journey in which everyone must find their own way." Well, although these phrases are flowery, they certainly are not helpful. Therefore, I have asked myself over the years why so many people are afraid to generalize, even when generalization is helpful. I have even wondered if, maybe, they say such things because they have a rough intellectual grasp but lack a full practical understanding of how to actually put these lean principles into action. At any rate, I will describe several "how to" prescriptive methods coupled with an occasional word of caution.

The First Prescription—The Path to Lean

The first prescription is "How to apply the five strategies to become Lean." It is a process that can be used for *kaizen*, a term that means a process of continual improvement. The prescription can also be used for *point kaizen*—that is, some local situation that might be used to remove waste from a work station or a cell. Or this prescription can be used in *value stream kaizen* (sometimes called *flow kaizen*), where an entire value stream is addressed. This covers the diagnostic tools used and how they can assist you in your efforts to eliminate the seven wastes in the production process. Once the five strategies have been addressed, there is a comprehensive "to-do" list of *kaizen* activities designed to make your system Lean. This prescription, The Path to Lean, is described in detail in Chap. 14, and the application of this prescription is further clarified in three case studies in Chap. 22.

The Second Prescription—How to Implement Lean… The Prescription for the Lean Value Stream

The second prescription addresses how to take all the information and resources and apply them in a project format so your lean design will become a reality. This prescription starts with a complex evaluation of the overall manufacturing system, goes into the specifics of the value stream improve-

> **P**oint of Clarity   This book is about the applications of lean principles.

ments, and turns it all into an action plan. In fact, this prescription is more than an action plan. It is an eight-step process that results in a leaner process. This second prescription is a proven technique (described fully in Chap. 16), and its application is shown in three case studies in Chap. 22.

The Third Prescription—The Change Prescription

In most of the changes that are needed to affect a cultural change, often times the technical aspect is not the most challenging. Most often the real challenge is in accomplishing the change with and through an engaged workforce. This Change Prescription addresses those issues. It is a five-step proven technique to guide the leadership to accomplish the needed cultural changes. It is discussed and detailed in Chap. 4.

The Fourth Prescription—How to Sustain the Gains

Again, to make changes in a process from both a technical and a personnel standpoint is highly challenging. However, the even greater challenge is to create a system whereby these changes can be sustained and even built upon. This is accomplished by using a practical and proven five-step process. The prescription for how to sustain the gains is fully described in Chap. 15, Sustaining the Gains.

"How to" Become a Better Money-Making Machine, a More Secure Workplace for Your Employees, and the Supplier of Choice to Your Customers

Make no mistake; this is not a primer on "What is Lean" or "What is the TPS," although much of that is contained herein. It is not a book about theory; it is a book to guide actions and is about applications. It is about how you can apply the lean principles and lead your facility, making it:

- A better money-making machine
- A more secure workplace for your employees
- The supplier of choice to your customers

And to Those in Manufacturing Who Seek Huge Gains

Finally, this book is written for those in manufacturing who wish to excel in making their business healthier and more robust, as well as for those who are struggling to survive. It is one of the themes of this book that you can make huge early gains—for those in survival mode, this should be encouraging. As you read, pay particular attention to how quickly many of these companies achieved gains—and how large those gains were—by applying the prescriptions outlined here for you.

Although manufacturing is the focus, the lean principles, once understood, have a very broad application. For example, we have applied these principles to staff functions within manufacturing such as training, procurement, engineering, and testing laboratories to name just a few. In addition, we have applied these principles to doctors' offices, hospitals, hotels, private schools, restaurants, and other service functions. They have a broad application to a wide variety of functions and industries.

Chapter Summary

Five themes run throughout this book. First, this is a practical book about Lean Manufacturing that includes many illuminating examples. Second, the book has an engineering perspective. Third, there is a significant discussion of the five key cultural change leading indicators to the deep cultural changes that are needed to be successful. Fourth, this book takes a definite "how to" approach so as to show the applications of principles. Finally, this book is written to those in manufacturing who hope to make their business more secure and more profitable. For these individuals, there are often huge early gains to be made.

The Lean Killers and Roll Out Errors

Part One, focusing on preparation prior to any roll out efforts, has two chapters. Chapter 2 discusses key management behaviors and attitudinal issues that need to be addressed prior to starting a lean transformation. These are the 10 Lean Killers. Chapter 3 outlines the design of the initial roll out and explains how to avoid the 6 Roll Out Design Errors.

The "Killers" to a Lean Transformation

Aim of this chapter … to explain the various management systems, attitudes, and behaviors that, if not properly addressed and mitigated, will virtually guarantee that your lean transformation will fail. Also, for each Killer, countermeasures are suggested so you can begin to design out these Lean Killers.

Background to the Lean Killers

In my research on lean transformations that have been succeeding, and especially those that have failed to reach fruition, I have used my and others' data. In analyzing the successes and the failures and doing correlations on each, I found 10 issues that were so common that they needed to be addressed before any significant progress toward Lean could be made. Companies that successfully overcame these issues often succeeded, and those that did not, failed. These are the Lean Killers.

Of the 10, Lean Killers 1 and 2 are specific to the roll out of the lean transformation, and consequently they may be new to your manufacturing system. Of the other eight, if these Lean Killers exist in your system, and likely they do, they have been present to some degree for quite some time.

None of the 10 Lean Killers present much of a technical challenge to make the needed changes so you can transform your manufacturing system. Most of the countermeasures are known and proven. Rather, I find that first, they require an increased awareness on the part of management to recognize and accept them. This then places a sim-

> **P**oint of Clarity At the end of the day, there are just choices and consequences.

ple choice in the hands of management: Do we or do we not choose to fix this issue? As a result of this choice, there are consequences, and that is simply between success and failure. It is no more complicated than that.

Each of the items discussed as follows is a deficiency of such magnitude that if it will more than hinder the progress, it will do more than slow down the transformation; if not addressed and mitigated in time, it will completely derail the lean transformation … . such is the nature of these 10 Lean Killers.

Again, this is yet another reason to rely heavily on your *sensei* at this phase of lean implementation. He will be a far superior observer of the current state and a far more dispassionate judge of both the countermeasures needed and the successes attained.

Lean Killer No. 1—Lack of Lean Understanding by Top Management

The biggest failing of all is to have a management team that is not lean competent. Yes, the management at all levels, C-Suite included, must be lean competent.

Just what do we mean by "lean competent?" The key managerial tasks include the ability to plan and budget, staff and organize, as well as control and problem-solve **at their specific level**. These are the key skills of management and they apply to management of the lean function as well. These managers then must have the knowledge base to execute these skills and make the management decisions relative to the lean transformation, especially the cultural transformation that is so necessary. It is lost on many who are making a lean transformation that **they are changing their entire manufacturing system and to some extent all the support services associated with the manufacturing system**. They are not only changing the policy, practices, and skills needed, they are asking for a wholesale change in the attitudes and behavior of everyone working in the system; they are making massive changes in their culture. Everything will change.

Lean Competence . . . an Example

Let me give you an example. You are the plant manager for a large manufacturing firm, with several functional managers and supervisors who report to you. One is in charge of a group of engineers whose task is to support the lean implementation. This supervisor approaches you with a proposal to change from the age-old practice of producing large lots to producing small lots with more frequent changeovers. You and your supervisor discuss the idea and decide not only is it part of your lean implementation, but also it is the right time to implement this lean concept. To assist the team, you do a little brush clearing at the next level, modify your existing strategy, and get input and agreement from the lower levels in your organization of how this strategy would be executed. Once the actions and metrics around the strategy are agreed upon, the production manager communicates and cascades the deployment, routinely measuring the progress. (This strategy deployment process is described more extensively in Chap. 17.)

However, as the production manager proceeds with the implementation, he gets pushback from the maintenance and logistics folks. This "batch-busting" philosophy of small lot production is clearly in harmony with the plant's lean implementation and both the maintenance and the logistics managers agree with the principles, that is—until they have to change. Then the real culture-changing work begins.

The team in charge of this change is at a crossroad. If they need support from you, can you intercede appropriately and support your team? I have seen this played out dozens of times, and it only goes well if you know your stuff. Simply put, you must be "lean competent."

By "lean competent," in this case I mean you know the benefits of a small lot of production, you can articulate and demonstrate how it is done, and you can positively relate it to both the lean implementation strategy your organization has committed to and the current business condition of your facility. If you are not lean competent, your team likely loses this argument and your lean implementation loses steam.

So let's say you are "lean competent" enough to convince the maintenance and logistics managers and things proceed smoothly for a while. Soon your plant will need to supply incoming materials in smaller lots, more frequently, to further mature your lean system. However, when your team discusses this with the storehouse manager, he balks completely. He cites the need for a new $60,000 forklift and two additional materials handlers per shift to accommodate the more frequent deliveries.

You and your team are pretty sure he has not thought it through well enough. However, in the face of shrinking capital budgets and downward pressure on costs, adding facilities and adding manpower are formidable enemies to the lean implementation. Will you succumb to the pressures of budgets and manpower costs? Or will you summon the courage to push the lean implementation forward? Will you keep your eye on the long-term objectives or succumb to short-term thinking??

Do you have the skills and the understanding of Lean to stand up and argue, "Our stated objective is to reach the Ideal State of small lot production and that means small lots in the warehouse as well? It is part of our vision, it is built into our 'House of Lean,' and we have all received the training that describes the benefits. We all know there will be obstacles along the way. We cannot allow these obstacles to prevent us from trying to attain our Ideal State. Now, how can we assist you as we try to overcome these obstacles?"

That statement requires a deep and abiding understanding of Lean, and a good dose of both courage and character. That statement came from someone who can be trusted to guide, assist, and support his organization as it embarks on this journey into the unknown.

> **"S**upport of top management is not sufficient. It is not enough that top management commit themselves for life to quality and productivity. They must know what it is that they are committed to—that is, what they must do. These obligations cannot be delegated. **Support is not enough; action is required** (emphasis mine). (W. Edwards Deming, 'Out of Crisis,' MIT Center for Advanced Engineering Study, 1986). **"**

Why Lean Competent?

You must be lean competent to guide and assist your employees as they pursue a practical "next step" in taking the entire organization through the PDCA (Plan-Do-Check Act) cycle. You must be lean competent so your guidance and assistance allows those affected to "see" the next steps as you proceed toward the Ideal State. You must be lean competent so these steps are large enough to be useful and challenging, yet small enough to be attainable in a reasonable time frame. As you proceed into the action steps, you must be lean competent so you can provide the "attaboys" when they are appropriate and yet intervene and support when pushback comes.

This is a topic I have discussed in detail with my mentor, Toshi Amino who is the retired (1995) Executive Vice President of Honda America Mfg. and constructed the first full-size Japanese automobile plant built on U.S. soil (1981-1982); Honda's plant in Marysville Ohio. While discussing this with Toshi, his forehead got all wrinkled and with a certain amount of consternation he commented; "I just don't understand how these managers who do not know their production systems and are not familiar with its details can really help their people." That was Toshi's first thought....helping the people. To Toshi, it was not only about making money and it was not only about providing high quality product....his common denominator was how these managers were not able to support their people. That should give all companies an insight into why Honda is such a powerful business and outstanding manufacturer.

The Importance of "Knowing Your Stuff" and Trust

A wide range of personal values and traits are required of lean leaders. Among those are courage and character along with good listening skills and large doses both of curiosity and humility. In practice I have found that a large fraction of lean leaders, particularly top-level lean leaders, have a good complement of the personal values and traits required. However, too many do not want to learn the basics of their own production system. They simply do not want to delve into the technical details—the very technical details that make their production system work. Consequently, the organization does not have lean competency at the top levels, which causes their employees to hesitate before they act. It causes employees to be risk averse, and it also causes employees to not trust their leadership and their management. With this cultural paradigm, progress is a long way off.

Trust in any field of endeavor requires that you not only have strong personal characteristics such as honesty, but it requires you be competent in the topic of discussion. For example, would you trust an honest but incompetent doctor to operate on you? Presuming you have a strong set of shared values and the courage and character it takes, you can support your team. But, quite frankly, that means your employees can only "trust you to be" a high-priced lean-cheerleader. With only half the formula for trust in place, you don't yet cut the mustard to be a lean leader. These are necessary but not yet sufficient traits to become a lean leader. However, if you have those strong personal traits and are also lean competent, your employees now can count on you not only to supply the support but also to supply the guidance and assistance that they will need to succeed. Then they can trust you to be a lean leader.

Only then will the team be "en-couraged" to walk metaphorically half-naked into the wilderness of the unknown as they aggressively pursue changes that are both risky and necessary to make your business a better money-making machine, a more secure work place, and the supplier of choice to our customers.

It would seem incumbent on all management, all in leadership positions, to then understand the manufacturing system which is the very center of their corporation. Yet many managers view this transformation of their manufacturing system as rather superficial and spend very little time learning and studying it and even less time practicing it. Given the tests for commitment to this effort, you will find they will fail several, including Test No. 1 (see Chap. 24).

> **P**oint of Clarity   "Do not let what you cannot do, prevent you from doing what you can."
> John Wooden

The Lean Mantra, "Learn by Doing"

So from where does most of this lean competency originate? In reality, most of Lean is learned by doing. A certain amount—insufficient to be successful—can be learned from books, videos, e-Learning, blogs, site visits, and discussions. But "learning by doing" is the mode for acquiring the behavioral traits needed to execute a lean transformation. This is how the managers who work for Honda and Toyota, to name a few, learned the skills to execute a lean system. They were taught the lean skills as a young employee by their supervisor as part of their personal development. As they progressed to new and higher jobs, they were taught new and advanced skills. All they while they learned by doing. They spent a significant portion of their career becoming lean competent. As I have had the pleasure to work with many of these managers, even at the corporate VP level, I have marveled at the strength and depth of their technical and personnel skills. They are very good at the details and can ask penetrating questions borne of experience. This makes them superior supervisors, superior managers, and superior leaders. There simply is no substitute for this; your entire management team must be lean competent if you wish to be successful.

Let's shift our metaphors and suppose you aspire to be a professional basketball player. To acquire the requisite skills and behavioral traits to perform on the court, no one would propose that you read a few books, check out the Internet, watch a few videos, have a few "in depth discussions over a beer," and then go try out for a professional basketball team. **Intellectual understanding** of how to dribble, set a pick, shoot the ball, and rebound are not adequate for you to **learn the behavioral traits** to compete at virtually any level. You need to learn the skills, practice them, spend time on the court, and finally you need to play the game. Yet somehow a majority of managers who try to implement Lean expect this intellectual understanding alone—gained from books, videos, and the Internet—to be sufficient for them to be successful in this lean implementation. They are not only wrong, they are dead wrong, and they are wrong to the extent that they will kill the entire lean implementation.

So for those top-level managers who have never practiced Lean on their way to the top, what can they do? A Lot. Yes, there is a lot they can do.

- First, they can actively study about Lean. Books, professional organizations, and plant tours are three sources of good information they are normally already involved in.

- Second, they should change the way they do things, starting right now. They can model the behavior they wish to see in others. For example, they can problem-solve using the lean tools they expect others to use. They can ask the Five Questions of Continuous Improvement (see Chap. 7). They can practice 5S in their office and meetings and create their own leader's standard work (see Chap. 12). They can change their planning system to *Hoshin–Kanri* planning and implement catchball (see Chap. 17). They can practice the five motivational tools of intrinsic motivation (see Chap. 6) rather than using the time-worn methods of bonuses and salary. There are an infinite list of things they can do to be a lean teacher. As a manager if you are at this stage and wish to know what to do, simply ask your *sensei*. He will be able to give you invaluable advice.

- Third, they need to understand in a cold, dispassionate, introspective way what they really do not know. In these areas they need to seek help. That is precisely what their *sensei* exists for…to advise, teach, coach, train, and mentor. At the onset of a lean transformation, the top managers may not be able to fully grasp the concept of line balancing across an entire plant and they may not be able to do the rigorous calculations to make a *kanban* system work or to design the three-part inventory system, but that does not preclude them from doing what they can do to broaden their base of knowledge. In their jobs as managers they will have to make decisions, to do this, I would argue those decisions will be much better if they are based on the presence of knowledge, rather than the absence of knowledge.

Countermeasure to Killer No. 1

The first part is pretty straightforward—it is of course education. Unfortunately, education, at least what we call education (see Chap. 9, A Learning/Teaching/Experimenting Environment), is mostly auditory and visual learning and only promotes awareness at the best. The second part, to get to the action level, so the management team can exhibit the appropriate behavioral traits, practice is mandatory. The lean mantra is "learn by doing." All the people involved in a lean transformation must "learn their way to greater competency" and management is no exception. The top-level managers must be "doing Lean" or they are not engaged. If management fails to model the correct behaviors, this will certainly kill any lean transformation. Finally a major cause of this Lean Killer is what I refer to as "the MBA Concept Gone Awry." It is discussed in Chap. 5, and pay particular attention to that section.

> **P**oint of Clarity   Lean strategy, tactics, and skills are learned by doing...and in no other way.

Lean Killer No. 2—Have No Management Guidance Team

There needs to be a key group of functional managers at the corporate level who are **lean competent** and can help support and guide the change. You need a **critical mass of the key functional leaders**. Typically this group is called a *Steering Committee,* and their job is to guide the effort and provide support as needed. To do this, they need to be lean competent.

This critical mass of key functional leaders is absolutely essential because as the lean transformation is deployed, it will begin to change each and every aspect of the business. It must be recognized early on that the lean transformation is a revolutionary change in the way you not only handle the process but handle the people as well. You will have a significant impact on all aspects of management, including planning and budgeting, organizing and staffing, as well as controlling and problem solving. Since it touches all of management, the senior leadership needs to be acutely aware of exactly how they are "steering" this event. In the absence of a critical mass of lean-competent senior leadership, it is very possible to "steer" this effort in the wrong direction.

Countermeasure to Lean Killer No. 2

Create, at day 1, a Steering Committee of the key decision makers. Of course, their first task will be to become lean competent. They must commit to implementing Lean and performing as lean leaders. In the absence of this commitment, the lean transformation will lack the guidance it needs. It will lack the support it needs, and the leadership will be lacking in one absolutely crucial characteristic it must have—the top management must model the behavior they wish to see in others.

Lean Killer No. 3—Insufficient Leadership at Many Levels

If you do not have good leadership at the top-level management, the lean transformation will fail. Simple, short, and sweet—it is no more complicated than that. I cannot think of a more definitive litmus test.

Got leadership, you have a chance—weak or missing leadership, you are doomed.

Keep in mind that the first and most important requisite skill of leadership is to have a plan. That is why lean competency at the top-level management is so crucial. The top management team cannot put together a viable plan unless they are lean competent.

Normally, but not always, I find there is a plan at the corporate level to transform the corporation. Unfortunately as this cascades down to the plants and finally to the value streams, I find fewer and weaker plans and frequently none at all. More often than not, at the plant level I find only budgets to be met with no real lean transformation or even continuous improvement plans at all.

By a plan, I mean as a minimum list of measurable objectives with an action item list that answers "who is going to do what by when" in order to meet the measurable objectives. The plan needs routine follow-up as well. Without this plan and the necessary follow-up, there is no leadership except the type needed in firefighting, which is not a healthy long-term solution. In terms of overall leadership, we speak of a leadership footprint and we want it to be as large as possible. In a lean system everyone has the opportunity to lead, and the more leadership we have, the larger the leadership footprint becomes.

All too often in the mass production plant, the plant manager is viewed as THE leader. Frequently, his direct reports including the production manager, maintenance manager, and process manager, to name a few, are referred to as key leadership positions in the plant. As I am reviewing this topic with the plant manager (PM) and we are evaluating the **leadership footprint** prior to preparing the lean rollout plan, I will ask him what I call the "coaching question." And for each of these managers I ask, "so what is his plan?" to which the PM often replies, "what exactly do you mean?" And the answer to that I respond, "Each of your managers, each of your supervisors, each of your team leaders, and each of your floor workers needs to have a plan. Don't you find it strange that the cell worker has a very specific plan? He knows exactly what he is supposed to do every hour of every day, yet your team leaders, supervisors, and managers do not? They all need to have a plan for today, they need to have a plan for things to do this week, they need to have a monthly plan, and they need to have a plan for the year" ...I go on.... "that is a key difference between a typical plant and a lean plant. In a lean plant all activities have a plan, for everything there is a 'should be' condition. We try to leave nothing to chance. We strive for a highly predictable workplace where 'what should happen, does happen, every time, on time.' This can only happen with a high degree of planning and execution at all levels. And when we have good planning and good follow-up, very likely we have good leadership. Certainly with no definitive plan and either spotty or no follow up...leadership is lacking". (See Creating a Leadership Footprint in Chap. 5.)

> **P**oint of Clarity   As a lean *sensei* or lean leader, you need to develop the skill of asking the "Coaching Question."

Countermeasure to Killer No. 3

First and foremost, you must understand the role and duties of leadership as outlined in Chap. 5. Of course, the top leadership must model good leadership behavior or the entire effort will be compromised. Teach, train, and model good leadership. Make leadership and leadership training a priority. Have a plan to expand your **leadership footprint**. In a job appropriate fashion, include leadership training in all other training plans.

Lean Killer No. 4—Lack of Customer Focus

All lean transformations, all manufacturing plants for that matter require a focus on three groups. Stockholders who usually supply the capital, employees who execute the system, and the customer who sets the needed standards for the products and, not insignificantly, pays us for them. Most plants only focus intently on the bottom line and hence really only pay attention to one group, the stockholders. Too frequently the customer is treated like the enemy. We and they both scratch, claw, and fight over the price, quality, and delivery, and feel that if they gain, we must lose. Rather than work to supply higher quality and better delivery we fight over standards and money alike. A highly antagonistic relationship is created. If we are a tier 1 automobile supplier, it is often lost in the day-to-day activities that we must work jointly with our customer to supply the ultimate customer. If we do not, prices rise and both we as suppliers and our customer will lose competitive position. As a consultant, I have worked with Toyota and Honda as well as Ford, GM, and Chrysler, and I can tell you with certainty that both Honda and Toyota do a much better job at keeping this in perspective. Even with Honda and Toyota, many times life is both tense and combative between them and their suppliers. However, they are simply better at providing support and direction to their supply chain. It is no wonder they financially outperform their American counterparts. My experience with Honda and Toyota is common. If you meet their goals of price, delivery, and quality, and show continuous improvement, it is highly likely that they will source with you again. If you fail to meet their standards, they will very likely source elsewhere.

Countermeasure to Lean Killer No. 4

This countermeasure is almost purely a matter of management attitude. Your manufacturing system needs to do two things: deliver on time and provide high-quality goods. These are totally within the control of management, and if they wish this to happen, it can. If the top management is talking and walking like the customer is the enemy, the whole manufacturing system will respond accordingly and disharmony will be the result. If, on the other hand, the top management is interacting with the customer to get clear contracts and both measuring and supplying the resources to meet the needs of the customer, the customer will get what he is paying for, high-quality goods, delivered on time. Most often this keeps the customer happy…at least as happy as customer's get—and that is to give you repeat business.

Lean Killer No. 5—Make Lean a Workforce Reduction Mechanism

Many managers view Lean as the holy grail of workforce reduction. Although a totally incorrect concept, there is a lot of history behind this. I will try to summarize it here. When Lean was becoming popular in the United States following the publication of the book, "The Machine That Changed the World" (Rawson associates, 1990) consultants came out of the woodwork to try to emulate this. Soon a number of books were on the market mostly based on work done by academics who studied and viewed the Toyota Production System (TPS) from the outside. Lean efforts boomed and there were soon many lean success stories further fueling this lean feeding frenzy. Unfortunately a large portion of these success stories were based solely on manpower reductions. This provided a completely warped, although popular perspective of what the TPS really was. Only later in 2001, did Toyota publish its own document describing itself. This book was affectionately called the Green Book and is a 13-page document entitled "The Toyota Way, 2001" (The Toyota Motor Corporation, 2001). In it, Toyota clearly states that the Toyota Way, which is described as "the company's fundamental DNA" is built on two pillars: Continuous Improvement and Respect for People. In the feeding frenzy from 1990 until 2001, some practitioners already understood this emphasis on respect for people but way too many consultants and managers actually believed that they were implementing the TPS when in fact they were focusing heavily only on the continuous improvement portion; respect for people was not even on the radar of many of these "lean conversions." Consequently a large fraction of those continuous improvement gains turned out to be improvements directly related to manpower reductions. Actually improving labor productivity through an active, engaged, empowered workforce is one of the strong points of Lean. However, if the workforce is then reduced and people are laid off as a result of productivity gains, the active, engaged empowered workforce will then become very inactive and highly disengaged. Consequently their productivity improvement efforts and gains will come to a screeching halt as they learn that they are working very hard to eliminate their own jobs. Unfortunately this is news to a whole slew of managers. Where companies embarked on a lean transformation attempting to emulate the TPS and used only the continuous improvement pillar, they normally failed and this fueled the fire that "Lean just doesn't work here."

Well, the fact is that any good idea, if it is executed poorly—will likely fail.

Regarding "respect for people" there are only two mental models you can hold. First is that people are an **asset to be utilized** and there mind and there body hold equal importance toward the success of the enterprise. Second, you can believe that people are simply an **expense to be minimized** like any other operating expense. After reading "The Toyota Way, 2001," you will have a clear understanding that the TPS and by extension Lean is obviously based on the first belief system.

In addition, although the typical lean transformation will allow large improvements in workforce productivity, the bottom line productivity gains will not be realized until people are released or overtime is reduced. Releasing the very people who created the productivity improvements is equivalent to killing the goose who laid the golden egg, so that is not a moral or a lean option. Reduction of overtime, is a possibility. Of course, the best choice is for new work to be acquired to utilize the freed up workforce and do more work with less. Barring that, put people to work on *kaizen* activities and make improvements.

Countermeasure to Lean Killer No. 5

The countermeasures depend on the current situation. First, if the perspective of management is that labor is viewed as a variable cost to be reduced, you have a monstrous task on your hands. This attitude will contaminate virtually all you do and will need to be corrected before you will make progress in a lean transformation. So let's assume you view your employees as an asset. First, create a policy statement that no workforce reductions will result due to lean improvement activities…and then stick by your word. Second, this means you need to have a plan to use the people you will free up as you reduce the seven wastes until the new work arrives. So have a list of *kaizen* activities ready to be worked on. In addition, there is almost always the opportunity to do training and cross-training.

Lean Killer No. 6—Have a Short-Term View of Success, Focused Narrowly on Financials

Many plants focus intently on the short term, at the expense of the long term. Nearly always this focus is on financials alone. This is anathema to lean success. There must be an eye to the long-term gains and there must be a balanced scoreboard. These firms intent on short-term successes are obvious by their intense focus on the monthly financial results, and their knee-jerk and almost instinctive reaction to minor deviations. Training, especially of management and staff is largely ignored. There are seldom employee development and job succession planning systems in place. Planning is seldom a priority and firefighting is the primary mode of problem resolution. These facilities always make time to do it over, but seldom have the time or the discipline to do it right the first time. Often during problem solving, issues of efficiency will take precedent over the issue of effectiveness. Frequently issues of maintenance will be inappropriately deferred and temporary employees will be used to avoid the cost of benefits—both of these causing serious quality issues. Almost without exception, their plant is highly chaotic with no sense of flow or direction either in the process, the product, the people, or the information. They do all this in the name of cost reductions, which nearly always end up costing more in the long run. Other management means used to support this Lean Killer include management's lack of reflection on prior efforts, *hansei*; lack of catchball in goal deployment; management's critical reviews of monthly performance, divorced from the long-term situation; an unbalanced scoreboard only focusing on financials; management overreaction to situations that are and are not problems; and the management reward system specifically designed to reward short-term performance.

Countermeasure to Lean Killer No. 6
This is simply one of the most difficult of the killers to address and change. Almost always the management system has built into itself so many means of enforcing short-term financial gains that this is a major undertaking and can only be solved by a concerted and introspective systemwide review of management.

The first and most egregious problem that must be resolved is that the entire management team must understand and teach the difference between special cause variation and random cause variation. Without this understanding, literally thousands of decisions made each day are totally or partially incorrect. **Even worse, without this understanding, the management team will routinely blame the workers for problems that are system generated.** This problem cannot be overstated. This not only leads to poor decisions, it leads to huge morale problems as well. I attended a class taught by Dr. Deming in 1985 and he was asked, "If you could teach two things to managers, what would that be?" and his immediate reply was, "Management must understand the difference between special cause and random cause variation and the use of operational definitions." Dr. Deming's writings are filled with examples of the damage done when managers do not distinguish between special and random causes, and I have also seen the fallout all too often.

Of course, *hansei* (a Japanese practice that means reflection) would shed a new light on nearly all these issues. In addition, the implementation of *Hoshin–Kanri* planning (see Chap. 17) with its emphasis on balanced, even daily planning and catchball would be highly corrective. The use of data and analytical especially statistical techniques go a long way toward making correct decisions.

Lean Killer No. 7—Have in Place a Financial Rewards System for Individuals That Is Not Supportive of Lean
It is commonplace in many businesses for senior managers to have a bonus system that is a substantial portion of their total annual remuneration. When this bonus is based on

the monthly financial performance, or any short-term gain, those will be the actions management will favor. No matter how much you might like them to be a team player, with a bonus system that favors short-term gains, that is exactly what you will get in the long run. In virtually all manufacturing systems, individual goals and rewards will trump group goals and rewards.

Countermeasure to Lean Killer No. 7

Do away with the bonus system. I know this is radical. And I do not say this lightly. However, the bonus system to accomplish the goal of higher performance must be based on the following logic:

- First, your managers are not properly incentivized and you must do something.
- Second, they are not giving their best efforts and have more to give to the job.
- Third, they have within their control all the means to make the company prosper.
- Fourth, more money will make them work both harder and smarter.

That is, your managers are not working as hard as they might, they can control the situation better if they so wished, and hence need to be bribed to do better work.

Really????? Does anyone really believe this logic?

I doubt it; yet we persist to have the bonus system built primarily on attaining these short-term gains.

In Chap. 6 we address each of these premises as each is either totally or partially false. Unfortunately in today's business environment, this is too radical a change for most management systems. Inherent in their total compensation system is the concept of bonuses. It might be hard to recruit and retain top managers because you might be the only firm doing this.

So what could you do? Well, first of all, do not confuse motivation with pay; they are two different dynamics that are addressed in Chap. 6 on motivation. However, if you have a motivational problem with management, attack that with vigor with the tools addressed in Chap. 6. Second, if you still feel you need a bonus as part of your compensation package, to remain competitive, your first move should be to shrink the bonus portion of the compensation package of top managers. Increase base pay if you need to. Third, make the bonus system based on long-term, not on short-term gains. Fourth, make sure that whatever basis the bonus system is built upon is based on objective measures that can be controlled by the management team. Fifth, make sure that the calculations are not based on random variations but can be shown to be statistically significant to some level, say 95 percent. And finally, work to develop a management compensation system that is supportive of creating and sustaining a lean culture.

Lean Killer No. 8—Unstable Process Flow

I cannot name even one client of mine in the last 25 years who had, upon initial evaluation, a stable process flow system. All have had to work on this issue.

There must be a stable flow from your shipping dock to the customer based on his/her needs. Then there must be a stable flow from the values stream to the warehouse. For this to happen, there must be a stable flow from each cell. And finally, each worker in each cell must perform to stable flow, and all this must be synchro-nized to process cycle time that supplies a good product to your customer, on time, every time.

In fact, the most common attribute of a non-lean facility is that the current mode of problem solving is euphemistically called *firefighting*. First, it is not problem solv-ing at all. Seldom, if ever, in this chaotic environment are root causes found and removed. True corrective actions are few and far between. And the entire workforce is left with the knowledge that they have fixed nothing. Rather, they are now anxious since they know that the "problem" will appear again, probably sooner, rather than later. Second and maybe worse, in this unstable environment, the entire sense of con-trol is lost by the workforce. There is no predictability and stress is very high. It is more appropriate to say that the unstable processes are controlling the workforce, rather than vice versa.

Unfortunately very few existing manufacturing systems have the high levels of process capability necessary to support a lean transformation and most are still very marginal to even meet current customer needs. Few things are more basic to lean success than stable flow at takt, yet this is often overlooked in a lean transformation. Unfortunately this is not highlighted in a lot of lean literature, mostly because stable flow at takt was achieved at Toyota long before much of what is known about the TPS was published (see Chap. 10). It would be good for everyone to pay attention to some parting words from James Womack in his book, *Gemba Walks*, in which in the chapter entitled "Hopeful Hansei," he says …

> For 20 years I have had a T-shirt on the wall above my desk in my writer's nook. On the front is a line drawing of Taichi Ohno staring down at me with the admonition, "Where there is no standard, there can be no kaizen." Yet somehow I kept thinking smooth flow and steady pull could be created first, with basic stability as an afterthought—or that maybe it would just emerge automatically. In retrospect, *it's like believing that a building can be built without first laying the foundation. I now see my error. I only wish I had realized this sooner (emphasis mine).* (Womack, James P. 2011-02-21, *Gemba Walks*, p. 328, Lean Enterprise Institute, Inc. Kindle Edition)

I too wished he had learned this sooner. This lack of understanding, of this most basic point, which was at the very heart of the teachings of Dr. Deming, has caused untold amounts of damage to those trying to implement a lean transformation. With stable flow, success is possible; without stable flow, no lean effort will be successful.

Countermeasure to Lean Killer No. 8
Attack the Five Precursors (see Chap. 24) with a vigor unseen in your company. The gains will be huge and early. Aggressively teach statistical process control and critical problem solving to all decision makers. Treat each aspect of unstable flow whether it is a quantity or a quality issue as a crisis…it is.

Lean Killer No. 9—Have a "Closed" Culture
Closed cultures have several characteristics. First, they have poor communications channels. Open, honest, frank, and balanced discussions are not encouraged. These cultures are typified by lots of "happy talk" where all successes are recognized, even "almost and impending successes" are reveled in—yet problems are kept out of sight

and frequently even swept under the rug. Problems and the opportunities they present are avoided like a cancer. Sometimes the bosses are not available to talk; more often there are a whole lot of topics that simply cannot be discussed. Usually these are the issues that the management team simply either does not see or sometimes does not want to see. Even if they do see it, they do not want to deal with the topic. All too often it is "that uncomfortable elephant sitting in the middle of the room" that no one wants to acknowledge. This is the topic that Chris Argyris (*Flawed Advice and the Management Trap*, Oxford Press, 2000) named "the undiscussables." And in fact, as he points out, more often than not the "undiscussables are themselves undiscussable." Dr. Deming referred to these as cultures of fear. His point no. 8 addresses this directly, "Drive our fear, so everyone may work effectively for the company" (*The Deming Route to Quality and Productivity*, by William Sherkenbach, ASQC Quality Press, 1988). Second, they have unclear objectives or the objectives are mismatched with the resources, and there are really no means to resolve this problem. Third, they do not use data and sound analytical techniques; rather they use emotions to make the majority of their decisions and value judgments. Fourth, problems are viewed as annoyances to be avoided rather than opportunities to be exploited to make the system better. Hence many problems stay "hidden." Fifth, all closed cultures are risk averse and hence action averse. Sixth, all closed cultures have high levels of "aversion to responsibility." They are cultures that are virtually expert at using the skills of avoidance, denial, rationalizations, and seek to place blame on others. Listen to their talk; it is filled with negatives, frustrations, and disempowerments. It is the "can't, won't, don't couldn't shouldn't wouldn't" culture. Seventh, many human resource functions such as appraisals, personal development plans, and job succession planning are executed in a very perfunctory fashion or often, not at all. Eighth, it is ironic that closed cultures are characterized by having lots and lots of meetings. You would think that meetings would facilitate communications. Unfortunately, in a closed culture open and honest communications are stifled and what comes out of the meeting is that, "after all is said and done, a lot more is said than done." Most of the meetings in a closed culture are pure *muda*.

Countermeasure to Lean Killer No. 9

Of course the answer is to open up the culture. Start with frank, honest communications based on the real needs and the real situation at hand. Learn to listen and listen well, that cannot be overemphasized. Create a workplace that is "transparent." Exploit all the aspects of visual management and share all the information as widely as possible.

Don't avoid problems. They are the vitality that, when resolved, make the manufacturing system better; that is your objective. Yet do not fail to celebrate your successes; this too must be done. See Chap. 7 on Problem Solving.

Make sure that goals and objectives are clear and properly deployed and that all resources are made available to execute the goals. In a phrase, make sure the results and the means to those results are properly addressed. See Chap. 8 on Engagement and Chap. 17 on Planning.

Create a culture that is willing to accept occasional failings and is not risk averse. There will be little or no innovation in a risk-averse culture...and not much action. Create a culture that is based on "learn-do-reflect." See Chap. 9 on a Learning/Teaching/Experimenting Environment.

Before any meeting make sure the objectives of the meeting are clear, and before you leave the meeting, confirm that you can answer, "who is going to do what by when" for each meeting topic. Get much better at designing and managing meetings.

Teaching and modeling the Skills of Lean Leadership described in Chap. 5 is a strong countermeasure to opening up the culture.

Deming wrote of the huge negative impact of a culture build on fear in both of his books as did Sherkenbach. However, neither author provides much guidance on how to really manage it better.

On the other hand, Chris Argyris attacks this directly in his book referenced in the text, "*Flawed Advice and the Management Trap*." "*Driving Fear Out of the Workplace*" by Ryan and Oestreich (Jossey Bass, 1991) is a good primer and the chapter entitled "Discuss the Undiscussable" is all by itself worth the price of the book. However, the best treatment is done by neither a businessman nor an academic, rather a doctor of psychiatry. M. Scott Peck's "*A World Waiting to be Born*" (Bantam Books, 1993) should be required reading for all managers, everywhere.

> **A**bout a culture of fear... William Sherkenbach in *The Deming Route to Quality and Productivity* wrote, "Fear is one of those highly leveraged qualities. You only need to kill a messenger once and word gets around. . . . Dr. Deming has found that the removal or reduction of fear should be one of the first of his fourteen obligations which top management starts to implement, because it affects nine of his other points. Without an atmosphere of mutual respect...no management system will work... ."

Lean Killer No. 10—Tolerate Certain Levels of Irresponsibility in Your Culture

Responsibility versus Accountability

Check a dictionary, especially one published in the last 20 years and you will often find that responsibility and accountability are synonyms. Not so in my 1946 Webster's that makes a clear distinction between responsibility and accountability. Specifically, accountability is just that—the "ability to account for." It means you are supposed to "know about it, and you can answer for it." As distinct from responsibility, which is—the "ability to respond." Hence you not only can answer for it, but you can change it. **You are able to respond**. Contained within the concept of responsibility is accountability, but the opposite is not true. In addition, even in my 1946 dictionary, responsibility has trustworthy as a synonym. So responsibility carries with it some moral implications as well.

Take, for example, the concept of plant profitability. The plant controller is certainly accountable for the plants profitability. He must know about it and be able to answer for it clearly. Yet the controller can do little to change it, at least not while following GAAP and some type of a moral compass. Hence, although he can "account for," he cannot "respond to" low levels of profitability and change them. However, on the other hand, the plant manager is both accountable to know the profitability and is also responsible to reach the best levels possible. Accountability has to do with knowing; responsibility is that plus the ability to change things. I know the distinction is often not made this clearly, but I wish to stress this, for more than just editorial reasons; I believe we should retain this distinction as it is of crucial importance both in business and in our everyday lives as well. You may agree or disagree, but to understand what follows, that distinction is critical.

The Nature of Responsibility

If you are a manager in a plant, to be held responsible for something, some elements must be in place. To have the "ability to respond," you first must have access to the relevant information needed, you certainly need to know. Second, you need the skill inventory to understand and take action on these data. Third, you must have the needed resources. Fourth, you also need the power to cause things to happen. Finally, you need the moral values to respond in the correct way; for you to act in the wrong way may be "irresponsible." The correct way is usually defined by one of three standards: the laws and regulations we are bound to follow, a code of morals or honor, or the rules of our business. So as a manager, you might violate some environmental law by inappropriately dumping environmental wastes to avoid the costs of compliance. This might be called legal irresponsibility. Or maybe, you might make an agreement with an employee that if he completes a certain project on time and on budget, he will get a raise, and then you renege on this promise. If you did that, many people would consider that you were exhibiting moral irresponsibility. Or maybe, as the plant manager, you fail to reach the plant earnings objectives due to poor decisions on your part; then your superiors might decide you were fiscally irresponsible.

> I give you this information on responsibility and accountability, as I, for one, believe that the basic concept of responsibility, both personal and corporate, is eroding, hence the reference to my 1946 dictionary. And I believe that our language is mirroring what is happening in society. As responsibility is becoming the same as accountability in people's mind, the language changes to explain the "new norm." I also believe that this erosion of responsibility and the "new norm" is deleterious to our society as a whole; and certainly to a lean transformation.

So to be responsible, you must have five things: relevant information, skills, resources, power, and a standard of some kind. I want to make a point clear. I seldom find a really, really irresponsible company. If I did, I would not work for them. Quite frankly, what I find is that all companies have "areas of opportunity," and the truly responsible companies are recognizing and working on their weaknesses.

That said, what do I see as examples of these areas of opportunity? Since the behavior of a company is a direct function of its management, we will focus on management.

Relevant information is one of the key functions needed by management to make decisions. Unfortunately the information they have is often inadequate. Today, with sophisticated information gathering system and huge data bases, the data we use to evaluate costs and determine product quality are almost always adequate for the typical manufacturing plant. Unfortunately as these data are manipulated into metrics for higher-level purposes, such as evaluating the performance of a process, a person, plant, or division, these data often get "pencil whipped" and often are distorted to say what someone "wishes" they would say. I have in my notes from a class I took from Dr. Deming and a quote he made regarding data. He said, "There are three things you can do with data. Use it properly, distort it or discard it." I find data often distorted and sometimes discarded. When the management distorts or discards the data, or they either support or accept this behavior, they are hindering their "ability to respond." This is a sign of some level of management irresponsibility.

Regarding skills, I find two major problems. First is the manner in which many managers get promoted. The criterion is too often arbitrary, sometimes based on personality rather than skills and some unskilled, but highly personable managers reach very high levels of impact. Second, and a problem of epidemic proportions, is the lack of management training. Some companies must certainly believe that by publishing someone's promotion in the company newsletter—and that act alone—somehow and miraculously, immediately increases their skill inventory because they do nothing else. In my former life in an international oil company, we used to say the difference between a mechanic and a maintenance supervisor was a tie—all too true and all too common. Expecting people to perform at a higher level without providing the training they need is hindering their "ability to respond." This is a clear sign a company is acting irresponsibly.

Regarding resources, this is most often the first thing a manager wants—more resources and normally he means more people. Most often I find this not to be true. Rather my experience is that the vast majority of skills needed to vault your enterprise into becoming a lean enterprise are already inherent within your organization. Specifically I find they need greater guidance, a better plan, and generally some skill-specific training. But the personnel on staff are generally adequate in intelligence, motivation, and number to effect the cultural transformation. When a company fails to support this precious resource, their people, they are hindering their "ability to respond." This becomes a very personal type of irresponsibility. To not fully use the talents we have on staff is so basic and such a huge loss; many have classified this as the Eighth Waste.

Power much like resources is something managers I have spoken to, would like more of. However, for the majority of the work done, I find that they generally have plenty of power. Regarding power, I routinely see two large problems. First, in many companies there are some strange imbalances in the delegation of authorities. Often in the area of capital, companies are very restrictive, ridiculously so. For example, I have seen several companies that if an outside contractor must be brought in for secondary inspection, the plant manager can readily make a decision worth say $20,000 per month. Yet this same manager cannot, without some higher level of approval, acquire a $2000 PC for one of his staff. This is often only a minor annoyance, delaying the purchase. On the other hand it makes little sense. The bigger and even more common problem is the way power is stolen from a supervisor or manager, by his superiors. More often than not, this power is stripped unconsciously and comes about by poor alignment and focus on objectives placing the manager in a double bind. For example, due to the increasing costs of overtime, an objective comes down from on high to reduce overtime. Unfortunately the high overtime was a response in itself, a planned one no less. You see, earlier that there had been an exorbitant increase in expediting costs and the only tool left to the manager was overtime. And guess what? The expediting was a response to poor on-time delivery performance. Well now the plant manager is powerless. The two tools he has at his disposal have all been removed and yet he still has the poor delivery performance to address. Because this person's upper management is reacting in this way, this plant manager has had his "ability to respond" severely compromised. He has been disempowered and his company has several other responsibility issues to address as well. This type of disempowerment is very common when the planning system is **management by objectives**. It is made worse when the objectives are largely sent down from the top in a heavy handed, one-way fashion without much dialogue. This basic problem creates a misalignment of the "goals" with the "means" and hinders the managers' "ability to respond" properly, making it very difficult for him to actually be responsible.

Standards is a minefield of major proportions. First, there are the legal issues. I find such things as minor forms of discrimination, but the major obstacles in a lean transformation are not legal in nature. But there certainly are issues of morals. The most common one is to try to disguise your workforce reduction effort as a lean transformation. Soon after this hits the shop floor it is shown for what it is and the "advertised lean transformation" is DOA. There are a litany of both major and minor moral transgressions in many companies yet most of them have nothing to do with the lean transformation directly; most likely they were present in the existing culture long before the lean transformation was initiated. The problem, however, is that lean transformations emphasize openness, honesty, and data-driven, dispassionate decision making. This transparency causes all kinds of issues to surface, and moral issues are no exception. The areas of standards that most frequently hinder our "ability to respond" has to do with standards of two types: those that are needed but don't yet exist, and standards that conflict, one with the other.

The first type, when we are missing a standard, creates an issue that makes it difficult to respond. For example, if our goal is to have "stable processes," we need some standard to define "stable." For example, we might then define this as a "process, that when placed on a standard Shewhart Control chart exhibits statistical stability." Without such a clear standard, it is not possible to chart our path to the destination so we simply do not have the "ability to respond" as we try to achieve this new directive. This is both debilitating as we try to make progress and is a clear function of our upper management to resolve.

The second type is a bit more insidious. Let's say we have the problem we discussed earlier of the need to achieve high levels of on-time delivery; and we must do it with no expediting and no overtime. Each of these is achievable in isolation. However, depending on your manufacturing system, achieving all three elements simultaneously may not be achievable today, or ever. Consequently, the very management system that is giving us guidance and evaluating our performance is greatly hindering our "ability to respond." Literally, **the system** is making us irresponsible.

This is a very common problem I see in industry. Not only this "high on-time delivery with no overtime and no expending" **crazy-making triangle**, but the entire concept of conflicting goals. It comes about by some simple logic, **taken to an extreme**. The logic is:

- If something is important, measure it.
- If we want to control it, make it a goal.
- If we then can't achieve our objectives, make it part of the individual performance appraisal system.

This logic is not so bad on the surface: however, it has definite limits. The purpose of goals is to guide behavior (see Chap. 17). That's OK so far. The purpose of putting goals into a plan is to align and focus the organization. Still OK. The problem I see is that this logic will work for one or two or maybe even eight or nine things; we can focus on "some" things. But when we have 40 important things, our focus falls apart because it is impossible to focus on 40 things. In fact, **looking at "everything" is the antithesis of focus**. (See insert on enantiodromia) So management has created an environment where it is impossible to focus on even one thing...and then this very system hinders our "ability to respond."

Countermeasure to Lean Killer No. 10

When managers tolerate a certain level of irresponsibility on their part or on the part of their employees, the problems are often very complex and the countermeasures can be very complicated. However, I have found two countermeasures to be very valuable in mitigating problems in all of the five elements required to be responsible.

The first countermeasure is awareness, that is, awareness and recognition by management that the problems exist. Until the problem is surfaced, it is not possible to address and solve it. However, a problem once recognized is half-solved.

The second countermeasure it to avoid the **Big 5 Responsibility Runarounds** (Big 5 RRs). They are sometimes subtle, sometime not-so-subtle, techniques used to avoid facing all kinds of responsibility. With some of their characteristic traits, they are:

> The great psychologist, C. G. Jung popularized this concept and gave it the old Greek name of enantiodromia. This means that any concept, taken to its extreme, tends to create the opposite of the desired effect. The classic example is freedom. If everyone is really totally free to do what they want, then idiots and bullies would rule the streets and people would be afraid to leave their homes and in this extreme, they would not be free to do what they want.

- *Denial*—When people are in denial, you will hear talk such as "the upset was not really an upset at all"; "it was not really our fault"; or "it could not have been avoided." Inattentional blindness and other phenomena feed our ability to deny the existence of problems—even in the face of overwhelming evidence. Never underestimate the power of the human ego to use denial. This is the most powerful and most pervasive technique of the Big 5:

- *Avoidance*—This is used when you have a known problem that is largely ignored or grossly minimized in importance. Avoidance is the skill employed to completely dodge dealing with the "elephant in the middle of the room" no one wants to acknowledge. Using this technique, problems are simply swept under the rug.

- *Rationalization*—When we rationalize, we only create "rational lies." This occurs when the words used to explain the problem away defy logic. Yet these same words seem to sound good to those who are rationalizing. There is no end to the possibilities with rationalizations; we are only limited by our imagination in the ways we can come up with excuses to "rationally lie" our way out of accepting responsibility. I believe Benjamin Franklin said, "He that is good for making excuses is seldom good for anything else." Such is the case with the rational liar. Frequently those with a little distance from the problem can readily spot the rationalizations.

- *Projection*—This is blaming someone else for your deficiencies. It is always "them other guys." The problems are "out there," never "in here."

- *Idealization*—This is the problem of being so close to a situation and so invested in it that you cannot see any of its flaws or failings—much like the idealizations you might have of your children. Idealization is used to convince ourselves and others that "we could not possibly create this problem. Our systems are just too robust, and after all, this has never happened before."

These techniques are used for some responsibility runaround, some "lack of acceptance" of responsibility. What makes them so dangerous is that they often are unconsciously used. Consequently, they can be used with a high degree of passion and even a high degree of sincerity. But make no mistake about it: Being sincere and being passionate about a topic does not make it correct. When people use any of the Big 5 RRs, they simply are dodging their responsibilities.

The Big 5 RRs are ubiquitous. Even the most dispassionate and analytical of us uses them from time to time, sometimes consciously and sometimes unconsciously. In the hands of someone with no power, these techniques are not too damaging to others. They simply allow people who use them to "save face," sometime just with themselves but other times with their co-workers and superiors.

Unfortunately, when the Big 5 RRs are used by those in power, entire organizations get completely upset and cease to function normally. When the powerful folks (leadership and management) avoid their responsibilities, someone else must fill the insidious responsibility void that is left behind or nothing gets done. Unfortunately this, by definition, places someone with the responsibility—but not the power—to resolve the problem.

Responsibility without attendant power is a perfect prescription for failure. First, there will be a failure of the issue at hand; later there will be long-term failure as it becomes less and less clear who is really in charge. If these responsibility runarounds persist, it becomes less and less clear how people are supposed to behave. The entire organization takes on an uncertainty both in decision making and in action that is always damaging and possibly crippling.

Any acceptance of, or fostering of, the Big 5 RRs has a deleterious effect on any organization's "ability to respond" and promotes corporate irresponsibility.

Chapter Summary

The 10 Lean Killers are management problems of such proportion that failing to address and mitigating them properly will virtually guarantee that your lean effort will fail. Lean Killers 1 and 2 are directly related to the management choice to make the transformation. The other eight were very likely present in your existing management system. All need to be addressed. The countermeasures suggested are a good starting point for your corrective actions.

CHAPTER 3

How to Design a Lean Transformation so Failure Is Guaranteed

Aim of this chapter ... to highlight six major pitfalls that are commonly made during the design phase of a lean transformation. These pitfalls are surprisingly common and each is consequential enough to place the entire success of the transformation in question.

A Way Too Typical Scenario

Let me share with you a scenario that I have seen played out way too often. Someone from the C-Suite like the CEO, makes a visit to a nearby facility that claims it is "Lean by every measure." He is completely wowed by the neatness, with a place for everything and everything in its place. He is amazed by the smooth flow of the product and the smooth flow of both the people and the information they need. He is further impressed with the "visual factory" depicting a high degree of control, resulting in excellent on-time performance with low levels of scrap and rework. Furthermore, he sees a workforce that is producing independently at a pace his facility cannot even approach, yet it is working with both a high degree of comfort and confidence—almost like a stress-free environment. A quick look at some of the production-by-hour boards and other metrics on the floor convinces him they are not only very productive, but they also have high levels of both internal and external quality as well as excellent delivery performance. In addition, the facility is performing at this high level with some surprisingly unimpressive machinery.

And he is sold. Absolutely totally sold on this "Lean thing."

He promptly gathers the rest of the C-Suite, and with conviction and passion declares that "this Lean thing is exactly what we need." With animation like he has never displayed, the CEO relays all the wonderful things he has just seen. Soon he is surrounded by a group of impassioned followers. Quickly they devise a plan by appointing Juan as the lean leader. He is a mid-level manager who is proficient in many of these techniques and has been a vocal advocate of Lean for years. Next, they appoint three others to work with Juan—the lean implementation team—and announce that Juan will unveil the lean implementation plan in 30 days. The team, on schedule, publishes the implementation plan. They want to get everyone involved so their year-one objectives are to implement 5S and standard work across the entire corporation of seven facilities. Juan and his team teach all the facilities and spend a great deal of time traveling to each one. Juan and his team not only introduce the transformation but also teach the tools. They then are required to be the support team and follow up with the various locations as those workers implement the lean tools.

Does this sound good to you? Well . . . don't be fooled. This is the absolutely perfect formula for failure!

So what's wrong? This all sounds so good. We have a jazzed-up top management. We have dedicated expert resources to train and support. We have a published plan. Everyone appears to be on board. Excitement and anticipation are high. Doesn't it sound like success is right around the corner?

Well, the lean answer is "no." The not-so-lean answer is "hell no."

And Failure Is Right Around the Corner

Culture-changing efforts designed like this are failing by the droves. And it makes no difference if it is a lean transformation, reengineering, something smaller like a total productive maintenance push or an effort to implement Six Sigma. They will fail if they are designed in this manner. Again I say, "So what's so bad about this program design?"

Well, it is filled with things that sound and feel good, but just won't work. They sound so good; it is amazing to most that they simply do not work, but they do not. Some are counterintuitive and normally all are countercultural. Of these major roll out failings, there are six, at least. These errors are so large, that if not addressed at the design phase of the roll out, it will guarantee failure, either in part or more likely total failure. The result will be the conclusion that many have erroneously made, "this lean stuff just does not work in our environment"... a rationalization of ignominious proportion. So let's discuss the Big Six.

Roll Out Design Error No. 1—Proceed without a Factual Review of the Situation

All too often these transformations are seen as some "silver bullet" and only a cursory review of the factual situation is performed. The decision to proceed, although deeply felt and sincerely believed, is a passionate one made with great emotion and with great hope. There is absolutely nothing wrong with being passionate, quite frankly in many areas, we need much more of it.

However, it is imperative that a dispassionate evaluation of the corporate competitive position be performed, listing the key issues to address. A solid analysis taking into account both internal and external environmental issues will be necessary. You will need to consider: your competitive advantages and weaknesses, present and future customer expectations, regulatory and legal impacts, competitive positioning, and business trends, to name a few. Then a second critical evaluation should be completed to ensure that a lean transformation is an appropriate response to these competitive needs of the business. Only through this type of analytical approach can a case be made and **sustained** for the changes that will be necessary.

Highly charged emotional decisions often can appear successful in the short run, and this energy may be necessary to catalyze a change. However, in the long run what will be sustained will be the things that make the business ... better, faster, and cheaper. Your lean transformation will test the steel of every man, woman, and child as you roll it out. As you change and ask others to change, you will need solid, business-based facts to guide the transformation. But even more importantly you will need these data to remind the employees, time and again, why the changes must be made. You will be tested repeatedly on your decision to implement Lean. And if all you have is an impassioned initial plea, that will not last very long if the business does not make improvements in the needed areas.

Even if you cannot show how the lean transformation will improve your business, over the short haul, you still may be able to sustain the change just because you, as the CEO, said so. Position power wielded by the CEO is a formidable force and many efforts have lasted for some time on this basis. However, if a good business case does not support your lean transformation, in the fullness of time the change transformation will slowly lose its energy and like a balloon with a slow leak, it will soon collapse.

Rah-Rah Speeches Are Not Enough to Keep the Facility Motivation High

Our mythical but all too real CEO got all "jazzed up" about what he saw, and I am sure he was sincere in his desire to improve his business. Again that "sounds" good, but the rank and file—the folks with their hand on the tiller—needs to know each and every day that what they are doing is necessary. Nothing will catalyze this better than if they can see daily that they are making a difference to what really matters in the business. They need to feel both a sense of accomplishment at work and a sense of urgency to stay focused. The CEO may convey his passion in his periodic speeches, but each and every hour of each and every day the rank and file must be reminded by this sense of urgency to stay focused. With it they can see their contribution and sense their individual importance toward the betterment of the facility. The point is that the motivation of the workers cannot come in fits and starts from the passionate speeches of the leadership. It must be present, with the worker, on the floor, continually reminding and reinforcing his actions. There simply is no substitute for this.

Roll Out Design Error No. 2—Pick a Mid-Level Manager Who Is a Lean Advocate as the Implementation Leader

Every company has a few of these. They are usually intelligent, better read than most, have dabbled in "lean stuff" and found some success on a smaller scale. They can talk the jargon and are usually far more versed in the lean tools than the corporate management. In addition, although they have a reasonable set of "lean skills" they are well versed enough through both reading and experience to know that "they don't know it all" and probably don't know it well enough to create a successful implementation. In short, having actually done some things, they know the power of Lean, the complexity of Lean and either know or strongly suspect that "there is a lot more to this lean thing than I know" and are somewhat apprehensive in their new role. But buoyed by the promise of corporate support, and very likely with a keen eye on success, even if they have concerns … onward they go.

This is OK, as far as it goes but at this point, the corporate management then tells this mid-level manager to "transform this corporation." He simply cannot do that. He is not able to affect a transformation without the requisite position power. The effort **must** be led by the CEO, for this there is no substitute. In any cultural transformation, it must be led by the top of the structure. In this case, the CEO would need to rely heavily on Juan, and his *sensei*, for advice; the CEO simply and definitively lacks the experience base to make many of the key decisions. (See Lean Killer No 1, in Chap. 2.) However, in the end, the decisions need to come from the CEO. People will follow him; they will not necessarily follow Juan. Juan does not have the influence or the power to make thing happen … unless everyone wants to naturally do this … and that is never the case.

By a "line" as contrasted to a "staff" person or group I mean this. The "line" organization starts at the floor worker who has his hands on the parts. These people are the ones who transform the raw materials into products and create the value stream. The "line" organization then proceeds up the structure to the worker's boss, maybe a team leader. Then to the group leader, production manager, plant manager, the division manager, chief operating office, and CEO. That is the "line" organization. The staff people are everyone else. The line organization makes the product. The staff organizations such as IT, maintenance, process and designs engineering, accounting ... are there to support the line organization. In leanspeak, the line organization is the customer of the staff organization. Unfortunately, this reality is lost in a lot of businesses.

In addition as you progress down the power structure from the C-Suite, the head of each corporate department must lead that department. The head of each division must lead that division, the head of each business unit must lead that business unit, the manager of each plant must lead his plant, the manager of each value stream must lead his value stream, and the group leader must lead his group. No staff person, either at this facility or from the home office, can substitute for this. The implementation must be management led by the line organization.

Roll Out Design Error No. 3—Have a Staff-Led Transformation

This error really gets to the point of precisely who is implementing the transformation and who will both learn from and lead the waste elimination efforts. When the staff is involved beyond (1) basic design support and advice, (2) initial training, and (3) specialized subject matter support, untold amounts of damage are done. The transformation absolutely must be line driven. The key change agents must be the top management in the line organization and they alone must lead this culture-changing transformation. The staff can provide support to the leadership, but **the staff cannot lead**. I do not mean they should not lead, although that also is true, I mean they cannot; not if you wish to be successful. At the end of the day, the facility manager is going to do what he believes will be in the best interests of the value stream, and if he needs to choose something that does not directly support the lean transformation at that moment, he will and should do just that.

Line Management Must Be Lean Competent

Hence to have the correct leadership for the lean transformation, line management must know their production system, they must be lean competent. There is no substitute for this, **None...Zip...Nada.** (See Lean Killer No. 1 in Chap. 2.) If they are not lean competent, they cannot lead the lean transformation. They cannot put together a plan: be it a daily plan, weekly plan, or a long-term plan that works. Furthermore, they do not have skills and knowledge to sustain that plan. **Lean leadership is not a spectator sport,** and all lean leaders must be able to "walk the talk." A lean transformation, like any culture changing process, must be led from the top down—through the line organization—with no layers in the organizational structure missed at all. Again, there is no substitute for this.

Don't Be Blinded by the Short-Term Gains of Staff-Led Efforts

Because the staff support organization initially has more lean experience, it may seem advantageous to have them lead and even execute much of the lean conversion. However, when the line organization relies heavily on the staff to execute waste countermeasures, you get three very undesirable effects. First, a great deal of effort is improperly directed as they are not the experts in your value stream and they often come up with inferior solutions. Second, knowledge transfer is missed, and although the staff support groups learn more, the value stream line organization learns markedly less. And third, lean leadership of the plant is completely lost to the staff functions and the line management has not gained from the effort. Therefore, the benefits may look good in the short term, but in the long term it guarantees failure. Once the staff people leave, they take with them the skills, knowledge, and leadership gained from the process improvement efforts. The staff team leaves and the improvement at the plant will predictably stop. The plant can make no more progress as the skill base experience and lean leadership have left; hence they are stuck. It is an effort marked by fits and starts. With a staff-led effort, continuous improvement is no longer continuous, because progress is made only when the staff group makes their periodic visits. Worse yet, the earlier gains will wither in the fullness of time as entropy takes its course and it will all look like a long-term failure. In the long run, no one is happy and progress is unlikely. This is a losing strategy.

Lean Is Not a New "Thing"...It Is a "New Way"

It must be recognized that **"Lean is not a 'new thing' but rather a new way to do the 'things' we need to do anyway."** It goes well beyond just the "tools" of Lean. It requires a new style of management, a new style of leadership, and not the least of which is a new style of dealing with all your people. It is a total cultural transformation. And those things go far beyond what a group of specialists can show you on some occasional visits. The skills of running the organization must be embellished by the "lean tools," and the staff organization can facilitate this by training. As they train and teach you how to apply all the lean tools as you problem-solve your way to the Ideal State, you can become lean competent. The actions of the staff support organization should be a manifestation of the overused but still valuable phrase "Give a man a fish, and you feed him for a day; show him how to catch fish, and you feed him for a lifetime."

> **P**oint of Clarity   Lean is not a "new thing" but rather a new way to do the "things" we need to do anyway.

It is imperative that you begin the transition of weaving the lean transformation into the everyday activities needed in the value stream. It must become "just how we do things around here," it must become part of your culture, and this can only occur if it is supported every day by everyone in everything we do. Only the line management can make that happen.

> **P**oint of Clarity   Lean activities must become, "just how we do things around here," to be successful in the long term.

Roll Out Design Error No. 4—Believe That the "Tools of Lean" Are the Very Heart of a Lean Implementation

Many times people talk about the "tools" of Lean Manufacturing, citing such countermeasures as standard work, SMED, *kaizen*, *heijunka*, 5S, and value stream mapping, to name a few (all described in Chap. 12). Then an effort is made to integrate these tools into the business culture like Juan and his team tried. This, too, "sounds" very logical. The people are then taught the theory and techniques on "what" these tools are, but all too often are left to their own inexperience on "how" to apply these tools.

In effect the implementation team is saying, "Here is a tool, now go apply it." It is very important to learn that all the "tools" of Lean are countermeasures (solutions) designed to mitigate some type of waste. So in leanspeak, when we use this "tools" approach, we are effectively saying, "Here is a solution, now go find a problem to use it on." If you had an auto shop and were training a mechanic, would you hand him a box end wrench and tell him to now go find a problem he can fix with this tool? As strange as that may sound, that is all too often the approach used in many lean transformations.

To the contrary, to properly root out waste and improve on a daily basis, we must challenge ourselves to develop:

- An understanding of the present state
- An understanding of the desired future state
- An explanation of what is preventing us from reaching the future state
- A plan for the next steps, the countermeasures we must take to achieve the desired state

This questioning approach then leads to a selection of countermeasures that are used. So, in the end, tools are selected. But they are selected based on the needs of the facility, not some arbitrary corporate-level selection process. When the problem-solving approach is used, which is the correct lean approach, tools are "pulled" based on the needs of the facility rather than "pushed" to those on the floor so they can be force-fit into some situation with the expectation that they are expected to be utilized.

> **P**oint of Clarity   The "Means to Lean" is problem solving to reach the Ideal State.

Roll Out Design Error No. 5—Plan the Implementation across Multiple Value Streams or Worse Yet, across Multiple Plants

To install a Lean Manufacturing system for all value streams in a large complex plant simultaneously is very difficult. Although this "clean sweep" approach is surprisingly popular, it is a grossly inferior technique to one-value-stream-at-a-time implementation. Like the other errors, this is a built-in failure mechanism.

There are at least four very large reasons to implement one value stream at a time.

- The initial learning curve is very steep, and no matter how well prepared you are, there are many unforeseen results. Sometimes they are positive results, other times negative, but it is best to experience these issues on a smaller scale and learn from them. This learning from the first line is invaluable in executing the conversion on the subsequent lines.
- There is the matter of resources, specifically trained eyes and ears to evaluate implementation, evaluate progress, and solve problems on the value streams being converted. No one has a plethora of such people and they are invaluable. These few people can move from line to line with the implementation, and

provide expert advice and training to the leadership team so the gains may be sustained.

- Each change has an effect on subsequent and past changes. It is not uncommon to lean out line A and then move on to line B for its implementation, and then find some aspect of line B implementation that has an unexpected effect on line A. Sometimes the effect is a positive one; other times it is detrimental, but either way this approach allows you to better understand these effects so they may be exploited more fully on all lines.

- I have seen a number of value streams by value-stream efforts progress quite nicely, but all the global implementation efforts I have seen have failed to meet their objectives. Some have "pockets of success," but none have worked globally. A very common postmortem comment, or mid course correction, on a global implementation effort is: "In hindsight, we should have done this one value stream at a time."

To attempt an implementation across multiple values streams is not recommended. Even worse is to attempt to implement in several plants simultaneously. This is a disaster just waiting to happen. Not only do you have the issue of resources and learning as mentioned above but now you have the issue of correcting mistakes. When a global mistake is made across several plants, now there is not one item to correct, but there are several. It is far better to learn from an Alpha Site and then apply your knowledge and experience. I know of no consultant that recommends anything other than "start small, learn, and then expand on that" when it comes to making a lean transformation.

However many companies, usually under the guise of consistency, try to proceed at multiple sites simultaneously. It is because the management totally underestimates both the power of Lean and the huge cultural transformation that occurs during a lean transformation. I can say that those who make this mistake do so on their first effort. Given a second chance, they never make this error again. But like my grandmother used to tell me in her thick Italian accent, "Lonnie, we get too soon old and too late smart."

Roll Out Design Error No. 6—Have No Plan or a Lousy Plan

In our not-so-fictional case, Juan and his team unveiled a plan. That is a step many others miss. However, without looking at it, I can tell you that Juan's plan will not transform his facility. It will not work because it is not the plan for the plant, nor is it the plan for the value stream. It is a plan to implement Lean like it was a new software program. Lean is no such thing; rather it is a new way to do business. It involves not only a plethora of new technical tools, but it also involves a great deal of behavioral change; in a phrase it is a huge cultural change.

In Juan's plan, they are using a "tools" approach and are going to teach one very popular leading tool, 5S; unfortunately they see 5S as just an organizing and neatness thing. They have missed the point that 5S is a countermeasure to a problem. However, upon careful scrutiny we find that of the seven plants in Juan's corporation, three of them are so neat and clean that you could eat off the floor, two have clear opportunities, and two other could best be described as pig sties. So for the three that don't really need the neatness and cleanup effort (i.e., their version of 5S), it will be some extra work. And in their effort toward corporate consistency (they also mislabel that as "standardization") they will also need to train all their folks in the same version of 5S and implement a corporate-created layered management audit system ... all to solve a problem they don't even have. Sound strange? It surely is. Unfortunately it is also, way too often, true.

What Is the Root Cause of These Six Roll Out Design Errors?

I am not sure I can definitively answer that question, although it certainly is a question worth answering. To me it is a bit like the post mortem that goes on after the Super Bowl when the losing team does an analysis of what they need to do to succeed next year or better yet what the winning team must do to repeat. After the game the answers seem very clear to many, especially the talking-head pundits. Unfortunately, the answers given, often with passion and seeming clarity, must not be too good as all too often the Super Bowl winner does not even make the playoffs the next year.

Although I cannot definitively answer this question, I am equally sure that I can give you some advice which will greatly improve your chances of future success in lean transformations.

The first and most common cause I find is that almost everyone, I can think of no exceptions, **underestimates both the power of Lean and the impact it will have when properly executed**. Only through the use and repeated application of the lean philosophy, lean strategies, and the lean tactics is an in-depth understanding achieved. Lean is seemingly simple and quite frankly appears to be like many things we do anyway. Take planning, for example. I cannot name a client that does not have a strategic plan that is deployed down to the plant level. So when we introduce *Hoshin–Kanri* (HK) planning (see Chap. 16) to them, they usually reply that they are doing all of that already. That may be, but in the next breath they will likely tell you they are well behind on their goals and objectives and cannot really understand why. Compared to the typical Management by Objectives methods of planning, I cannot give you even one example of a firm that gets the type and level of "buy in," of true engagement, that is achieved through HK methods when catchball is properly deployed. At the same time, firms using HK planning usually do an excellent job of on-time goal attainment, and they can explain with specificity why they are successful.

A second core problem is that they are often **very defensive or otherwise in denial** (see Lean Killer No. 10 in Chap. 2) about the gap between a truly Lean Manufacturing system and their current system. They simply will not see or will not want to see the differences involved in the two processes. These differences in planning may seem simple and superficial, yet they are anything but simple nor are they superficial. In fact, they go to the very heart of what motivates us and are extremely consequential in the long run. This is but one example. As you read the case studies, you will find this issue of denial repeated often.

Therefore, a lack of appreciation for the magnitude of the differences between a classic and a Lean Manufacturing system, coupled with denial in all its forms, is as close as I can come to a root cause. I know that may not be a very satisfying answer to many, but I am hopeful it will cause some **to pause, reflect, and learn more** ... as they embark on their lean transformation.

Chapter Summary

The Six Roll Out Errors are major pitfalls that can seriously endanger the success of a lean transformation. They include proceeding without a sound business case, not leading the transformation from the CEO, using the staff rather than the line organization to drive the transformation, using a "tools" rather than a problem-solving approach, making the initial transformation at multiple value streams or locations, simultaneously, and doing so without a specific plan. These pitfalls are an outcome of poor understanding of Lean by the top management along with poor awareness of the deficiencies in their current manufacturing system. If each of these Six Roll Out Errors is not properly addressed, very likely this will yield an unsuccessful transformation.

PART II

The Issue of Culture and the Five Cultural Change Leading Indicators

Part Two focuses on the absolutely critical cultural change that is needed to make a true lean transformation and sustain any gains achieved. There are six chapters. Chapter 4 covers cultural basics, including a new "prescription on how to change your culture." Chapters 5 to 9 address each of the five key cultural change leading indicators to success which are leadership, motivation, problem solving, total facility engagement, and learning/teaching/experimenting.

CHAPTER 4

Cultures

Aim of this chapter ... to cover two major topics on cultures. First, we will offer a brief introduction into cultures—business cultures in particular. We will just skim the surface on the topic of cultures, but our objective is to give you enough information so that the complexity and depth of the Toyota culture can be appreciated, as well as the depth of effort needed to truly transform yours into a lean culture. Second, we will introduce the five cultural change leading indicators.

The Importance of Culture

This chapter was included because the defining aspects of the Toyota Production System (TPS) are its cultural elements. As an extension of that thought, some of the truly unique aspects of the TPS were consciously and painstakingly developed over an extended period of time; hence, they are not easily imitated by others. More importantly, no real transformation occurs until the culture has changed. Until that time, changes are highly transitory and have no sustaining power; they lack "stickability." That is, things will change—but they just don't stay changed. It is hard to overstate this issue. There are literally thousands of references in the literature of businesses that have been transformed and improved .... quite frankly that is not too difficult. The really hard part is to sustain those changes. To sustain the improvements you must effect a cultural change and that is not only hard, it is a frequently misunderstood and often a neglected aspect of sustaining the gains you have made in your transformation. To express the importance that we at Quality Consultants place on understanding and managing your culture, our mantra is "Culture Trumps Everything."

> **"...M**anagement must awaken to the challenge, must learn their responsibilities, and take on Leadership for Change... .**"**
> W. Edwards Deming

What Is a Culture?

We define a culture as "the combined actions, thoughts, values, beliefs, artifacts, and language of any group of people." It could be the culture of the Catholic Church, the culture of AARP, the culture of the South, the culture of Toyota, the culture of the New York Yankees, the culture of your plant, or any group of people. The people within these groups think, talk, and behave within predictable patterns of behavior. These thoughts, language, and behaviors then identify them to be a member of that culture. Often, to clarify their inclusivity in that culture, they will have specific artifacts that help identify them as part of a culture. These artifacts may include such things as symbolic clothing, jewelry, and even uniforms.

However, simply put, a culture is "just how we do things around here."

For example, when I started as an engineer in the oil industry in 1970, all engineers wore a dress shirt and tie, at a minimum. Generally, even first-line supervisors wore a suit or at least a sport coat. I worked in Southern California, the heart of the "take-it-easy-and-let's-go-to-the-beach" culture, yet the dress was very formal. When we asked why, they would say, "That's just how we do it around here." A second example at my employer was that there was no formal program to indoctrinate engineers. For the most part, we were given work and expected to find out what we needed to do to perform. A middle manager once said, "Our engineer training program is like asking the engineers to put on ankle weights, throwing them out into the middle of a pond, and telling them to swim to shore. Along the way we lose a lot, but the ones who make it to shore are real strong." That too was an aspect of their culture—that is, engineer training was not very highly valued, yet we were still expected to perform. A less obvious, although equally strong, aspect of that culture was the intolerance for failures. Failures were not easily forgotten. If an engineer completed 10 projects, for example, and 9 were very successful, with one having some problems, he was often chastised for the failure and it was seldom forgotten. Should another engineer complete maybe only three projects and all three were successful, frequently he would be viewed as a superior engineer. It was the "It only takes one ah-shit to cancel 100 attaboys" aspect of that culture. It was joked about, at least behind closed doors. But in the end it caused engineers to hesitate before taking even the slightest chances. Consequently, qualities such as imagination, creativity, and innovation were repressed. As you might expect, when these qualities became repressed, other qualities would rise as being important. In this culture, company politics became a dominant quality that helped in salary increases and promotions.

These qualities—dress, training, and development—are all important cultural aspects. They need to be understood and managed just as costs, profits, and customer satisfaction are understood and managed.

Why Is It Important to Understand Culture?

There are at least five reasons that make it worthwhile to consciously create and sustain a healthy culture. First, goal alignment is much easier. Second, workers are more highly motivated in a predictable culture and will not only take action with more confidence, they will also be more likely to initiate new methods. Third, a culture can provide some needed structure and avoid the bureaucracy that dampens enthusiasm, innovation, and actions in general. Fourth, until the changes made as part of the lean transformation become part of your culture, the gains will be not be sustained over time. Your lean transformation will lack "stickability." Fifth, even if you make no conscious effort to create a culture, one will "occur" spontaneously. This **organic** culture may either help or hurt your business. However, it is much better to intentionally design and create the culture you need, rather than having it develop organically as a result of the lean transformation.

How Are Cultures Developed?

There are three major contributors to the culture of any business: the external business environment, the few key managers at the top, and the history of the business.

Most important is the business environment. This is the single largest driving factor. The area of the country you work in, the labor market, the products you make, the degree of government control, and a litany of other external factors all impact your culture. If you are making food products, the FDA will play a major role in how you run your business. If you are tier-1 auto supplier, your sense of urgency is much greater than if you are a tier 3. If you produce a medical product, your business will need very specific quality behaviors, dictated by government regulations to survive and prosper. The business environment is the number 1 factor in culture creation. Second is the few powerful people at the top. It is their actions, thoughts, values, beliefs, artifacts, and language that demonstrate to the rest of the organization what is desired and what is acceptable. They set the cultural rules whether they appreciate it or not. Some are set consciously, yet others are set unconsciously; that is, the top managers are not really aware of the large cultural impact they are making. Unfortunately most cultures are developed unconsciously. They are developed **organically**. By this I mean the external environment dictates almost totally what is done in the business and everyone in the business simply responds. These are often what I call "survival cultures." Usually the dominant motivation that exists inside this type of culture is "pain avoidance." No one really has a sense of direction they just do what the previous guy did and hope they don't get into trouble. That leads us to the third element that determines the culture—the history of the organization. In the absence of a conscious plan to design and manage the culture that is what perpetuates the culture, its history. People simply look around, see what other people do, evaluate the consequences of their actions, and respond accordingly. This is more often than not, the norm in culture perpetuation. Finally all healthy cultures are developed consciously. They are not a random happening, they are designed. The external business environment is evaluated and a culture is selected, designed, and created that allows your organization to survive and prosper. This culture is largely created when the key top managers define, explain, teach, support, reward, and model the values and beliefs needed for the business to survive and prosper.

Regardless of whether we are conscious of how we develop the culture or not; we define that certain people will do certain things in certain ways. As a manager we try to explain the behaviors we want to see from everyone. For example, we have schedules, work rules, and business practices. It is this behavior that then begins to create the culture. Take a business—a restaurant, for example. In our restaurant, we are in business to make money. We decide we want to cater to upscale patrons with the upscale prices they are willing to pay. However, we must also develop a group of chefs and waiters who are compatible with our upscale business. Let's discuss the waiters, for example. In our upscale restaurant, we will need waiters with significant social skills, like the ability to carry on a conversation—something we would not need, nor even want, if our restaurant was a fast-food business. At our upscale restaurant, we might even want them to get some training in handling customers, or we ourselves might even give it to them. As soon as we start to define the behaviors we want, we start to define our culture. Now, as we proceed to develop our business, we further define the behaviors—the skills—we need. This then goes further toward developing the culture we will have. The more aware we are of the behaviors we seek and the impacts of these behaviors, the more we understand our culture. Hence, the culture is developed based on the required action of the personnel, which is dictated by the needs of the business. Pretty simple, huh?

How Are the Cultural Rules Set?

Cultures are created and sustained by way of three major factors: the business environment you are working in, the history of the group, and the actions of the few top people. These few top people are the ones who set the cultural rules. They are the culture creators. Whether the rules are stated or silent, they are made and enforced by the top few people. Most cultural rules are not stated, and when we are unaware of them, this makes them potentially very dangerous.

The Most Common Rules: The Silent Rules   This seemingly odd aspect of cultures, the silent rules, creates a major cultural problem. It is often the reason why those within the culture, especially the rule makers, simply do not see what is happening within their own culture. For example, in the culture of my early engineering job, it was not unusual for some high-level manager to comment and even criticize the organization for being so "close to the vest"—for not being willing to step out and be innovative. He might say something like, "Why is it that the new ideas always come from our competition?" or "Where is the entrepreneurial spirit we so often talk about?" or maybe, "Where is the rugged individualism we Americans are so proud of?". In fact, he might say this just minutes after having reprimanded someone for some mistake that was made. Once you become aware of what is happening within your culture, some of these comments are almost laughable—that is, if they were not so tragic.

> "In studying the history of the human mind, one is impressed again and again by the fact that the growth of the mind is the widening of the range of consciousness, and that each step forward has been a most painful and laborious achievement... . Ask those who have tried to introduce a new idea!"
>
> Carl Jung

The silent rules would not be so damaging, except (more often than not) we are not consciously aware of them. **And if we are not conscious of them, they control us; we do not control them**. This is not only dangerous; it leads us into all kinds of aberrations.

Healthy Cultures

Cultures can be either healthy or unhealthy. A healthy culture is one which fully supports the long and short-term needs of the business. It is the combined actions, thoughts, values, beliefs, artifacts, and language that are needed so the business can survive and prosper. For a business culture to be "healthy," it must have three basic qualities. It must be strong, flexible and appropriate.

Cultural Strength

By strong we mean that all the people embrace the same thoughts, beliefs, and values. The measure of a strong culture is that given a set of circumstances, you can predict with relative certainty that people would react similarly. To be strong, a culture must have the two characteristics:

1. Its thoughts, beliefs, values, and actions must be widely accepted, acknowledged, and practiced across all levels and functions of the group. The thoughts, values, and beliefs of sales must match those of engineering and production, for example.

2. Its thoughts, beliefs, values, and actions must be in harmony with one another. There needs to be a logical consistency. For example, for a business that needs a highly structured environment, like that necessary for firefighters, it is not possible to have an "everyone can do it his own way" attitude and still expect things to get done. Those behaviors are not consistent.

Simply put, to have a strong culture, the thoughts, beliefs, values, and actions must be consistent with each other, as well as vertically and horizontally integrated throughout the group.

It is a certainty that weak cultures will lead to weak performance since everyone will be doing their own thing. There will be lousy alignment and likely very little focus. It was once believed that strong cultures were, by definition, healthy cultures. Unfortunately there are many examples of strong cultures causing businesses to fail...precisely because they had strong culture. While strong cultures are often very comfortable for those within the culture, if they are successful businesses over a long period, they can become very arrogant, with its attendant problems of a strong inward focus, becoming both very political and very bureaucratic. Unfortunately these qualities then lead to an inability to recognize and respond to changes in their business environment (see Lean Killers Nos. 9 and 10 in Chap. 2). This inability to recognize the changing external environment can cause a severe drop in business performance. Once very strong firms such as Sears, J. C. Penny, K Mart, and General Motors are examples.

Cultural Flexibility

The second quality of a healthy culture is that it must be flexible. Now on the surface this sounds contradictory to being strong. But strong alone, as described previously, can be damaging in a changing environment. Those firms that are flexible have a very strong attachment to their basic values and beliefs but are very flexible on the strategies and tactics to achieve those values. They have the ability to retain their basic identify while adapting to the changing business environment.

Appropriateness of the Culture

The third quality of a healthy culture is that it must be appropriate for the needs of the employees, the stockholders, and of the customers. For example, a culture that is appropriate for a professional football team would not be appropriate for a business such as Disney, and may not be appropriate for your manufacturing facility. This concept of an appropriate culture is missed by most. Often, a culture that has demonstrated success is one that many will want to copy. Take the successful football team that demonstrates their competence by winning the Super Bowl, for example. The coach is often next seen on the motivational speaker's circuit explaining the "Road to Success" or some such thing. Altogether too often, many business managers flock to these meetings trying to get the most recent success formula, believing fully that if they could embody the principles of the football team, they too could be successful. Well, it just doesn't work that way.

Appropriate Cultures: An Example A key cultural characteristic is leadership style. This characteristic then dictates a whole set of behaviors by both the leaders and the followers within the culture. Now back to our football metaphor. For example, on game day, the quarterback is the offensive leader. It is his responsibility to call the plays and audibles. In his role as the "on-field leader," he will perform based on his judgment alone. In so doing, he does not consult with those affected. His actions are very autocratic. In fact, he acts in a very dictatorial fashion.

In Super Bowl I, Bart Starr was the quarterback of the winning Green Bay Packers, and Forrest Gregg was his all-pro offensive tackle. While he was in the huddle, calling a play, do you think he might have bent over to Forrest and said something like:

> "Well Forrest, I know you have blocked that big guy across from you for some time now, and I know you could use some rest. I really appreciate your efforts and am pleased to play with you. But would it be all right if we ran over your position just one more time? We really do need the yardage to get the first down, and I firmly believe this will be the best thing for the team. But before we do anything that might affect you, we wanted to solicit your opinion. So do you think you could support that and knock that big guy on his butt—just one more time?"

Well, for a quarterback to say such a thing in a football game would be patently ridiculous. The quarterback, the team leader on the field, is a dictator—there is no better word to describe his style. He is not seeking input or agreement, nor is he trying to create relationships. He tells people exactly what to do, when to do it, and how to do it, and if they do not do it, he has them replaced—they have no options. Worker freedom, on this occasion, is nonexistent. The quarterback is under extreme time pressure and *his necessary style of leadership* is dictatorial. Yet no one finds this odd. It is the way it has to be for that business. It is *appropriate* for that business, at that time.

So when the plant manager looks for the answers to his cultural problems in the football team metaphor, he is often looking within the wrong type of a culture. What is appropriate for a football team is not necessarily appropriate for a manufacturing plant.

How Do We Design a Cultural Change When One Is Needed?

In most cases, when some top manager decides his culture needs to be changed, he is aware of some weaknesses that he would like to see corrected. It could be something huge, such as changing from a system of "firefighting" to a real-time problem solving. Or something really emotionally charged such as changing your system of rewards. It might even be something as mechanical as learning how to manage meetings better. Maybe he thinks he needs to modify some aspect of his culture such as his system of recognition and rewards or his change mechanism. Or maybe he wants to make a full cultural transformation and create a culture like that of Toyota, for instance. In either case, he would follow the pattern of the Five Questions of Continuous Improvement (detailed in Chap. 5). First, he would need to assess the present situation. Hence he would need to perform a cultural evaluation. This is a tool that allows an evaluation of the present situation of the culture—how it is structured, right now. For a large change we do this using a variety of techniques, including a formal written survey, individual and group interviews, and observations of the culture in action. These data are compiled and compared. This will give a good picture of what the culture is like now. Smaller issues can be handled less formally.

Next, he would need to decide what is the "desired" culture he wishes to have. For a large project, this can be done in a facilitated workshop, and will determine what type and structure of a culture will be needed in the future. Likewise with smaller projects this can also be done less formally. With these data in hand, it is now possible to create a gap analysis and ask questions 3 and 4 so the obstacles can be defined and the "next steps" also defined.

Following the cultural evaluations of the present state and the possible future state, improvement and growth then focuses on defining and solving problems (closing the gaps found) in three major areas of opportunity. These are as follows:

- Are the thoughts, beliefs, values, and actions appropriate for the business and meeting the needs of the customers? (Appropriateness of the Culture)
- Are the thoughts, beliefs, values, and actions in harmony, one with the other? (Harmony of the Culture—one part of cultural strength)
- Are the thoughts, beliefs, values, and actions disintegrated either vertically or laterally in the facility? (Integration of the Culture—the second part of cultural strength)
- Does our culture have a change mechanism, with adequate support to be flexible enough to accommodate the changes in the external and internal environment? (Flexibility of the Culture)

This becomes a project and is managed like most projects with one huge exception. The desired cultural changes must be prioritized and the largest cultural changes must be done first. Recall that every cultural change will affect all other aspects of the culture; so once the first major change is accomplished, a minor reevaluation is in order. It sounds complicated, but in the hands of a cultural expert, such as your *sensei*, many of the interdependent changes are predictable.

How Do We Then Modify the Culture?

At this point, it is worth discussing how we might proceed to make the changes. Let's say we want to implement a full-blown lean transformation. There are two schools of thought on how to go about changing the culture. They center on the question, "Should we change a man's attitude so his behavior will change (first point), or should we change their behavior and expect their attitude to follow (second point)?"

I have seen much of this "rah-rah" positive mental approach stuff (the first point), which in the end fails. It fails because it is not supported by the needed actions, especially the actions of management. Consequently, it dies of its own weight and does not sustain itself.

If you want sustained change, you must change their behavior, and their attitude will follow. This works because it is based on action, not words, which yields results, and that in turn feeds all the intrinsic motivators (see Chap. 6) especially the sense of control and the sense of accomplishment.

So how do we proceed to change the culture?

With leadership, of course.

The Change Prescription

1. The facility leadership must announce and "sell" the change. Utilize the Leadership for Change paradigm (see Chap. 5) and clearly explain, in a formal way, the three aspects of change. Specifically ask and answer:

   A. What is wrong with what we are doing?
   B. What do we intend to change to?
   C. How do we intend to go about executing that change?

2. Teach the change.

3. Model the needed behaviors.

4. Support the needed behaviors.

5. Reward the behaviors.

To make this cultural modification, we need strong leadership. The zeroth step is to put together a plan that is action based. Do this with a lot of help; remember most of the culture is unconscious to those within the culture, so it is almost guaranteed that to properly modify your culture, help is needed from someone who has not been "contaminated" by your culture. On this issue, clearly seek the advice of your *sensei*. Once a plan has been developed, the first action of the leadership must be to advertise and sell the plan by explaining the three aspects of Leadership for Change (Chap. 5). It is important that this be explained well. It must be both understood and accepted before proceeding to the action phase. Take the time to let it sink in, take the time to allow those who need to change to ask questions, and get concerns addressed. Take the time for everyone to "reflect" on the proposed changes. Second, perform the teaching that is required to affect the necessary behavioral changes. Make sure through validation that the training was effective. Third, as a leader, you must model the correct behaviors so others will know you are serious. If your cultural change involves 5S for instance, the first thing you should do is 5S your area and make sure your direct reports do the same. Fourth, you must have in place a support system to assist in the cultural change. Finally, you need to have a system of recognition and rewards that is supportive of the cultural change you wish to see.

The Toyota Production System and Its Culture

After having studied the culture of Toyota, and having worked with many Toyota facilities and suppliers, I have come to not only appreciate but quite frankly admire the Toyota culture. It is the healthiest culture of any business I have ever studied. It is a culture that is strong, flexible, and appropriate.

Their culture is absolutely appropriate for their business. First, while using their culture as a strong tool to improve their business, they have grown from a small manufacturing firm making a few thousand small trucks per year to a manufacturing giant. In today's environment, while Chrysler and General Motors have needed government subsidies to survive, Toyota continues to prosper. All of this is largely due to the culture within Toyota.

The culture is strong. The beliefs, thoughts, and actions are consistent throughout the business. These attributes are vertically integrated as well as laterally integrated throughout the entire Toyota business, suppliers included. They are so consistent that it is almost boring. Furthermore, it is such a strong culture that it has developed a language of its own with new words like *kanban* and "autonomation," and old words like "leveling" that have an entirely new meaning.

Yet the culture is flexible enough to even go from one country to another and yet retain its basic values and business position.

A Culture of Management Responsibility

A facet of Toyota that has always stood out to me was their unrelenting respect for people. Even more so is their thorough acceptance of responsibility—especially the responsibilities of management. Think about their policy of no layoffs. This has been their policy since before the days of Kiichiro Toyoda, who resigned as president in disgrace in 1948. His responsibility was manifest when he resigned because forced layoffs were required to avoid bankruptcy—all to save the company. Toyota has maintained this policy of no layoffs, even until today. It is clearly a part of their culture. A part that is neither well understood nor appreciated by Western managers.

Today, a few companies have agreements like this with employees. These companies are not the norm, but some do exist. However, I can provide no examples where, when the company suffered financially, the President stood up, took responsibility for the problems, and resigned. Quite the opposite is the norm. The stories are legion of those CEOs whose companies failed, but they dropped out to safe landings with their golden parachutes—but back to Mr. Toyoda for a second. Can you imagine what impact it has—on the culture, on the entire organization, and on the managers and workers alike—to know that this level of responsibility is not only expected of all, but is also practiced by the highest echelon of management? Somewhere on and beyond huge is my observation.

What Toyota is saying through their actions is what any responsible company would say, that is, "If the company fails, it is because management has failed." Across the business world, unfortunately, what is seen differs greatly from what Toyota says. Rather, almost without exception, management will take credit for the successes, but not for the failures. As soon as a failure appears, it is amazing how management will work to find both circumstances and "others" to blame. Yet at the first blush of success, they will stand in line to not only collect the accolades but the bonuses as well.

What kind of logic supports that? It is said in a folksy way that, "Success has many fathers, but failure is an orphan."

Well, some management teams will take responsibility for their failures … or sort of. They will talk about it and say nice things like, "We take full responsibility" or "the buck stops here," or some nice-sounding phrase—and they might even feel bad about it. But that is not the measure of what true responsibility is really about—it is about RESPONSE-ability (see Chap. 2, Lean Killer No. 10). Responsibility is simply the ability to respond. So, just how do they respond? Clearly Kiichiro Toyoda not only felt bad, not only did he say he took responsibility, but his actions were unmistakably supportive of what he said. This man, and hence this company, "walks the talk." He set an example for what is expected of each Toyota Manager, what is expected of each Toyota Supervisor, and what is expected of each Toyota employee. Consequently, they have a culture that expects responsibility, and they work hard to maintain that. They are responsible.

A Culture of Worker Responsibility: *Jidoka* and Line Shutdowns

The second example, also focusing on responsibility, but closer to the shop floor, is the *jidoka* principle. It is the concept that they place an unwavering emphasis on never passing anything but the highest-quality product, and this is supported by a culture of empowered problem solvers. In a Toyota facility, upon finding a defect, the line may be shut down and not started until the problem has been resolved. This line stoppage is done by the worker or anyone who finds a defect. It is not only their right to do this, it is their responsibility. Toyota considers this a good, if not mandatory, business practice and they have given it a name, it is called "pulling the cord."

Most American plants would consider this like turning the asylum over to the inmates. At the typical American plant, this idea is close to the truth. Not because the workers are insane, but often because they do not have what I call "the context of the problem." Consequently, when they find a defect, they cannot make a good business decision to shut down or keep producing. They do not understand enough about the problems and the consequences of shutting down the line, or the consequences of *not* shutting down the line.

What does this say for the Toyota culture? Let me just mention three aspects:

- It shows a great respect for the decision-making ability of everyone. What this means culturally, is that they demand—and expect—respect from their people. It is a culture of individual respect.

- They have confidence in the training of these people. They not only train their people but expect them to use this training. It is a culture that values knowledge and training.

- It shows, beyond a doubt, that the worker is fully capable of producing 100 percent good product. It also says that if we cannot produce good product, we will produce no product. **It is the ultimate statement of the importance of quality**.

Again, how does your manufacturing system and your management measure up to this?

The first concern I hear from many managers is that this concept will not work in their facility. Well, maybe in a continuously operating facility such as a petroleum refinery or a chemical plant, shutdowns by "pulling the cord" are not practical. However, I ask these managers to look deeper into the concept behind this "pulling the cord" behavior. Look at the basics. Can they create an environment where only high quality product is sent forward and do so using empowered problem solvers to respond to problems in real time? Every facility can create this in their culture—even if they cannot "pull the cord" like Toyota. Hence every organization, if it chooses to, can create a culture of *jidoka*.

> **P**oint of Clarity *Jidoka* is the concept that only high-quality product advances in the process, and the process is supported by empowered problem solvers that can respond in real time.

The Five Cultural Change Leading Indicators

There are many cultural characteristics to measure and evaluate as we proceed through any cultural change. Our Cultural Matrix (not included here) has 14. Some are lagging indicators such as achieved quality, whereas others are leading indicators. Of these, there are five leading indicators that are so critical to success; if you fail to successfully implement any one of these, your lean transformation will fail by some amount and possible totally fail. Stated in question format, they are as follows:

- Do we have the leadership to make this a success?
- Do we have the motivation to make this a success?
- Do we have the necessary problem solvers in place to make this a success?
- Do we have the full engagement of the entire workforce?
- Do we have a learning, teaching, and experimenting work environment?

Although I found these five leading indicators empirically, the logic as to their importance is equally compelling. First, if you do not have great leadership and a significant "leadership footprint," you simply will not be able to set the direction and make the changes necessary. Leadership is all about guiding change. Next, you need the initiative to make the changes. If you do not have a properly motivated workforce with the supervisory skills to not demotivate the people, failure is assured. And next is the question of "change to what?" and "how do we change?" which begs the question of problem solving. If you do not have the problem solvers and the problem-solving mentality, there is no way to actualize the "Means to Lean," which is problem solving our way to the Ideal State. Then the question is "who is to initiate these changes and the answer becomes 'everyone.' Therefore, if you do not have full engagement of the workforce, led by the total engagement of the top management, then failure is knocking at the door. And finally if you do not have a learning/teaching/experimenting mentality, there is no hope to implement the scientific method and hypothesis testing, which is at the very heart of problem solving and the means to perpetuate the improvements. Each of these is a cultural element of such magnitude that these indicators must be understood and managed well to be successful. And each is a leading indicator of the cultural change that must occur if you wish to sustain the gains achieved. Each of these five key cultural change leading indicators is more fully discussed in its own chapter.

> **P**oint of Clarity   The "Means to Lean" is to problem solve your way to the Ideal State.

An Example Application of the Five Cultural Change Leading Indicators

These five leading indicators to success are described well in Chap. 20, The Story of the Alpha Line. This story illustrates in graphic detail how these issues were so wonderfully managed. As a result of the proper handling of these five issues, the plant succeeded, grew, and prospered beyond all expectations. On the other hand, Chap. 23, The Precursors to Lean—Not Handled Well, showcases an example of what happens when the five leading indicators of cultural change are not managed well. I suggest you skip ahead and read these descriptive stories before continuing with this chapter.

> **"...M**anagement must awaken to the challenge, must learn their responsibilities, and take on Leadership for Change... **"**
>
> W. Edwards Deming

Some Cultural Aspects of a Lean Implementation Worthy of Further Thought

A lean implementation has several unique aspects. None of the items listed next are unique in concept, but all are unique in the intensity in which each subject must be approached when compared to a typical project. These three aspects are as follows:

1. The interdependence of activities
2. The emphasis on foundational issues and basics
3. The implementation of *jidoka*

Some people will see this list and say, "Well we're aware of that" and just move on. If that is your attitude, I can almost guarantee your transformation will fail. On the other hand, if you are one of those enlightened managers that exhibit humility, curiosity, and insight, and you spend time observing, understanding, and acting on these three items, you will obtain long-term gains you had not anticipated. These three issues, or more importantly, the depth to which these three issues are addressed, are often driven by some deep cultural issues, including the following:

- Respect for the workforce
- The natural maturation process in going from dependence to independence, and then ultimately to interdependence
- The need to avoid the "convenience of compartmentalization" and "simplistic thinking"
- Fundamental business and managerial humility
- The need to properly balance the long- and the short-term needs of the business and the culture
- An awareness of human needs, system needs, and business needs—to name just a few

The Level of Interdependence

When we are little children, we are totally dependent on our parents. This dependence, coupled with our survival instincts, goes a long way toward shaping our personality. However, as we grow and get older, we are expected to become far more independent— being able to dress ourselves, later keep our rooms clean, and still later do some work around the house. Becoming independent is also a sign of maturity and is often equated with maturity. We are so enamored by independence, we coined the phrase, "rugged individualism" to somehow capture the American spirit. Well some of us just don't buy into this as the highest of ideals.

In fact, I for one do not even think it is an accurate description of reality. A far more accurate representation of reality is the concept of *inter*dependence. It is the concept that gives recognition to the reality that all things are intertwined, and if one aspect of an entity is changed, sympathetic changes occur in other aspects without fail. It is a key aspect of "systems thinking." For example, at the human level, if you change your exercise habits, your patterns of eating and sleeping will naturally change because your body is an interdependent system. At the family level, if one person gets sick, frequently all are affected, because the family is an interdependent system. At the business level, production lead times cannot change without a resultant change in inventory, overtime, or delivery performance. Thus, your business is an interdependent system as well.

Some call this "systems theory," as I mentioned, and parts of it are taught in engineering, business, and medical schools. Systems theory implies that whenever one part of a system changes, other parts must also be able to adjust or the entire system will breakdown. This adjustment requires several characteristics:

- The system must be able to recognize that a change is occurring, and it must have a conscious awareness of its state

- It must be flexible enough to make the change
- The system must be responsive

This is true of all systems, including human systems and certainly in Lean Manufacturing systems.

I still find a large number of managers who either do not understand this or do not believe in it. For example, I still see managers deciding that to improve bottom-line profit, all they need to do is cut labor, as if there is no impact other than to reduce the overall costs.

- What happens to the overall skill level? Is it affected?
- What about the morale? Will it cause a reduction in productivity when they see the layoffs?
- What about the effect of working as a team now that some members are gone? Certainly this has an effect.

But it is easier for the manager—*not better, just easier*—to ignore the impacts and do some simple straight-line mathematics, as if that was somehow an accurate representation of reality.

We see the same thing with expedited freight, for example. I have seen many managers who decided that they needed to reduce this cost (they must think that their personnel are incompetent and so expedited the shipments, even though they didn't have to). When the manager imposes limits on their ability to expedite the shipments, what happens to on-time delivery or overtime? This will impact the system somewhere, but where? Or just maybe their people *are* incompetent or poorly trained, or maybe they don't understand the goals and objectives. If they are any of these, these managers have some failing in their management system, which also needs changing. Yes, this system concept is very much intertwined.

> "The key to the Toyota Way and what makes Toyota stand out is not any of the individual elements... . But what is important is having all the elements together as a system. It must be practiced every day in a very consistent manner... not in spurts."
>
> Taiichi Ohno

It is clear Ohno recognized the issues of systems and the concept of interdependence.

The degree of interdependence of activities is a truly amazing phenomenon in a lean implementation. Most people would like to think the systems of the world work independently and are simple and linear in nature. It makes them easier to understand. Seldom is this true. More often than not the systems of the world interact with interdependence rather than independence. For example, as you begin to reduce the variation in production rate, implementing such lean techniques as load leveling using *heijunka* boards (both terms are described in Chap. 12), many other aspects of production will sympathetically change. As the variation in cycle time and workload is reduced, employees have a more repeatable pattern and a more even pace that produces fewer mistakes. Consequently, scrap is reduced, buffers reduce in size, and the workload variation and the production rate variation are further reduced.

The changes resulting from interdependent causes are often large and hard to foresee by the novice, and are sometimes unique to your circumstances. They are an area where the best advice I can give you is twofold:

- First, look for them, so when they appear you will be prepared.
- Second, on these topics listen very carefully and defer to your *sensei* as many of these interactions are not only counterintuitive, but they are paradoxical also.

For example, we have already discussed the paradox of change in Lean. In addition, there is the paradox of *jidoka*, which is "We shut down the system so the system can run continuously."

The beauty of Lean is that it recognizes these concepts of systems and interdependence. For example, the concept of transparency is prevalent, so we can understand and become conscious of the workings of the system, can delegate deeper into the organization, and can foster a greater leadership footprint as we become better, faster, and cheaper suppliers. These concepts were understood early on during the formative phases of the TPS. Ohno is quoted as saying: "The key to the Toyota Way and what makes Toyota stand out is not any of the individual elements…. But what is important is having all the elements together as a system. It must be practiced every day in a very consistent manner….not in spurts." Taiichi Ohno.

The Emphasis on Foundational Issues and Basics

Support by top management on the foundational issues (see A House of Lean, Chap. 24) must be applied to a degree seen by no other initiative. There are few things a manager does that are so culture changing as the implementation of Lean, and when the culture needs to change, management must lead the way. I can say with certainty that, next to inadequate leadership, the most common reason facilities do not reach their goals is a marked weakness in addressing the foundational issues.

- If you find a weakness in one of the foundational issues, immediately fix it. Don't ponder, don't budget, don't meet to discuss it, and don't organize—**just fix it**.
- If you have concerns that one of the foundational issues may be weak, immediately improve it. Again, don't ponder, don't budget, don't meet to discuss it, and don't organize—**just fix it**.
- If you see a problem appear that should not be present and if this problem is related to a foundational issue, attack it with a 24/7 approach until it is fully understood and then aggressively implement the corrective actions necessary.

Spare no effort on this topic.

All deficient foundational issues are crises, requiring immediate attention, and need to be managed as such. They are emergencies and do not lend themselves to the time-consuming decision making of optional activities. They are not optional; they are crises. In effect, the patient is bleeding uncontrollably, and the bleeding needs to be stopped. In such cases, we don't seek out advice from others, we don't request permission to proceed, and we don't think about the cost of the bandaging—**we just do it**. This is the attitude that must be taken with foundational issues. I call it the Nike attitude—**just do it**.

The Implementation of *Jidoka* Is Always a Fundamental Weakness

I have yet to see a lean transformation where *jidoka* was the leading strategy used. It always lags behind other strategies in its implementation, and lags behind other strategies such as JIT (just in time) in both application and depth of application. There are a variety of reasons for *jidoka* lagging behind rather than leading the effort; none of them good. The best I can come up with is that it is hard to do, that it's hard to see progress, and that it takes a lot of time and effort to do it well. Or to put it in more flowery terms, it is not quite as sexy as JIT, TPM (total productive maintenance), or *kanban*—or in more practical terms, it is not seen as important by management.

Many companies put in lots of *andons* and other tools and refer to that as *Jidoka*. Often, they also have the inappropriate belief that "Lean *is* the tools of Lean." Yet even more important than the tools is the concept behind the tools. And for *jidoka* the key concept is the unwavering emphasis on never passing anything downstream except high-quality product, and this is supported by a culture of empowered problem solvers.

> **P**oint of Clarity Toyota started with *jidoka* before they started with either SPC (statistical process control) or DOE (designs of experiments). More accurately, they started with *jidoka* before either SPC or DOE *had been invented*! So it's not surprising they are among the world leaders in *jidoka* implementation and that the rest of us simply need to catch up.

An example of the Five "Whys" technique:
Jidoka is a weakness in lean implementations.

1. **Why** is *jidoka* a weakness in lean implementations?
   - Because quality problem solving, an integral part of *jidoka*, is hard to do.
2. **Why** is quality problem solving hard to do?
   - Because we avoid these problems rather than embrace them
   - Because we do not have a sound continuous improvement philosophy
   - Because we are not skilled at problem solving
   - Because we do not have time to work on problem solving
   - Because our data are not good enough for solving quality problems
3. **Why** are our data not good enough to solve quality problems?
   - Because the data are total reject data only, not stratified by type of defects.
4. **Why** are the data total reject data only and not stratified by type of defects?
   - Because we only need total reject data to answer the questions we are asked by management.

5. **Why** are the total reject data adequate to answer the questions from our management?

- Because the questions we get from our management focus on production and yield and not quality.

6. **Why** does our management ask questions about yield and production and not quality?

- Because management is more interested in yield and production than they are in quality (solution statement).

A couple of points about the Five "Whys":

- First, the Five "Whys" technique is seldom a straight-line linear process. For example, you will note that the second "Why?" has five possible answers and we only addressed one of them. We have two options when we have multiple causes.

- We could answer why to each of these, which would give us a very branched problem solution. Quite frankly, all five branches might converge on the same conclusion.

- However, the normal advice is to quantify the issues and follow the branch with the largest impact.

- Second, as is the case here, it is arguable if you have reached a root cause, which is the objective of problem solving. For example, why is management more interested in production than quality? Maybe it's because of their bonus structure, or just maybe it's the best thing for the company today. Nevertheless, with problem solving of human systems, often the root cause is not really found, and sometimes it isn't necessary. What is necessary is to find an **"actionable cause that reasonable people agree should be changed."**

Let's check this possible solution statement with the "Therefore" technique. If "management is more interested in yield and production than they are in quality" is a good answer to our problem, then the "Therefore" technique will logically connect the "Whys" in reverse order, starting with the solution statement.

An example of the "Therefore" technique, which is a check of the Five "Whys."
Management is more interested in yield and production than quality (the solution statement).

1. **Therefore**, the questions we get from management focus on production and yield, not quality.
2. **Therefore**, we only need total reject data to answer these questions.
3. **Therefore**, the data are total reject data only and not stratified by defects.

4. **Therefore**, the data are not good enough for solving quality problems.

5. **Therefore**, quality problem solving, an integral part of *jidoka*, is hard to do.

6. **Therefore**, *jidoka* is a weakness in lean implementations.

And it works, which tends to confirm the validity of our solution statement.

Jidoka is hard to do, but the reason it is hard to do is caused by two other issues.

- First, management does not emphasize it.
- Second, as shown in the preceding example, a number of foundational issues must be in place to make *jidoka* work. These include a quality test and inspection system, a quality data system, and root cause problem solving by all, to name just a few. And why aren't these foundational issues taken care of? Just go back to reason one—that is, it was not emphasized by management in the past.

What Happens When *Jidoka* Is Not Done Well?

When *jidoka* is not done well, production problems appear. Most typically, defects advance in the system, causing quality production problems. Almost always, the source of the problem is the failure to properly implement some foundational issue. For example refer to the Five "Whys" exercise done earlier. However, now you have two problems: the production problem and the poorly executed foundational issue. Consequently, we must go back and repair the foundational issue before we can improve on the *jidoka* system.

This is like finding that the roof on your house is not level and you search for the problem and find the concrete foundation is not level. To correct the roof issue, you must first correct the foundation issue. However, as soon as you modify the foundation, you find that it affects other things like the walls and floor. The same is true of your lean system. Once you go back and clean up your operational definitions, for example, you need to clean up your work instructions, your training instructions, your training matrix, your visual displays on the floor, and the list goes on. All of this is rework, wasteful, and is only necessary because the foundational issues were not handled well. All of this is waste of some of the most precious commodity we have: the thinking doing problem solvers in the business.

Recalling *Jidoka*

Do not forget about the purpose of *jidoka*: It is there to prevent defects from advancing in the production system, and it is a continuous improvement tool. Because both of these tactics are crucial to lean implementation, it is clear we must have some kind of *jidoka*, even if we cannot capture the most mature forms of *jidoka*, which include line shutdowns by operating personnel followed by "Rapid-Response PDCA (Plan-Do-Check-Act)" problem solving.

We must keep in mind the definition of *jidoka*:

- It is a 100 percent inspection technique that will prevent defects from advancing in the production process.
- It is done by machines not men.
- It uses techniques such as *poka-yoke* (error proofing).

- It will prevent defects from advancing in the system by:
  - Isolating bad materials
  - Implementing line shutdowns
- It is a continuous improvement tool because as soon as a defect is found, immediate problem solving is initiated, which is designed to find and remove the root cause of the problem.
- In the design case, the line does not return to normal operation until it has totally eliminated the defect-propagating situation.

When Line Shutdowns Are Not Practical

First, and without delay, we need to work on all the foundational issues mentioned earlier. They need to be analyzed, prioritized, and aggressively brought to maturity. If this is not done, there is no hope for real progress.

Second, while working on those foundational issues, we need to do some other things.

1. Change the attitude toward defective materials. Make sure the concept of defective material advancing in the system is unacceptable and that it must be reduced. Make sure "continuous improvement" is understood by all.

   A. Teach it, preach it, and do it.

   B. Measure it and post it.

   C. Make sure that all know we cannot send scrap to the customer; the customer is the next step.

2. Improve quality responsiveness by reducing first-part lead time (see Chap. 13).

   A. Implement all seven techniques to reduce lead time.

   B. The most powerful initial activity will be to reduce lot sizes to a minimum. Very often this is easily done. If this cannot be done everywhere, do it where you can.

   C. The next most powerful activity is inventory reductions of all kinds.

   D. Usually, the third most powerful tool is the reduction in changeover times.

3. Emphasize JIT problem solving: Our objective is Rapid-Response PDCA. This, very likely, will require a huge cultural change, but you should do some of the following:

   A. For example, if your engineers are the problem solvers, move them to the production line.

   B. Post a flipchart at the line and list the problems to resolve. Keep the flipchart up to date.

   C. Get the managers to spend time on the floor, evaluating, observing, and just being available.

   D. Have a top manager be at the daily production meeting where the problem solving is reviewed.

E. Be brief, but make problem solving a part of the meeting, which should only be a few minutes if it is run well, even with a problem-solving review added.

F. The point is to be proactive in problem solving: "Do something."

4. For all process cells and lines that are highly people dependent for quality results, begin the process of line shutdowns for quality issues. If you can't shut down for all quality problems, shut down for some problems and prioritize your activities. Start this practice, and then expand it. There is no substitute for it.

5. Processes that are highly machine dependent, such as high-speed pick-and-place machines, ovens, and other continuous process equipment are a little more difficult to manage in this way.

A. If you can't shut them down, use *andons* to signal abnormalities and make sure the *andons* are responded to.

B. Make those responsible for the machinery also responsible for the defects the machinery creates.

6. Aggressively categorize all defects as to the most likely source. Assign the appropriate groups to take on the issue of problem solving for defect reduction. Support them; hold them accountable.

7. Become more introspective in your problem solutions.
   A. Cease completely the philosophy of using inspection as the means to achieve quality of product and process.

      i. Eliminate visual or human inspection as an acceptable quality-control technique; emphasize *poka-yokes*.

      ii. Institute *poka-yokes* widely; emphasize process improvements.

   B. Eliminate improved procedures—done by humans—as an acceptable quality improvement technique. Change the process design so it can't be done wrong; *poka-yoke* the process.

> **"Do** not let what you cannot do, prevent you from doing what you can do... ."
>
> John Wooden—Former basketball coach at UCLA (Wooden won 10 National Championships as a coach, and one more as a player at Purdue University).

The point is that you can do a great deal. So start now! Show the commitment, and show the initiative.

Jidoka is a crucial element and is absolutely necessary to make Lean work.

Chapter Summary

It is critical to a lean transformation that we pay particular attention to our culture. It is the key guiding force in our business success. We will want to work to create and sustain a healthy culture that is strong, flexible, and appropriate for our business. Because our lean transformation is a culture-changing event, we will need to learn the Change Prescription and become very skilled at Management for Change. The effectiveness of the cultural change is dependent on five internal skill sets we must cultivate. These are the five cultural change leading indicators: leadership, motivation, problem solving, whole facility engagement, and a learning/teaching/experimenting environment.

In addition to the five cultural change leading indicators, we will need to pay particular attention to three other cultural characteristics. First is the degree of interdependence that all the changes will have. As we modify one aspect of our culture—one aspect of our business—other parts will change sympathetically. These are pronounced in a lean transformation and may be changed for the good or the detriment of the transformation. Second, we cannot stray from our commitment to emphasize the foundational issues. Finally, the key issue that is likely to get less attention than needed is *jidoka*. It will be a key to our future vitality and we will need to implement it aggressively, starting now.

CHAPTER 5

Cultural Change Leading Indicator No. 1, Leadership

Aim of this chapter ... to elaborate on the skills of leadership and its critical nature in a lean transformation, and we will also distinguish the skills of leadership from the skills of management. We will explore the new leadership style that is required, lean leadership, and how to enlarge your leadership footprint. Finally we will explain the 4 "ah ha" experiences.

Leadership Basic

Just what is a leader? Simple: It is someone who has a following. Often his followers are willing to do what he says, just because HE says so. They may follow him because of his character or they may follow because of his position, or maybe it is his personality, or his competence, or just maybe a critical combination of these four aspects of who he is. But the critical factor is that they are willing to follow him, they are willing to act, and many times they are willing to undergo severe hardship and sacrifice for this leader. If he says to do something, they do it. They need nothing more.

How does this leader get his following to do these things? There is a great deal in the literature on leadership, and some of it is outstanding—Robert Greenleaf's being some of the best. Some of the literature puts the topic of leadership completely out of context, yet others touch on all the major points but do not emphasize the very critical few necessary characteristics. I do not intend to fully explore leadership. That is not necessary right now, but what we will discuss is the critical few characteristics that all leaders must have. They are also the skills that the leader of this lean transformation must have. Depending on the type of cultural change undertaken, or maybe where or when it happens, the leaders of these efforts may need other skills as well, but these are the critical few skills that are absolutely necessary for *any* leader. If he is lacking in any of these characteristics, the effort—be it a lean transformation or the overtaking of a nation—will almost surely fail.

Think of a leader, any leader. It could be a political great like Churchill, Martin Luther King, or

> **P**oint of Clarity   Lean leaders must be situation leaders and they can lead through:
>
> - Competence
> - Character
> - Position
> - Personality
>
> Our lean leader must lead primarily through competence with character a close second.

even John F. Kennedy. It could be a sports leader like Vince Lombardi or may be a spiritual leader like Gandhi, Mother Teresa, or Christ. All these people were great leaders. Their individual fields of endeavor were different, yet they shared some common abilities that led to their success. That said, what are the common characteristics shared by these leaders? By all leaders?

First, all leaders have a vision. Martin Luther King's plan was immortalized by his "I have a dream" speech. Even today, some 40 years later, many people quote his words, "...one day live in a nation where they will not be judged by the color of their skin, but by the content of their character." What about Gandhi's vision of passive resistance? Maybe the grandest of all was the vision of Christ. Even having died nearly 2000 years ago, his following still grows. They all had a vision.

> **P**oint of Clarity   It is not possible to lead without a vision.

Your lean transformation leader must also have a vision and a plan. Lacking this person or his plan, my advice is strong and clear. *Don't even start.* If you attempt to undertake this transformation without strong leadership and a strong plan, you will fail. Furthermore, you will have raised the expectations of your employees, and their expectations will be crushed. This only makes the next effort more difficult. Certainly, do not rush into this effort. Find a good leader and develop a solid plan before you start.

The second necessary characteristic these men all shared was the ability to articulate their vision so people could understand and buy into it. They have the ability to get people engaged. It is not coincidental that all these examples were also great orators. I'm not so sure about Gandhi and Mother Teresa, but the others could mesmerize a group with their speeches. About Vince Lombardi, his great middle linebacker Ray Nietzsche once said, "When coach Lombardi said to sit down, I didn't even look for a chair." Their speeches were an extremely strong tool that allowed them to attract, motivate, and activate a following. Often, the content of their speeches was very motivating in its own right—but make no mistake: The choice of words and the type of delivery made a huge difference. The leader of this lean transformation must have this skill also. They must be able to make the lean transformation come alive to those involved. Not only must they encourage the hands and feet of their followers to move in the right direction, they must also motivate the people and engage their hearts as well.

Some examples exist of successful transformations where the person who put together the plan was not the person to articulate it. This type of brains and mouthpiece combination will not work in our lean transformation, however. Both skills must be resident in the same person as they will have to be on the floor on a daily basis— observing, dealing with problems, and interacting with all the people in the facility. The leader must be able to handle questions and problems quickly—in short, they must be able to lead from competence.

> **P**oint of Clarity   The leader must be able to translate their concepts into behavioral traits that will support the execution of those concepts.

The third, and final, requisite characteristic of a leader is the willingness to act on the plan, at the exclusion of all else. He must exhibit the skill of not losing sight of the goal and stay the course in spite of roadblocks, obstacles, and resistance. He must be willing to act, at all costs, to reach the necessary objectives. This ability to act is a huge need. Many great plans have failed not because the plan was not needed or the plan was not good enough, but because the leadership did not have the courage and character to make the difficult decisions. In the end, this compromised the entire effort, and the plan, simultaneously. We have seen this all too often.

> **P**oint of Clarity It is not possible to lead unless one is willing to act, especially in the face of adversity. This sounds almost trivial, but it is a key failing of leadership.

To act properly, they must first be able to recognize exactly what the situation is, and they must be acutely aware of happenings in the facility. Second, they themselves must be excellent problem solvers. They must be able to sort through the options and properly apply the values needed. Finally, they must have the wisdom, courage, and character to act when action is required, and conversely, should use those same skills to hold back when thoughtful inaction is the appropriate behavior.

So the leader of your lean transformation—be it you or someone else—must have these three requisite skills:

- The ability to develop a vision-based plan
- The ability to articulate this plan and engage others
- The ability to act on the plan

Carefully choose your leader, and if you do not have someone with these skills, find a person who does. There is no substitute for leadership. Any compromise on this topic will guarantee failure.

Our leader and all the key persons in the lean transformation will be well served if they have an abundance of some of the blessings we humans are endowed with. These include the blessings of:

- **Awareness** The ability to fully see and appreciate the reality of all that is happening around us.
- **Imagination** The ability to see things that may not, but could be there, and the ability to conjure up options that translate into opportunities.
- **Conscience-driven values and principles** These are the things, ideas, and ideals that we consider important. They become the basis for prioritizing our options.
- **Choice** The independent ability to apply our priorities and consciously decide what to both do and not do.

Management versus Leadership

There is a great deal of difference between the skills it take to lead and the skills it take to manage. However, because the supervision and management of a facility are often the leaders as well, we frequently mistake the skills of leadership with the skills of management. They are different and the distinctions are worth codifying. Arguably the best effort at that to date has been done by John Kotter in his writing, *A Force for Change: How Leadership Differs from Management* by John Kotter, (Free Press, 1990) and summarized in the following table:

Comparison of Management and Leadership		
Needs	Management Skills	Leadership Skills
Desired outcomes	Producing a predictable, orderly outcome of key results expected by key stakeholders, e.g., on time, on budget, high quality.	Producing change, particularly change that is needed to respond to the changing environment of the business.
Creating an agenda	**Planning and budgeting.** Establishing the detailed steps, timetables, and resources needed to accomplish the goals. Answering the question, "who, is to do what, by when?"	**Establishing a vision.** A realistic, convincing, and attractive depiction of the future. Also included in a vision are the strategies for producing the needed changes.
Developing a human network to accomplish the agenda	**Organizing and staffing.** Establishing a structure of people, policies, and procedures to execute the plans.	**Aligning people by communication and deeds.**
Execution of the agenda	**Controlling and problem solving.** Monitoring the planned results in detail, identifying deviations, and taking corrective or remedial actions.	**Motivating and inspiring.** Energizing people to overcome inertial barriers to change largely using intrinsic motivators.

Much of what you find in the literature tends to downplay the role of management and focus only on the critical need for leadership. Although this is explainable, it arguably represents an imbalance of what is needed.

For example, Kotter in his writings frequently refers to many companies as "being over-managed yet under-led." My observations more than validate his value judgment. For a long time, the external environment was not rapidly changing and hence the more needed skill was that of management. This probably explains why we have an MBA and not an MBL (Masters of Business Leadership). Excellent management, by its very nature, is somewhat conservative; methodically incremental; filled with bureaucracy, reports, and follow-ups; and is focused on the short term. As a result, the very best management simply cannot produce major change. Only with leadership does one get the boldness, the courage, the long-term vision, and the energy needed to activate the workforce and create changes.

Kotter will say in almost his next breath, that both skills are needed for a business to be successful. The current major emphasis on leadership is largely because most people believe we simply need more of that now; that weaknesses in leadership are the bottleneck that is holding back business success. There is no doubt that we need more and better leadership, but that does not mean the skills of leadership will replace the skills of management. Quite the contrary—both are needed. Our businesses need to provide predictability for the business through good management yet navigate the changes needed to survive and prosper through good leadership.

Situational Leadership

So much has been written about "command and control" style of leadership that it is considered a bad leadership style. Rather than say it is a bad leadership style, I say it is leadership style that is way overused and often abused, but still needed at times.

Unfortunately, for many leaders, it is all they really know. For others, although they know other styles, command and control is the one they use, regardless of circumstance. Both of these uses of "command and control" leadership create problems of execution and morale.

Let me share an illustrative example. As a Six Sigma trainer I have taught many black belts. As part of the curriculum, we always discuss leadership and leadership styles. Often, we will start with a brainstorm on "the qualities of a good leader." Normally the first 10 items or so, focus on a leader that is compassionate, empowering, thoughtful, and good listener to name a few. In short they want a kind, caring, compassionate leader—or at least they are heading that way. Unbeknownst to the class, I will have prearranged for the security guard to come to the training room, knock on the door, and pretend to have a discussion with me. When he arrives, I will immediately turn to the class and say, "we have a situation, we need everyone to exit the back door immediately." And they always comply and do so quickly. While outside, in July in El Paso it is very hot and soon enough someone will say, "hey why isn't everyone else out here?" or some such observation. We will then go inside and I will ask them "what just happened?". Soon enough it will come to light; that in a heavy-handed, dictatorial fashion I had sent everyone outside, yet no one felt it was inappropriate leadership, at the time. For example, I did not seek their input, I did not explain to them in an interactive fashion how I wanted and needed their help. I did not explain to them how they could be affected by this…I simply ordered them to go outside…they complied and yet they did not feel abused.

We then go into the topic of situational leadership. It simply means that all managers must have the skills needed to lead in a variety of circumstances, and when there are time constraints or severe consequences to be suffered if a quick decision is not made, maybe a more direct style of leadership is needed. Some of the best material I have read include *The Situational Leader* by Paul Hersey (Warner Books, 1984) and *Leadership and Decision Making* by Vroom and Yetton (University of Pittsburgh Press, 1973).

The purpose of Vroom and Yetton was to find the proper role of the manager in making decisions such that these decisions:

- Would be a high-quality rational decisions
- Would engender the support of those who must execute it
- Would be reached in the least possible time

Vroom and Yetton go on to say, "No one leadership style is the most effective. Each is probably effective under a given set of circumstances. Consequently, I suggest that effective leaders are those who are capable of behaving in many different leadership styles, depending on the requirements of reality as they and other perceive it."

Their book is then a treatment of this concept of situational leadership and just how the manager should act in decision making, depending on the reality at the moment. The really strong point of their treatment is the aspect of engendering support from those who are affected. They make a compelling argument for this point. I strongly recommend it.

To be effective as a lean leader, you need to be a situational leader.

Leadership for Change

The key distinction between management and leadership is that leadership is all about change; whereas management is all about predictability. So to clarify and support the role of the leader, the change agent, we teach them "Leadership for Change."

There are three simple basic principles that must be addressed by the leader if he expects to change something in his company and expects to be successful with the change. They are as follows:

- The present case must be unsatisfactory.
- There must be a better place to go to.
- There must be a path to the new destination.

The lean leader keeps these three aspects of change management in mind at all times. Because if he wants people to change, to **behaviorally relocate**, he must make the current situation bad; provide a better place to go to; and provide a path to the new location.

Lean Leadership

As I have worked with companies that are very Lean and especially with my Japanese clients, I have noticed a dramatically different style of leadership. I have observed that there are a number of things they do, as a routine, that typical mass production managers don't do nearly enough of, if at all. Earlier, in the preface, I boldly stated that:

> The time for excuses is over, it is time that U.S. managers, and managers all over the world accept the real reason the Japanese have prospered. (Page xiv).

So, how did we get to this inferior position? The genesis of our problem starts with the basic concept of an MBA and the teachings of the business schools, coupled with the penchant of American management to take the MBA concept to an extreme.

What's Wrong with the Concept of the MBA?

Nothing really, until these tenets get taken to an extreme. (See Lean Killer No. 10, especially the insert on enantiodromia.)

The basic concept underlying the theory of management, and the MBA in particular, is that there are more commonalities in this field than there are differences. Probably true. However, extend that a little and you soon come to believe that nearly all business are so common as to be indistinguishable such that an MBA can run a business—any business, anywhere. And you reach the point in your beliefs that a successful manager can run a bank, a petroleum refinery as well as a hardware store—which is a commonly held belief underlying the current failings of American management today. Taken to its extreme, I call this "the concept of the MBA...gone wrong."

This type of thinking leads to four very unsound concepts that have crept into the skill set of many American managers. They are so common that they are accepted as "the way things should be done around here." Unfortunately these four items are so damaging as to place American manufacturing in a followership role rather than being the leader of the pack. The scorecard has clearly shown this.

- The first unsound concept is that all the important aspects of a business can be expressed in numerical terms, primarily financial terms. Everything can be reduced to numbers such as cash flow and IRR and RONA. Conversely, if they cannot be quantified in numerical terms, they simply are not important and hence not worthy of management's attention. What about morale? What about the cost of an unhappy customer? The extension is that since these are not easily, if at all, expressed in quantifiable terms, then they are not important enough to demand management attention. Who really believes that? But what do most managers do? They try to manage the money, the results, and hope all else will fall in line. Dr. W. Edwards Deming felt this concept was so important he listed it as "Deadly Disease No. 5: Running the company on visible figures alone."

- The second unsound concept is that since everything can be reduced to numbers—mostly financials—almost anything can be understood, evaluated, and corrected by evaluating documents called spreadsheets, monthly balance sheets, and quarterly financial reports, to name a few.

- The third unsound concept is that since only the financials matter and they can be expressed in spreadsheets, there is no need to leave your PC and certainly not your office to make the decisions needed to be a good leader or a good manager.

- The fourth unsound concept is that because you can now do "managing and leading" from your office, there is no need to interact with the action on the floor. Consequently, the machines become a fungible piece of capital that needs none of management's attention. But worse by far is the concept that the people on the floor are another piece of fungible expense—not even capital—**but an expense that we are actively trying to minimize to improve all those financial metrics we are using to drive this business. This is the worst of the unsound concepts—with far-reaching and deleterious effects.**

In *Confronting Managerialism: How the Business Elite and Their Schools Threw Our Lives Out of Balance*, (Locke, Robert R.; Spender, J.C., 2011). (Kindle Locations 926-931). Zed Books
Locke writes,

"...managers were already familiar with the concrete details of the operations they managed, no matter how complicated and confused those operations became. Such individuals, prevalent in top management ranks before 1970, had a clear sense of the difference between "the map" created by abstract computer calculations and "the territory" that people inhabited in the workplace. Increasingly after 1970, however, managers lacking in shop floor experience or in engineering training, often trained in graduate business schools, came to dominate American and European manufacturing establishments. In their hands the "map was the territory." In other words, they considered reality to be the abstract quantitative models, the management accounting reports, and the computer scheduling algorithms."

Does this not sound just like Lean Killer No. 1??
Following this model of "the MBA gone wrong," we then created over time, what I refer to as managers who "execute at the next level down." I am not sure if this effect is because the managers did not want to deal with these details, they did not know how to deal with these details or if they just felt it was beneath then to get involved in the details; I suspect it is a bit of all three. For example, discuss with them the need for a better scheduling program and, not wanting, or not seeing any need, to get directly

involved in the details they will merrily tell you "To work with my Logistics Manager." Never mind that the program once installed in one facility may cost upward of $500,000 initially, with huge training support and organizational structural issues that will cost a small fortune long term. It is no wonder I see huge, expensive scheduling programs going largely unused. An even larger negative effect occurs when any culture-changing issue is implemented. Your classic "let the next level execute that" mentality then causes several if not all of the 10 Lean Killers to be fully activated, as well as probably most of the six Roll Out Design Errors.

> **Can You Pass the Legal Test for Lean Leadership??**
> Ask yourself, "If lean leadership were an illegal activity, could someone find enough evidence to convict me?"

The Solution: First, Awareness and Then Acceptance of the Problem

Management, which is also the leadership in most firms, must realize they need to change.

Unfortunately, this is not happening at any reasonable rate. What is happening is that plants are closing, businesses are shutting down, and entire industries are being taken over by those who are better managers. We are losing the battle to keep manufacturing in the United States.

If you hadn't noticed, the little island nation that was devastated about 70 years ago has become a world manufacturing power, and about the only natural resource it had in abundance was fish.

No natural resources??? Unless you count the people and their ability to work and think, and their ability to manage and lead—then they have what has been definitively proven to be some really powerful resources. That ought to make all of American management not only think but wake up and take notice. Unfortunately, way too many are still asleep.

Once we overcome the first hurdle—a monstrous problem in and of itself—the next question is...

What Model of Leadership Should We Change to?

First, leadership is not some manager sitting in his office presuming that he knows it all, and "all of it" can be expressed in a spreadsheet that he can receive on his PC and then, from the comfort, quietude, and solitude of his office analyze this information and make a cold, calculated decision that will drive his business to prosperity. This model must be discarded. It simply does not work and it will never work.

I call the new model of leadership—well, it is not really new; the Japanese have been using it for 70 years that I know of. It is called lean leadership. It has six basic qualities, which are as follows:

- *Leaders as superior observers*: They go to the action—they call it the *Gemba*—to observe not only the machines and the products but also to spend significant time with the employees. They strive to be aware of not only the products and the processes but more importantly the people. They also are in contact with their customers. A much overlooked leadership skill they have in abundance is the ability to be an empathetic listener.

- *Leaders as learners*: They do not assume they know it all. Rather, they go to the floor to learn. They are in "lifelong" learning mode. They are masters of the scientific method. They learn by observing and doing, but most importantly they are superior at asking questions, they learn by questioning.

- *Leaders as change agents*: They plan, they articulate, and sell their plans, and they act on their plans. They are not risk averse, yet they are not cavalier. They do not like to, but are not afraid to make mistakes.

- *Leaders as teachers*: They are "lifelong" teachers. When something goes wrong, their first thought is not "Who fouled up?" but "Why did if fail?" and "How can I use this as a teaching opportunity?". They teach through the use of questioning rather than just instructing.

- *Leaders as role models*: They walk the talk. They are lean competent. They know what to do and they know how to do all of Lean that is job specific to their current function in the organization. There is no substitute for this. NONE.

- *Leaders as supporters*: They recognize they mainly get work done through others, so they have mastered the skills of "servant leadership."

So, how can the United States win back the manufacturing capacity it has lost? The formula is rather simple:

- Recognize and accept there is a better way to manage.
- Start by employing the Six Skills of Lean leadership.

"Put everybody in the company to work to accomplish the transformation. The transformation is everyone's job….Deming Point No. 14.

Leading by Using the Five Questions of Continuous Improvement

The power of leading by questioning is not well understood by most Western managers. Like I said earlier, all too often, the leadership style is leading by command and control.

Let me explain the power of questioning. The other day I had my morning planned. I was going to write an article for a magazine after I took the dogs for a walk down by the river. I was looking forward to it as I had just returned from a week-long trip to Mexico. It was going to be fun. As I was putting on my shorts and sneakers to go on the walk, my wife said, "Oh dear, I forgot to mention but the garbage disposal is not working correctly, could you look at it?" At that very moment my entire day changed **because of one simple question from the right person**. Rather than go on a leisurely walk, I was now scrunched up under the sink, with some water pump pliers gouging me in the ribs, working on the disposal. It was no fun, but that is not the point. The point is that most managers don't really understand the power of a question. If, in the middle of the day, you get a call and the boss says, "could you take a look at the budget for the Roto-Ruback Line, compare it to the Fritzengrabber cell and tell me why the per-unit conversion costs are so different?", just how will you respond? Of course, with the right question by the right person, you immediately drop all you are doing to answer their questions. Most managers really fail to understand how impactful a simple question might be. Whether it is a question about costs like in this example or questioning as a teaching method, both are extremely powerful. Socrates knew how to teach by asking questions as did the best managers I have worked with, but much of that skill has gotten lost in Western society.

Consequently, a strong skill to add to your leadership toolbox is the skill of asking good questions. And in particular how to ask the Five Questions of

Continuous Improvement (covered in greater detail in Chap. 7). These questions are as follows:

1. What is the present state or condition?
2. What is the desired future condition?
3. What is preventing us from reaching the desired condition?
4. What is something we can do now to get closer to the desired condition?
5. What is our expectation when we do this?
   A. What will happen?
   B. How much of it will happen?
   C. When will it happen so we can "go see"?

Avoid Non-questions

Non-questions are of the variety, "Don't you think that John could have done a much better job on the project?" You are not so much looking for discovery as confirmation of your bias, as you speak with John's supervisor. Your "question" clearly indicates that you thought John had tanked it. Rather, you should state what you know and ask a genuine question. For example, "John missed the due date by 3 months and the customer is not happy with our quality." What can you tell me about that?"

Other Questioning Skills

There is a plethora of information on asking good question, so I do not wish to make this a treatise on good questioning. Rather I wish to elaborate on the most common of the "next questions" that come up when you receive an answer to one of the Five Questions of Continuous Improvement listed above.

Open-Ended Questions   First, learn how to ask "open-ended" questions. Closed-ended questions can be answered with a "yes," "no," or a simple fact. For example, when your son comes home from the Friday night football game, you could ask him, "how was the game?" or "did you have fun?". My son would respectively answer "OK" and "Yes." And the conversation would end there. Those closed-ended questions do not cause new information to surface, nor do they further the discovery you are hoping for in the discussion. As you lead by questioning, often this is what you want: new information and discovery. On the other hand, you could say, "Tell me about the game" or "How did you feel that you attended the game alone this time?". Open-ended questions like these elicit longer answers. They usually begin with "what", "why" and "how". An open-ended question asks the respondent for his or her knowledge, opinion, or feelings. "Tell me" and "describe" can also be used in the same way as open questions.

Follow-Up Questions   Second, learn how to ask the "follow-up" questions. Frequently, especially in a closed culture, respondents will not be very likely to share details. Yet you may both need and want the details. Simple follow-up questions are, "could you be more specific?", the catchall follow-up is, "can you say some more about that?". If you may doubt what they say, or need more specificity, you can ask the two Management Follow-Up Questions. In order, they are as follows:

1. How do you know?
2. What data do you have to support that?

For example you ask, "How is the new press performing." He replies, "just fine." Since other similar presses had initial problems, suspecting an "ole" type answer, you would then ask, "How do you know?" and follow that with, "and what data do you have to support that?"—pretty universal and good questions to gain more specificity in your dialogue.

Be Persistent in Your Questioning . . . .   Many senior managers ask "how is the plant running today?" and accept the "ole" answer of "OK or just fine" without digging much deeper. That is not enough. If, for a question like this, you do not get a response with some facts such as "we are shipping to schedule, had a quality problem but contained it and had a safe day today," you are getting no information. If you don't want information, don't ask the question. If you do want some, don't fail to get it. Persist in your questioning until you get the specificity you need.

> **P**oint of Clarity If your boss isn't asking about "it," then "it's" just not important to him.

Questions Help Clarify Your Intent . . . .   More often than not, you will ask questions about what is important to you. Consequently it is a way to express and clarify the importance of goals and objectives.

Creating a Leadership Footprint

In our lean transformation we often refer to creating a larger "leadership footprint." Basically it means that "all can and should lead." Yes, we can get, and we need, leadership from the cell worker to the CEO and everyone in between. As you read in Chap. 15, you will see that one of the first evaluations made is a leadership assessment. When completing this assessment in the plant, and when we truly and objectively assess the actions of the managers, more often than not we find that rather than preparing and executing plans, they are responding to the instabilities in their plants as they embark on firefighting activities. They have no plan because they know soon enough that there will be an upset or a crisis to which they must respond. In effect the plants are leading them, rather than they being the leaders in the plant. Yet there are many ways in which they can and should lead. A great tool to foster this is the use of *Hoshin–Kanri* planning described in Chap. 16.

We need to foster and support leadership at all levels, including the line worker. As he is preparing a *kaizen* idea, he has the opportunity to create a plan, sell the plan, and execute the plan…this absolutely must be management supported and management encouraged. Leadership at all levels is a real key to success of the transformation and to the vitality of our company.

Training to Be Leaders

I simply do not understand the controversy on "leaders are born versus leaders are made." I can give you literally thousands of examples of training people who learned to be leaders. Maybe great leaders like Lincoln and Kennedy have some genetic material the rest of us are missing but make no mistake about this; the skills to lead can be trained such that almost everyone, in the right environment, can lead in some fashion. What I find hindering people's ability to lead is first and foremost the culture of the business. I have frequently witnessed first-hand, people who have been deemed unqualified to lead inside their workplace, yet they will exhibit strong leadership skills outside the workplace. They do this in a variety of circumstances. Yet when they go to work, the culture does not allow it to happen. This is a sad state of affairs and further supports those folks who call this the Eighth Waste. Second, leadership is seldom taught. Companies need to take the responsibility to teach leadership. Quite frankly, if they valued it, they probably would. Leadership is not only teachable; it is very teachable. The best document I can recommend on that topic is the *U.S. Army Leadership Manual*, FM 6-22, available online for public use. Another great book is *The Bass Handbook of Leadership* (Simon and Schuster, 1974). Normally you need a hand truck to carry this incredibly comprehensive and wonderful book, but luckily now you can get the Kindle version. I find the works of Robert Greenleaf to be very valuable; he is best known for his book *Servant Leadership* (Paulist Press, 1977). Finally a great and also funny book is *Managing Management Time* (Prentice Hall, 1984) by William Oncken Jr. As a young supervisor I took a seminar from William Oncken, long before he wrote his book. Without a doubt it was the best lesson in management and leadership I was ever given. When his book became available, I quickly bought it and have read it several times since. It is a classic, and I recommend it to all.

Screw-ups, Mistakes, Errors … and the Teaching Opportunity … a Paradigm Shift for Lean Leaders

Whenever there is a problem in the value stream and it appears to be a screw-up, a mistake, or an error, there are at several approaches that you, as a lean leader, can take toward creating a countermeasure. Most of the time the first response a leader has, "ok, who screwed up?" and start by finding the guilty parties so they can apportion blame and then begin to fix their screw-ups. This is extremely counterproductive.

Early in a lean transformation we teach all the leadership that at least 90% of the problems are system problems and not people problems. Soon enough, with only a little coaching these lean leaders come to understand this and change their focus from "fixing" the people to making the process more robust. And since the system is largely a management-created entity, almost always buried in this "screw up, mistake, or error" is a management problem. Hand in hand with that management problem is another issue. The people executing this system that failed were not taught properly. Hence, there is a technique we teach which is called … "every problem is a teaching opportunity." So when the next mistake occurs, think of turning it, **from a mistake into a teaching opportunity**. Now you not only have removed the specter of blame from the situation, but you have created an environment where people can learn and be part of the solution…and that can be nothing short of healthy.

Often when I get involved in problem solving, the first and most predictable response I get from the management team is, "but we told them exactly what to do, they have the tools to do it and it simply did not get done." And on the surface, that may seem to be true. Although as you dig deeper, "using the Five Whys" and reflecting deeply while being both open and honest, almost always you find there is more to the story. I often find poor training; missing, inadequate, and conflicting instructions; conflicting and irreconcilable priorities; equipment in poor condition; inadequate raw materials; and a litany of issues that prevent the worker, supervisor, or engineer from completing his assigned tasks. Like I have said many times, 90 percent of the problems we encounter in a manufacturing facility are system problems. That is specifically why the "concept of the MBA gone wrong" is so deleterious. It locks the manager up in his office and he will not get to the floor and observe the actual situation…and consequently "his system" will not get fixed. However, using the model of lean leadership problems get surfaced and real root causes can be found and corrected and the old management paradigm of "but we told them exactly what to do, they have the tools to do it and it simply did not get done" dies an early death…as it should.

Your Plant Has Lots of Leadership Raw Material

I find it amazing that when you ask the floor workers what they do for fun, you will find they are involved in a large variety of activities and not surprisingly, many of them are in key leadership roles **outside of work**. This happens when they get involved in their children's school activities by helping on the PTA and even being involved in the PTA leadership. I find line workers in key leadership and teaching positions in their church, synagogue, or mosque. I find them very often involved in coaching and in the league administration for their children whether it is sports, computer club, cheerleading, karate, or dance. Outside of work, they seem to be able to lead and manage; yet when they cross the threshold and enter their work building, these skills can seemingly no longer be found. If you are the least bit curious, you might ask, "just why is that?".

And guess what, it does not take you too long to come to the conclusion that, once again, it is the Pogo Syndrome… "we have encountered the enemy—and they are us?"

So Let's Teach Them the Four Leadership Work Rules

So how do you activate this resource of leadership, that you desperately want from your workforce? It seems to be plentiful outside of work, but you cannot seem to activate it inside your organization.

Well it is a skill set that is not so complicated, but it is so rare as to qualify as a precious gem. Once again it is a matter of teaching with the appropriate follow-up and support. Do the following.

- Make sure your line employees, all employees for that matter, know that the only reason we exist as a company is to supply value to our customer. Furthermore, for the most part, the only people who are adding value are the line workers. Therefore, we cannot waste this precious commodity; the time of each worker is simply too valuable. There is no better measure of our vitality than measuring how much of the worker's time goes to value-added work. In fact, their time is so valuable, everyone else has, as part of their job, the task to support this line worker. So they cannot be waiting; they should be adding value.
  - Leadership work rule number 1…No one can **wait until told**…all must be working.
- Furthermore, all these workers are experts in the work they are to do. We have trained them to be experts and as such they must follow:
  - Leadership work rule number 2…No one can **ask what to do**.
- Well it seems we have painted these workers in a box. If they can do nothing… and they cannot ask what to do, two former activities that were totally acceptable, then just what do they do? Well they must now exhibit, inside workplace, the very skills they exhibit outside the workplace. They must lead. However, keep in mind that Pogo taught you that you are the problem here…

well luckily the flipside of being the problem is that now you can become the solution. In fact, no one else, only you, can affect this solution. So now, I am sure you are going to say with a balance of both exasperation and curiosity, "Ok, just exactly how do I get this guy to lead?" Some Harvard MBA would tell you to get real smart, I am going to tell you to get real dumb. Yeah, get dumb. Instead of showing the worker all you know about what needs to be done, how it needs to be done, and why it should be done this way—get him to tell you what you and what **HE KNOWS** about this situation; after all, recall that he has been trained to be the expert. So instead of putting your knowledge on display, draw out the knowledge he has and put it to use. I will guarantee you that he has all kinds of process wisdom that is not readily available to you. He is there all the time with the process and yet you only get to see it in bits in pieces. You have the snippets; his picture is broader and deeper. Yeah…put his mind in high gear…it is ready, willing, and able. And, guess what? You do that with questioning!! But before we go into the questioning aspect, let's forge ahead and cover work rules numbers 3 and 4.

- Leadership work rule number 3….the worker must state what he **"intends to do"** and
- Leadership work rule number 4….the **intent must be accompanied with sufficient detail** to explain to the supervisor, so the supervisor can answer with a pithy sentence such as:
  - "Perfect"… (that is my favorite answer).
  - "Thanks, that is a great idea."
  - "Yes, that is exactly what we need."
  - Or "sounds like you have thought this through, that'll work well."

An Example of Leadership and Leadership Training on the Floor

Pay particular attention to how questioning was the primary leadership technique that was used and also the technique that was taught in this real-life experience.

So let's say you, as the plant manager, are walking around the floor and see Gerhardt, your very best machinist, doing precisely nothing. Well he is really not doing nothing…he is fidgeting because first and foremost he does not like doing nothing. Like all of your workers, he is self-motivated.

Since you initiated the Four Leadership Work Rules just recently, you see this as a teaching opportunity. Six months ago you would have stormed over to Gerhardt and told him that "we're not paying you $45/hr. to stand around, I can get minimum wage earners to do that; so either find something to do or clock out." At which time, Gerhardt would have walked to the tool room and back a couple of times, looking occupied, because that used to qualify as work in this organization. Alternatively, you resist this former behavior and ask Gerhardt. "Is it accurate that you are doing nothing?" "Yeah," he says, "but I am **waiting** on Martin to give me the go ahead on my next job." Seeing that he is violating rule 1, he is "waiting to be told," but at the same time recognizing from his tone and his body language that he knows he is "waiting" and he knows this is wrong and he is currently unable to correct it. You then ask, "Well, if you owned this company, what would you do?" So Gerhardt, without a moment's hesitation, says, "I would set up and run Job 4867. I have the steel, all the tools are here and so is the drawing." To which you reply, "Gerhardt, you are way too valuable to be waiting. Your plan sounds good, so go ahead and I will find out what crisis is holding up Martin and have him here within the hour. Will that work?" Gerhardt smiles and nods OK.

Immediately I find Martin and politely and calmly inquire as to why Gerhardt was "waiting to be told?" Martin informs me that the clerk who updates the routings does not publish them until 9 and he did not want Gerhardt to start a job until he was sure it was consistent with the plan. Earlier, in a lesser state of emotional maturity, I would have dressed Martin down right then and there. However, I now see yet another "teaching opportunity." I guess my vision has improved as well. Anyway I approach him and after inquiring "how's it going?" and getting no feedback on the fact that Gerhardt was waiting around, I engage Martin thusly. "Martin, while walking round this morning, I found Gerhardt doing nothing. I spoke to him and he said he was waiting on a go-ahead from you. I then asked him 'if you owned the company what would you do?', to which he replied 'he would start on job 4867' and he also gave me the supporting details. Sounded like the right thing to do. So first, I told Gerhardt that he was too valuable to be 'not working.' Next I told him to start on job 4867 and you would be by within the hour to confirm. However, you know that rule 1 is 'no one can wait until told,' so why were you waiting on that routing slip, holding up Gerhardt?" Martin replies, "Well the clerk does not update and distribute the routings until 9, and the workers are at the machines at 7:30." So very calmly, I ask, "Well, Martin what would you do about that if you owned this company?". To which Martin replied, "I would have the clerk come in at 6:30 so he could get the routings to the machines before the machinists arrive." To which I reply, "Ok, what is preventing us from doing that?" To which Martin replies…" yeah, I get it, that is Continuous Improvement Question No. 3, …and the answer to Continuous Improvement Question No. 4 is that I'll get with the office manager and have the clerk's schedule modified. I'll do that immediately after I check in with Gerhardt. Tomorrow we will have the routings out to the floor before the machinists arrive." "Excellent" I say. I go on, "since this will be done tomorrow, what would be a good time for you and me to do a quick audit of the routing slips?" To which Martin replies, "First item on my Leader Standard Work is to check with each machinist at shift start up. I am done with that by 8:30, so how about tomorrow at 8:30 we meet at the large boring mill and check to see that each workstation has the correct and current routing slips." By now I am feeling a little better about my teaching skills as Martin has also answered all three parts of Continuous Improvement Question No. 5.

- What will be done?… "the correct and current routing slips will be there."

- How much?... "at all workstations."

- When?…. "by 7:30 tomorrow morning."

That is an example of both teaching leadership and exhibiting leadership. At any point in the process I could have answered a variety of questions. And the problem of Gerhardt being busy—at that moment—would have been resolved. We would have fought that fire and put it out; a short-term win clearly. However, we still would have had the problem each morning when any routing needed to be changed. So instead of putting some temporary band aid on this problem, through the use of lean leadership and teaching leadership, we now have created a more robust process that will work better each and every day. Furthermore, we got it done working with the people by questioning and teaching each and every step of the way. And guess what??…they know all the answers and now since the answers are theirs, the "stickability" of this solution is dramatically improved. Finally both Martin and Gerhardt are one step closer to being strong leaders and we have improved our leadership footprint.

Three final comments are as follows:

First, in this example you will see all Six Skills of Lean Leadership exhibited. You saw:

1. *Leaders as superior observers*: I was on the floor, and not only did I observe, but I was listening, a much underrated type of observational skill.

2. *Leaders as learners*: I was on the floor and asked just a few questions.

3. *Leaders as change agents*: The action was immediate, not inventoried to handle in some future meeting. It was real-time PDCA (Plan-Do-Check-Act).

4. *Leaders as teachers*: They are "lifelong" teachers. When something goes wrong, my first thought was "how can I use this as a teaching opportunity?" And then used teaching through the use of questioning rather than just instructing. Furthermore by teaching and acting on the four work rules, we were teaching the others how to lead.

5. *Leaders as role models*: I have taught real-time PDCA as a way of life, so I too much act immediately.

6. *Leaders as supporters*: By acting in this way they had no higher-level obstacles; they now had total control over this corrective action and improvement activity.

Second, you saw all Four Leadership Work Rules executed

1. Leadership work rule number 1…No one can **wait until told**…all must be working

2. Leadership work rule number 2…No one can **ask what to do**

3. Leadership work rule number 3…The worker must state what he **"intends to do"**

4. Leadership work rule number 4…The **intent must be accompanied with sufficient detail** to explain to the supervisor so the supervisor can answer with a pithy sentence.

Third, you will find in this example the use of all Five Questions of Continuous Improvement.

More than any techniques I can think of when the lean leader models the Six Skills of Lean Leadership and particularly follows the Four Leadership Work Rules, guided by the Five Questions of Continuous Improvement, the leadership footprints will grow exponentially. This will be a major factor that will march your business directly into the world of continuous improvements. The benefits of that impact are both unknown and unknowable, but they will certainly be large and likely far more impressive than your wildest dreams.

Use them wisely.

This process of teaching leadership is often nothing more than making your teammates responsible for the things they are capable of doing. In creating a larger leadership footprint, the worker has a greater contribution and is usually intrinsically motivated to do more and better. He feels a lot better about his work. First, he is accomplishing something rather than just waiting; nobody likes to stand around and try to look busy. Second, by suggesting ideas he is exhibiting some control over the timing and content of his work. Third, he knows he is contributing to the profitability and contributing in a way to make his company a better supplier to their customers and doing it with you. Finally he is growing in his leadership skills. By using these techniques we have activated all five of the intrinsic motivators described in Chap. 6. This is truly the most highly leveraged work you can do as a supervisor or managers.

By enlarging the leadership footprint, there is yet another very powerful advantage. You free the supervisor and managers for other work. You not only have the worker, who is closer to the information, to make job appropriate decisions, but this frees up a lot of supervisory and management time. Then, they too can *kaizen* their work areas. Everybody wins.

Some Other Questions to Guide the Floor Worker Might Be:

- If you were me, what would you recommend?
- What is consistent with our plan?
- What would be best for the company?
- What would the customer want/need?
- What do you think I would do?
- What makes sense?

In getting this level of engagement at the cell worker level, I have found two techniques to be invaluable.

- First is the use of visual displays. The more you have a "visual factory," the more you can delegate these tasks. There are many times when the cell worker does not make a decision but seeks assistance. More often than not, I find they seek assistance, not because they lack the skill to make the decision, but rather the limitation is more often a lack of information. The classic here is the use of *kanban* (see Chap. 17, Planning and Goals) which can take production planning, not only out of the hands of the supervisor but out of the hands of the planner as well. There is an old lean adage, "all the information is in the parts." If we can get that info into the hands of the cell worker, often we cannot only get the work done faster and better, but we can also free up the supervisor and planner for other work.

- The second issue is a larger issue and a cultural issue. That is the topic of delegation. Specifically we must get the supervisor and the planner to "want to let go" of this task. On the surface, it would seem they would want to shed this work, for almost everyone is overworked. However, you are now dealing with a lot of what makes a supervisor a supervisor, in many companies. This is very sensitive, emotionally charged, and must be handled well. This is a major cultural change so plan this well, discuss it well, and get all the issues out before taking this action.

The Four "Ah Ha" Experiences ... or ... If I Am the Key Leader, How Do I Know If I Am on the Right Track?

As you progress in your lean transformation, there are certain times you will go through the "ah ha" experience. This happens the moment that light bulb goes on and you have that moment of clarity. In a moment of clarity, you become suddenly and deeply aware of some truth. Your vision becomes unclouded and sharply focused. There is a rush of understanding that appears in the form of an epiphany. You are simply amazed and new doors of understanding swing wide open. Things you had been previously told and maybe were skeptical about have demonstrated themselves in a clarity such that you once and for all "get it." The effect is not just intellectual; it is behavioral. You are effervescent and want to share this new-found understanding with anyone who will listen.

These "ah ha" experiences are not only enlightening, but they are energizing and liberating as well. You cannot only "see" with a newfound clarity; you have a new type of initiative and are not only driven, but compelled to do something about it. There is both a sense of being empowered by the truth and a compulsion to correct things you now see as deeply in need of improvement.

In a lean transformation there are four such moments for the leadership. The good news; in fact the great news is that each of these "ah ha" experiences is a strong sign that you are on the right path. These "ah ha" experiences are a special kind of feedback that says beyond any doubt… "you are really and truly transforming this place, you are on the right track, push on."

The First "Ah Ha"—Exhilaration

This experience normally comes very early in the transformation. It occurs when some part of your lean system works as designed and yields great results. For example, you have worked hard to install the three-part inventory system. You believe in it fully. But then you have the design case shutdown of some critical piece of equipment and productions stops. It stops cold. All your resources scramble to get back into production. The focus of your entire team is on fixing the issue and returning to normal production. Totally focused on the failure, you lose sight of the fact that the shipping department is still shipping. However, 16 hours later you are back on line producing at rate and you breathe a sigh of relief. Then it dawns on you—and with a natural sense of impending despair, you ask "What shipments did we miss, who did we short?" And to your amazement and total glee you learn that you have not missed one shipment! No shorts, no lates—100% full order, non-expedited on-time delivery! You are not only elated—you are totally amazed. This has never happened before, not even once. You and your team marvel at the fact that earlier you would have had to short 10 or 15 shipments but now with "our 3 part lean inventory" in place…we did not short a single customer. The light goes on, you experience an enlightenment that confirms beyond any doubt that…this lean system really works!!

Alternatively you are taking your daily *Gemba* walk and run across Ethyl, a normally contentious worker with over 30 years seniority. You ask her, "Ethyl are we winning today?". And she says, "let me tell you, since we implemented this lean stuff, I go home less stressed. My husband and I get along better, it has actually improved our marriage. Thanks." And you are completely blown away. The rest of your day you have a permanent smile carved into your face. You simply can't smile. You go home and explain this to your wife who totally and completely "gets it." You and she discuss it to the wee hours of the morning. Each of these examples could be an "ah ha" experience for you. Either way, you are amazed, exhilarated and walk away assured you are on the right track on your lean transformation and motivated to accelerate it even further.

The Second "Ah Ha"...Not So Much Fun This Time

The second "ah ha" experience is equally telling, but initially is not so much fun. It is simultaneously an "ah ha" and an "ah shit" moment...depending on how you view it. While reviewing your lean objectives, you notice a slowdown in improvements. Upon investigation you find that something in your manufacturing system that was not part of the initial lean transformation design is holding you back. It could be your management training program; an ineffective maintenance program; some failings in your existing systems of recognition and rewards or the support from your IT group. Whatever the problem was, it was in your existing manufacturing system, and you NOW discover it is hindering your progress. Initially you say "ah shit" because you did not foresee it as a problem. Then you recall the wise words of your *sensei* who told you earlier, "all of what you call the lean tools are things whose purpose is to cause problems to surface" and you also recall that he told you, "you gotta learn to love problems, they guide us to what we must improve upon." At first you may feel a little deflated but then you say, "ah ha." Now I get it! There are huge gaps in our existing system and we must review every aspect of our system with much greater scrutiny. Enlightened, energized, and empowered, you then begin a long and arduous review of your existing manufacturing systems. You then modify your plans, prioritize these issues, repair the deficiencies, and make your system even stronger. The "ah ha" experience drives you to act.

In addition with this big "ah ha," there are several smaller "ah has." Some other issues are likewise clearer. Specifically you are now a bit more circumspect about how much you really understand about this lean transformation. This awakening cautions you to open your eyes, your ears, and your heart more fully. It humbles you because you view this as a setback. But then you recall the words of your *sensei* who cautioned you that there would be periods like this. Specifically, you feel like you are going backward by repairing your existing system. However, all that really occurred was that you found a weakness because **now you are more aware**. The deficiency in your operating system was there all along. There was no backsliding except compared to your initial scale of evaluation; that scale was based on less knowledge and less awareness. You recall the words of your *sensei* who said, "we must learn our way to greater competency," and those words now have newer and deeper meaning.

Normally these first two "ah ha" experiences come early in the transformation— certainly less than 12 months from transformation launch and normally less than 6 months.

A Bit Later, the Third "Ah Ha"

Frequently, there is a decent gap before the third "ah ha." It too is preceded by an "ah shit" moment. This normally comes about when changes you made earlier do not stick. For example, after installing several *kanban* systems, you find they have problems losing cards and keeping above minimum inventory levels; or your TWI (Training within Industries) training program cannot keep up with the process changes made on the floor; your production-by-hour boards as part of your visual factory are not being acted upon; or the SPC (statistical process control) charts are not being maintained. Upon review you find that many people knew of these signs of deterioration but thought the management team wanted more "things" done. You investigate further and find this is not the only example of "favorable information" traveling fast but "bad news" not even making to the formal or informal communications channels. You find out we have too much "happy talk" and not enough dealing with reality. In a phrase we have a cultural problem. In your investigation you also find that people are not willing to "tell it like it is" to the management team and you find other elements of a closed culture. And you say, "ah shit" we have a cultural problem. However, you recall your second "ah ha" experience and at first you thought it was a negative, yet once the problems had been resolved, your lean system improved and with it all plant performance parameters likewise improved. There were no accidents, and on-time delivery improved six points and profits rose to record levels. Reflecting on your last "ah ha" you ponder…"maybe this problem is not a problem, rather it is an opportunity being serendipitously shown to us—so we can improve." You recall the words of your *sensei* when he said, "culture trumps everything." You thought you understood what he said earlier, but now it has a completely new and clearer meaning. At that moment you understand the deleterious effects of a closed culture and you get it, "culture does trump everything." The third "ah ha." You think back to all you had done to make those *kanban* work, to make the SPC work, and how you reveled in the early successes…only to have it unravel because we have cultural problems we had not addressed. You also recall the cautions your *sensei* had given you about the dangers of going "a mile wide and an inch deep" and that too has new meaning. It is all just so much clearer. This epiphany of understanding energizes and empowers you to action. Immediately you begin to codify your culture, decide on the culture that you will need, and begin the hard work to design the cultural changes that will be necessary.

The Fourth "Ah Ha"…and the Problem of Your Own Making

As you begin the work to redesign your culture or during the changes, you will experience the fourth "ah ha," which like the second and third "ah has"; is preceded by that uncomfortable predecessor, the "ah shit" moment. Which is this? Uncomfortably you realize that the problem with the culture is one of your own making. You see while trying to unravel the issue of "happy talk," you realize that the only reason it persists is that you have accepted it. Heck, you find you not only accept it, you have encouraged it—as has your whole staff. Wow! It is the Pogo concept, "we have encountered the enemy and he is us." You have an "ah ha" as you realize this. You then truly understand what your *sensei* told you long ago: that the culture was created at the top; that you will get the type of behavior you accept, not what you expect; and a million other pieces of wisdom. However, this time the words have a clarity that is cathartic; this time they have real meaning. You have clarity of understanding as well as clarity of responsibility. This is a truly significant moment in the entire lean transformation. The moment the very top guy, the CEO, reaches this epiphany, success is almost guaranteed. This epiphany is followed by an ongoing process of personal development and growth. Almost without exception, this particular "ah ha" experience vaults you into a process of ongoing questioning, reflection, and personal examination that is not only very healthy for you as an individual but healthy for the entire organization.

Armed with this new information you spring into action. You gather your team, with energy and conviction you explain to them your new found learnings; declare we must do this better; and solicit their help in how we do it better. This is the "teaching opportunity of all teaching opportunities." To get the top management to understand,

The Four "Ah Ha" Experiences

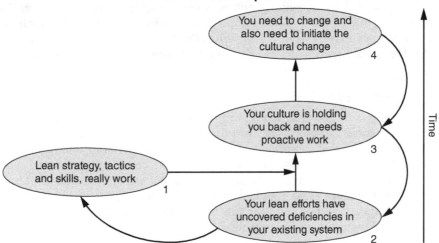

accept, and act on the principle that a consciously and properly managed culture is the responsibility of the top few managers; and that they are the only people who can orchestrate this is a huge step forward.

Later, maybe quite a bit later, someone comes to you with some kind of "bad news" that, even though it was critically important, it had previously been filtered by your culture bent on happy talk and you say, "ah ha, I am glad we attacked this issue of culture, now we can address this critical issue in a timely fashion" and you feel justified in all the work you had put into consciously designing and managing the culture. From that very moment, you begin to focus your attention on the culture and make it a priority. You make a conscious decision to design the culture, rather than to let it grow organically and force itself upon you. You also recognize, beyond any doubt, you are once again on the right path.

With this final "ah ha" the success of your lean transformation is virtually guaranteed. Once the top few people buy in and are truly engaged you are on your way. Happy improvings to all!!! (At this point I suggest you read the Fivefold Definition of Lean in Chap. 10.)

Two Cautionary Statements

First, this is not a one-way street. Once you get your first "ah ha" and take the leap to action, there is no physical or metaphysical law that says you cannot regress. Very likely you will. Watch out for deterioration in any system.

Second, almost any manager will tell you he has the understanding of the "ah ha" experience, even before the actual experience. He may actually believe this or he may just wish this to be true; but he is most often mistaken. How do you know if he has really captured the "ah ha" experience or just thinks he has? First, look in his eyes. It is unmistakable. Second, listen to his voice. Once you experience the "ah ha" experience, your voice takes on a new speed and a new energy. Unfortunately many have learned how to fake this aspect quite nicely and among us amateur behavioral scientists, many of us cannot really discern. In the presence of a real behavioral expert he will hear and see the difference. But the real acid test is this: look at what he does. Prior to the "ah ha" experience, actions are planned and prioritized with the "business as usual" approach. Following the "ah ha" experience, this is an activity we simply cannot fail to do. There is an urgency that transcends logic when the "ah ha" is truly and deeply experienced. You feel it once and for all, and all around you can tell that, "something has changed."

Chapter Summary

Leadership is the skill set that is required to effect change and is distinguishable from management. It is a crucial element that is needed to make a lean transformation and the new and necessary leadership style is situational in nature and called "lean leadership." It is a style that is easily distinguished from command and control styles. A key element of lean leadership is the ability to lead by asking questions and by the use of questioning you can also greatly increase your leadership footprint. Finally, by knowing and understanding the four "ah ha" experiences, you can confirm that you are properly transforming your organization.

CHAPTER 6

Cultural Change Leading Indicator No. 2, Motivation

Aim of this chapter … to discuss the motivation to execute a lean transformation; what typically drives it and what it takes to be successful. The second topic is the role of motivation in the cultural change transformation and in the daily activities within the plant. We will cover the myths and the reality of what it takes to sustain a highly motivated work environment.

Motivation to Implement a Lean Transformation

It has been our experience that most successful lean transformations are driven by one of two motivating factors. The first factor is evident when the company is looking survival square in the face, and is on the verge of going broke. Under these circumstances, it is easy to get people's attention. However, the most common situation occurs when your customer says you must implement Lean—that is, they say, "If you wish to continue doing business with us, you must implement a Lean Manufacturing system." In the end, both are about the same issue: survival.

> **"I**t is not the strongest of the species that survive, nor the most intelligent, but the one most responsive to change. **"**
> Charles Darwin

On some occasions, we run across companies that want to implement a lean transformation because some visionary has decreed it so. This choice is often an informed choice that is part of an overall business strategy. In my experience, these "optional" efforts have very mixed results. More often than not they proceed far slower and with much less success than those that are survival motivated. However, optional or not, those that proceed with careful attention to avoidance of the 10 Lean Killers and Six Roll Out Design Errors are among the most successful.

So in the end, if your concern is not that of your immediate survival, the issue is likely one of *long-term* survival. Because you see, you will learn that the competition is improving, and if *you* do not improve, you will not survive—so, in fact, regardless of your apparent motivation, in the end, it is always a survival issue. It will be important to carry this message to the entire facility so they will have the proper motivation to make this transformation a success. Do not be surprised if you find resistance to this issue, since what we are talking about is cultural change—and with cultural change, unless designed and managed properly, you *always* get resistance.

Although there is a lot in the literature about people having a natural resistance to change, I simply do not agree. I can look at the two biggest changes I made in my life: the decision to get married and the decision to have children. Each was a huge change. However, I did both of them quite willingly, actually with a great deal of zeal in each case.

On the other hand, that was my choice. I wanted to make that change. Human nature is such that we do not mind that kind of change. However, **most people do not really wish to be changed by others** and those type of changes engender resistance. Consequently, the leadership must develop a culture that is mature enough to recognize that change is needed, then people will, at some level, "want to" change. Furthermore, the leadership must then equip the culture with the necessary skills so they "can" change.

So in a nutshell to avoid the resistance to change associated with any culture-changing issue, get the workforce to "want to" change and equip them so they "can" change. That is no mean feat and the leadership will have to work hard to avoid the 10 Lean Killers, and they will need to design a transformation that avoids the Six Roll Out Design Errors and follow the implementation advice in Part Three—that is all!

As I said earlier, regardless of whether the change is forced or an optional matter, there is always a certain level of resistance to change or resistance to growth involved. The basic issue is that all entities, including cultures, seek stability.

This resistance to change or resistance to growth is also seen in the human body—something called *homeostasis*. It is the body's desire to seek a position of equilibrium. If we wish to change our body, we must force it to go beyond its limits. We must make it uncomfortable. Take, for example, someone who wants to become a great athlete—a soccer player, let's say. In addition to the many skills a soccer player must develop, he must have tremendous cardiovascular fitness—in a phrase, he must be able to run all day. To acquire this ability, he must push his body beyond its normal limits time and again; he must undergo discomfort and even pain; he also must exhibit the mental strength to fight through the physical pain and discomfort again and again, knowing well it will be painful once more. He must resist the desire to take a day off or go partying because, even though he would like to do those things, he knows it will not be in the best interest of his goal to become a great soccer player. With his eyes clearly focused on his goal, he can maintain his regimen in spite of what both his mind and his body tell him. To measure fitness levels, he can chart his 2-mile run time. When he sees this improving, he is further motivated by his mini success, and then when he sees how this better fitness translates into better play on the field, he feels justified in putting his body through the discomfort and pain he had to endure. Frequently, his feelings at this point go beyond justification. He is proud of his efforts and often extracts a serious and mature sense of accomplishment.

Point of Clarity It takes great courage and character to say "no" to the things we would like to do so we can focus on the things we must do in order to attain those goals worth achieving.

In spite of this, he will encounter other distractions and challenges. Others will tell him he is working too hard; he needs a day off or he is taking the project too seriously. Frequently, they tell him this not because it is in his best interest, but because it furthers some issue of theirs. Nonetheless, it sounds good and his inner voice says, "Well, I have been working hard. Maybe I do deserve a day off." Well, maybe he does and maybe he doesn't, but these distractions test his resolve and he must decide which the best course of action is. Those who want to be great will summon up their courage, fight through the challenges, make the sacrifices, and will not be deterred from their goals. Often, this makes them less popular or puts other social pressures on them, but those who are focused keep their eye on the goal and proceed.

Well, the changes we must undergo in our lean transformation are much like the changes an athlete must undertake. Like the athlete, we must:

> **"W**e are made to persist. That's how we find out who we are.**"**
>
> Tobias Wolff

- Have a clear goal in sight.
- Recognize that we need to change to reach the goal.
- Recognize that the changes will be uncomfortable and even painful at times.
- Recognize that the short-term problems and losses will be just that, short term; and we must keep our eye on the target, the long-term objectives.
- Recognize there will be forces within ourselves and outside ourselves, which are driven by different motivations. These forces will resist the necessary changes and often coax us in the wrong direction.

In spite of all this, we must focus on the goals and persist!

Here, in particular, leadership is crucial. The leader must know all this. Furthermore, they must prepare the people for the changes, and the constancy of the change that is to come. The leadership must realize and make all those involved understand that the entire manufacturing world is improving and that if we stay where we are, we are really regressing. If we wish to survive, we must improve, and to improve we must change. Change becomes a given, and in fact if we want to become more competitive, we must improve faster than our competition. So, to survive we must change, and to prosper we must increase our rate of change so it exceeds the rate of change of our competition. That is not an easy sell for any leader.

A Lean Paradox

In leanspeak, we talk of the removal of variation in many forms. One such technique to reduce variation is the standardization of work. We continually look to standardize things so they are all done the same; we do not want even one operator to vary from the other in work methods. We seek standardization. On the other hand, we tell everyone that the system must change, and we must foster creativity to improve. So we must remain the same, yet we must continually change. It is a major paradox worth thinking about.

So our leader has a daunting task in front of him. He feels he must create within the organization a willingness to change. Causing this change will always be a challenge for the leadership and the management. The desired state they must reach is similar to the mental state of the athlete. The leadership must develop a mind set, a culture that says, "We are willing to undergo the temporary discomfort of the change because we know that

> **"I**n times of change, learners inherit the Earth, while the learned find themselves beautifully equipped to deal with a world that no longer exists.**"**
>
> Eric Hoffer

this discomfort will go away, and only by going through this discomfort can we reach our goals."

Just How Does the Management Do All That, and Do It Long Term?

As a young manager I was taught by my mentor that my job was largely to "create an environment where it is possible for my employees to succeed, it was up to them to choose if they wanted to or not." I took this advice to heart and was incredibly successful in training and developing young men and women. Then as I got a little more seasoned, I was "taught" a whole new list of things I needed to do to get people to perform. These things included using money and perks to motivate along with a whole slew of what I would call "pop psychology." Techniques including such items as employee suggestion systems to encourage "involvement" and the guiding management principle I often saw in the 1980s, "I don't care how you get it done, I only care about the results," which was a guiding principle at my employer at the time. Many of these concepts I resisted but would get severely criticized when things did not go well. Fortunately, I found soon enough that the advice my mentor had given me earlier was still the best and I regressed to that earlier advice, so I might progress.

> **P**oint of Clarity   Great leadership will move us into and through the needed pain so we might change and improve. Sometimes the leadership technique used is one of nurturance; other times it is simply force.

I learned, relearned, practiced, learned more, practiced more, and came up with three basic principles of motivation. When I speak of my three basic rules, they have some limits. First, they apply to the geography where I have practiced. In Canada and the United States where incomes are high on a world scale, all this applies fully. I have worked for a few European companies from France, Germany, Sweden, and Italy and a handful of English companies as well, I can say these principles apply equally well there. In addition I have done a great deal of work in Mexico, Central and South America, and these principles still largely apply in that geography. I would not expect them to apply the same in really poor countries such as China, Bangladesh, or India where many people are at Maslow's first stage, trying to satisfy basic survival needs.

Rule Number 1—Don't Try to Motivate People, They Don't Need It

Rather I find that they come to work not only motivated but highly motivated. They not only want to learn about the business, they want to learn about their job, they want to excel at their jobs and more often than not, at day one, they have several ideas on improvements that can be made. And in fact, most efforts to directly motivate the workforce, although well-meaning and sounding good, are counterproductive and should be avoided.

Who Also Says This Is True?

This is premise you will find in Douglas McGregor's famous book (*The Human Side of Enterprise*, McGraw Hill, 1960) and also supported by the works of Frederick Herzberg (*The Motivation to Work*, Transaction Publishers, 1993) as well as the writings of Abraham Maslow (*Motivation and Personality*, Harper and Bros., 1954). In fact the great psychologist, Erik Erikson (*Identity and the Life Cycle*, International Universities Press, 1959) felt we had a natural innate motivation to perform and we would all, at a very early age, go through a series of "crises." Three of these "crises" were autonomy versus shame and doubt; initiative versus guilt; and industry versus inferiority. Furthermore he states that nearly all of us will have some success in overcoming these crises and this will give us, in order, a certain level of will; a certain degree of

> **P**oint of Clarity   Motivators come in two types; there are extrinsic motivators, which are external to the work; and there are intrinsic motivators, inherent in the work itself.

purpose; and a certain level of competence, respectively. All three of these personality traits: autonomy, initiative, and industry, then motivate, they drive us to do things. Hence we come to work, wanting to do a good job, wanting to improve the workplace; and this penchant came to us at a very, very early age, long before we went to work.

Rule Number Two—Don't Demotivate the Workforce

As a manager I can think of two ways in which you can do this. First, there are a "**thousand little things management does and says**" to demotivate the workforce. Second, you can demotivate employees by misusing the extrinsic and not using the intrinsic motivators.

A Thousand Little Things

Well some of these are not so little. But they are all things that management can do to demotivate the workforce. They can presume the workforce is not competent and must be overly controlled to get the work done. They can presume the workforce is lazy and needs to be watched very carefully. They can presume the workforce is not motivated and needs to be enticed with rewards or worse yet, punishment, if they do not "hop to." Most of these are attitudinal on the part of management and were addressed and debunked by Douglas McGregor in his seminal work (*The Human Side of Enterprise*, McGraw Hill, 1960). In summary McGregor says there are two types of managers—those that ascribe to Theory X and those that ascribe to Theory Y. Theory Y managers have been and are guided by a certain list of beliefs. Theory Y managers believe:

1. The expenditure of physical and mental effort in work is as natural as play or rest. The average human being does not inherently dislike work. Depending upon controllable conditions, work may be a source of satisfaction (and will be voluntarily performed) or a source of punishment (and will be avoided if possible).

2. External control and the threat of punishment are not the only means for bringing about effort toward organizational effectiveness. Man will exercise self-direction and self-control in the service of objectives to which he is committed.

3. Commitment to objectives is a function of the rewards associated with their achievement. The most significant of such rewards, for example, the satisfaction of ego and self-actualization needs, can be direct products of effort directed toward organizational objectives.

4. The average human being learns, under proper conditions, not only to accept but to seek responsibility. Avoidance of responsibility, lack of ambition, and emphasis on security are generally consequences of experience, not inherent human characteristics.

5. The capacity to exercise a relatively high degree of imagination, ingenuity, and creativity in the solution of organizational problems is widely, not narrowly, distributed in the population.

6. Under the conditions of modern industrial life, the intellectual potentialities of the average human being are only partially utilized.

Above all, the assumptions of Theory Y point up the fact that the limits on human collaboration in the organizational setting are not limits of human nature but rather they are limits on management's ingenuity in discovering how to realize the potential represented by its human resources.

Specifically, as a minimum, management needs to accept:

- That workers are self-motivated and thrive on responsibility; hence, management needs to involve the worker in a participative relationship, but management must still retain appropriate control.

- That workers have considerable skills they naturally wish to embellish, and hence managers should encourage skill development and personal growth and also encourage them to make suggestions and improvements.

- That workers consider work as a natural part of life and will solve problems naturally so management should teach them and include them in the activities to improve their workplace

By contrast, the Theory X managers believe:

- The average human being has an inherent dislike of work and will avoid it if he can…reflects an underlying belief that management must counteract an inherent human tendency to avoid work. The evidence for the correctness of this assumption would seem to most managers to be incontrovertible.

- Because of this human characteristic of dislike of work, most people must be coerced, controlled, directed, and threatened. The dislike of work is so strong that even the promise of rewards is not generally enough to overcome it. People will accept the rewards and demand continually higher ones, but these alone will not produce the expected results; only punishment and the threat of punishment is truly effective.

- The average human being prefers to be directed, wishes to avoid responsibility, has relatively little ambition, and wants security above all. McGregor, Douglas (2005-12-21). *The Human Side of Enterprise*, Douglas McGregor, Annotated Edition, pp. 43–44, McGraw-Hill, Kindle Edition.

Hence a manager driven by Theory X beliefs would best be described as a "command and control" type leader who uses that style at all times, regardless of circumstance. And while there is an occasional time and an occasional place for this style of management, with today's workforce this is largely an antiquated style that does not support Lean and hence is not an effective means to achieve a more effective and efficient workplace.

It is interesting to note that McGregor made quite a point that Theory X, while overused and misused, still had its place. Keep in mind McGregor wrote this in 1960 and things have changed. However, theoretically I agree. But you would need to have a workplace where the worker has no possibility to contribute to the betterment of the facility and a management and support staff that knows all the answers…to name just a few qualities to support a Theory X basic management style. I know of no such facility in the world of manufacturing, or the service sector for that matter, although there may be some, somewhere. Hence the countermeasure to "**these thousand little things that management does and says**" is to proclaim, teach, foster, model, and support a Theory Y management style.

Misusing the Extrinsic and Intrinsic Motivators

Herzberg was the first to point out the factors which create job satisfaction are a different list of factors than those that provide job dissatisfaction. Until his research it was largely believed that if the presence of a certain factor would motivate people, then conversely withdrawing or minimizing this same factor would demotivate the individual. His studies debunked this theory. He codified these factors into two groups. The first group he called motivators, the second group he named "hygiene factors." Hygiene factors are frequently called extrinsic motivators and really do not give positive job satisfaction in the long run. Hygiene factors are not really motivators at all. Rather they are factors, when supplied in sufficient quantity, they become non issues. However, these same "hygiene factors," if not supplied in sufficient quantity, then become demotivators, usually large demotivators. The classic "hygiene factors" are pay and benefits. However, Herzberg's list of "hygiene factors" went beyond pay and benefits to include many things including such items as company policies and supervisory actions; but all were extrinsic to the work itself. On the other hand, the motivators were all directly related to the work and included such items as achievement, personal growth, the work itself, and even the concept of job responsibility.

Herzberg was right. Unfortunately many businesses still do not accept this.

Consequently and consistent with Herzberg's theory, we must do two things. First, for these "hygiene factors," especially pay and benefits, we need to make sure they are satisfied to the level necessary so they do not become demotivating. In most cases this means you will need to match the pay and benefits to the competitive level in your area. My recommendation is to pay at a level slightly above the average, for example, at the 60th percentile. The objective is to turn these "lightning rod" issues into non issues. Second, the extension of this is to quit trying to use these hygiene factors as some form of motivation. Simply put—they do not work. In fact, very often they actually demotivate the workforce. Especially in work that is complex, requires creativity or involves problem solving, these short-term, extrinsic motivators actually reduce creativity, and foster very short-term thinking at the expense of long-term results. They extinguish intrinsic motivation, diminish performance, crush creativity, and crowd out good behavior. If that is not enough bad news, you will find that this misuse of these extrinsic motivators also encourages cheating, shortcuts, and unethical behavior. I know this flies in the face of a whole lot of prior and even planned management actions, but history has shown Herzberg to be right time and time again. If you wish to study this further, I recommend you read Herzberg as referenced earlier and also *Punished by Rewards* by Alfie Kohn (Houghton Mifflin Co., 1993).

How the Misuse of Extrinsic Motivators Actually Demotivates

This is a topic beyond the scope of this book but is very interesting nonetheless. Metaphorically think of being transported to the airport, first by your wife, next by a taxi driver. Whom should you tip?? If my wife dropped me off at the airport and upon arrival, I hopped out, said a perfunctory "thank you," and gave her a $15 tip, she would be very offended. Sometimes I am not very socially aware but in this case I would not even attempt this as an experiment. Yet if I took the taxi and did not tip the driver, he would be very offended and understandably so. You see the same action can create a wildly different response in different circumstances, and this goes directly to what motivates each of them. If you wish to understand this and learn many of the twists and turns of this dynamic, often called "market versus social conditions," you can read Deci and Ryan's work; it is amazing. On the other hand, there is a simplified version focusing on intrinsic motivation in an entertaining, although controversial book by Daniel Pink entitled, *Drive* (Riverhead Books). Alternatively you can go to YouTube, Google "Drive or RSA Animate-Drive," and watch the short video. It is entertaining and covers nearly all the relevant material in Pink's book. I also recommend, *Predictably Irrational* by Dan Ariely, where he discusses this and many other aspects of human behavior (Harper-Collins e-books, 2010).

Rule Number Three—Manage to Sustain This Motivation

So now we know some things we should not do; but what should we do to keep the workers motivated? Herzberg taught us there were "motivators." He included such topics as challenging work, recognition, personal growth, and responsibility. All these motivators give positive job satisfaction arising from "intrinsic" elements of the work itself. Later in the 1970s Edward Deci and Richard Ryan (*Intrinsic Motivation and Self-Determination in Human Behavior*, Plenum Press, 1985) did studies regarding worker engagement and further validated the concept of intrinsic motivation. However, the real proof to me that the use of these "intrinsic motivators" is the major tool to be used by the manager to keep worker motivated comes from my own personal experience, both as an employee and as a manager. Early in my management career I used these intrinsic motivators extensively. Then later, in one period of my managerial experience I was taught to use other motivational tools, thought to be more powerful. Unfortunately I found myself to be less successful in both my managerial and supervisory roles, and actually by a large margin. Having then learned from that experience and those failures, and after studying even more, reflecting on what I have done well and done not-so-well and focusing intently on those intrinsic motivators once again…I found they not only work, they are extremely powerful tools.

Furthermore, the power of these intrinsic motivators has been reconfirmed once again during my consultancy of 25 years. In that experience I have tried these principles and found them to be not just successful— but wildly successful as I have not only used them but taught them as well.

The five intrinsic motivators are as follows:

1. A sense of meaningfulness
2. A sense of control

3. A sense of accomplishment

4. A sense of growth

5. A sense of community

Think about each of these motivators:

1. **A sense of meaningfulness:** Do your workers show greater interest in the work when they understand they are working for a meaningful task or when they are serving a higher purpose? Do they understand the company mission and vision to represent a company that seeks to be competitive, thriving, growing, a company that not only makes money but gives back to the employees and is a good corporate citizen in the community? Or are your management actions solely focused on the goal of making money? Just how motivated do you get to know that you have improved the portfolio of some stockholder?

   If management's actions are heavily focused or solely focused on "the bottom line at all costs," your employees' sole focus will, predictably, be, "What's in it for me?" They won't "want to" work for the company, only for themselves— and they won't "want to" improve the workplace.

   Can your employees "see" that their contributions are not only necessary but significant? That their ideas are considered? Or is your entire workforce just another fungible piece of easily replaced hardware, or worse yet, are they treated as "an expense to be minimized"?

2. **A sense of control:** Do your workers have some way to get input into the things they can affect and the things they should affect? Are they being asked to participate in *kaizen* activities in their workplace? Are they being trained in and training others using the TWI (Training within Industries) methods? Is their input regarding their job redesign being sought on a regular basis? Do they have ways to control what and how they do things, or are they just following the instructions some engineer wrote from his desk away from the production floor? If it is a "my way or the highway" management style, employees will find the highway as soon as something slightly better appears.

3. **A sense of accomplishment:** Do your workers have ways to determine whether they have done a good job? Can they answer the question, "How did I (we) do today?" Can they go home knowing they did well? Or is "not getting your ass chewed out" the definition of a "good day"? Can they tell each hour of each day if they are doing their job well? Are their visual indicators in place? Can they codify and quantify their contribution? Can they give an honest and positive response to the question, "Are we winning today"?

4. **A sense of growth:** Do your workers have a way to contribute and grow as individuals? Can they improve their skills via cross-training and advancement? Is there a conscious effort to create "future opportunities," or does your company supply no sense of hope for the future of the individual? Can your company reward your employees, even the cell workers, with opportunities to exercise their demonstrated skills, such as writing new procedures or training other employees to their level of competence? Does each employee have a personal development plan?

5. **A sense of community:** Do your employees have a true sense of teamwork at work? Do your employees have reason to proudly wear the company logo on their shirts, or do they only proudly wear their "Big Strikers" shirt from the bowling league? Or worse yet, is there a sense of community focused on the union and not the company? Humans are a social animal; and if their sense of community is not fed at work, they will seek it out elsewhere.

Reflect and Do a "Reality Check"

You can perform a reality check on these five intrinsic motivators. Simply ask yourself, "How is it that volunteer organizations can function and persist year after year?" After all they are "volunteer" organizations and as such are often unable to "bribe you" with pay and perks. Their "motivational toolkit" is virtually devoid of the extrinsic motivators, and they are left with only the intrinsic motivators as arrows in their quiver to keep the staff motivated. Successful volunteer organizations understand how to utilize these five elements to acquire and retain their workers. Consequently, for-profits can learn a lot by studying how not-for-profits retain a highly engaged workforce.

Managers must learn that developing continued engagement—beyond day 1—among their employees is largely an issue of recognizing and feeding the five key elements. Some call this "motivating the employees." Since I believe they come to work "motivated," I have another name for it, which I will share with you at the end of this chapter.

Once recognized, what can managers do to feed the five key elements? Five suggestions cover most of the ground needed. They are as follows:

1. **Create and live—and I mean live—your company mission and vision.** Many create them; far fewer live them. Give your mission and vision the "tombstone test." That is: Is this a mission and vision that I would want on my tombstone to represent me and the type of company I work for?

2. **Create goals, and metrics for those goals, that properly reflect all of the vision and mission.** Make sure the goals are aligned and focused throughout the organization. *Hoshin–Kanri* planning is one method to get aligned and focused, and its use of "catchball" is a simple tool to assist this effort (see Chap. 16).

3. **Provide the support needed at each level** so everyone can contribute to the execution of the goals and, hence, the mission and vision. This includes providing clear work instructions, training in those instructions, and the support and resources to accomplish the goals.

4. **Make the goals and metrics "transparent"** at the floor level with the use of visual management tools. With such tools, everyone is clear on what is required and "how we're doing right now." Success is visible. Support these visual tools with management feedback systems that include going to the "gemba" and finding someone doing something right and recognizing that accomplishment.

5. **Make employee involvement a way of life on the floor.** Include them in all that they are inherently capable of doing. Stretch them; they are capable of a great deal. If they don't have the skills, train them. TWI-type training (see Chap. 12) and real-time problem solving are two very practical, powerful, and empowering tools to delegate to the floor worker.

If you find the preceding list of five suggestions not very earthshaking…neither do I. This
thing that many call motivation, I simply call
good management.

> **P**oint of Clarity It's all about good management. The rest is just details.

I Came Full Circle…

This brings me full circle to what my mentor told
me over 40 years ago, "Wilson, your job as a supervisor is to create an environment
where it is possible for your employees to succeed, it is up to them to choose if they
want to or not."

And just how do you "create this environment"? First of all, supply them with the
three elements of engagement (see Chap. 8) by teaching them what to do, how to do it,
and then giving them the resources to do it. Second, learn and execute the three rules of
employee motivation by: first, accepting they are naturally motivated; second, accepting that they are more often than not demotivated by mis management and avoid these
pitfalls; and third, use the intrinsic motivators in your daily activities to sustain the
motivation they brought to work at day 1 to keep them at high productivity with a
resultant high morale. **This is all done by implementing the Six Skills of Lean
Leadership as detailed in Chap. 5 which is fed by maintaining the mind set and
mental models of the Theory Y manager as outlined by Douglas McGregor over
50 years ago.**

Lean Manufacturing Is Not Only Built upon the Intrinsic Motivators but the Lean Manufacturing System Fuels These Same Intrinsic Motivators

Intrinsic Motivation Is Designed into Lean Manufacturing

When Toyota Corporation published "The Toyota Way 2001" (Toyota Motor Corporation, 2001), they also publicly shared what they referred to as the DNA of the now
famous Toyota Production System (TPS). "The Toyota Way" is built on two pillars. The first
is *Continuous Improvement* which was obvious to all. Less obvious, but quite frankly even
more important was the second pillar, *Respect for People*. As you read this amazing document, you see topics such as "challenge" and "promoting organizational learning" and
you immediately recognize these as the intrinsic motivator we call, "a sense of growth."
You read about "mutual trust and mutual responsibility," "sharing of goals," and "emphasis
on consensus" and you immediately can see they support the intrinsic motivator of "a
sense of community." The document is literally a huge lesson on how to design an organization that uses the intrinsic motivators and you can see how intrinsic motivation is built into
the DNA of the Toyota Way, the, TPS, and by its logical extension, Lean Manufacturing.

The Lean Manufacturing System Fuels These Same Intrinsic Motivators

Often, when describing the lean factory, we use the term of transparency and call our
floor condition "the visual factory." We told you that one of the primary functions of the
lean tools of *heijunka*, production by hour (PBH) boards and *andons* is to bring problems
to the surface in an early fashion so they can be recognized and solved. Well a second
factor of "the visual factory" is its motivational effect. Take the simple visual display of

the production by hour (PBH) board. I like white boards, manually filled out with columns for "Planned Production," "Actual Production," and a column for "Comments," as a minimum. In addition, abnormal deviations in production are quantified by $+/-$ X units/h and written in plain view on the PBH board. Planned production is entered on the board by the supervisor. The operators, during their shift, fill in actual production each hour. If the production is within normal limits, production proceeds normally. However, if an abnormal deviation is noted on any given hour, immediate problem solving is initiated, normally by the team leader with the help of the cell workers. It seems odd to many, but this simple manual, visual display fuels several of the intrinsic motivators. For example, by having it done manually, by the operators themselves, they are recording their production rate on their visual display. This feeds directly into a true "sense of accomplishment." Everyone cannot only see that they are producing but when they compare the "actual" to the "planned," they can see they are meeting the goals and "voilà" can see they are "winning." The cathartic effect of winning should never be underestimated. Not only that, but they get this feedback every hour. Imagine the effect of positive feedback every hour on your workforce. Don't forget that feedback, rather than Wheaties, is really the breakfast of champions. In addition, should a problem occur, the problem solving is done locally by the team in real time. In so doing they are exhibiting a "sense of control" over their work environment, using a team which feeds the "sense of community" and very likely by improving the work situation, they are feeding their "sense of accomplishment." **It is the very nature of a Lean Manufacturing system these intrinsic motivators are both designed into and brought out by the Lean Manufacturing system itself.**

Chapter Summary

Normally the driving force to implement a lean transformation is company survival. At the individual level of motivation, we must remember to not try to motivate our workforce, they come to work that way. We also must remember that most of our well-intended efforts are really counterproductive and demotivate the workforce when we use old, outdated strategies and paradigms. Rather we need to study and follow the writings of Herzberg, McGregor, Deming, and others. Largely, this is accomplished by learning and practicing the Six Skills of Lean Leadership and using the intrinsic motivators to keep your workforce at peak interest levels.

CHAPTER 7

Cultural Change Leading Indicator No. 3, Problem Solving

Aim of this chapter … to explain the importance of the presence of good problem solvers and good problem solving to a successful transformation. We will introduce the Five Questions of Continuous Improvement, the relationship of problem solving to standardization, the implications of personality on problem solving, and why group problem solving often works very well. We will introduce our basic principles of problem solving and why our mantra is "think small, think fast and think lots" of cycles through the Rapid Response PDCA (Plan, Do, Check, Act) process is important. We will also integrate the concept of hypothesis testing.

Talented Problem Solvers

The results of a lean transformation are a facility that will produce product with an ever-improving quality; at an ever-reducing cost; and at an ever-improving lead time. The "Means to Lean" is to problem solve our way to the Ideal State, consequently the third cultural change leading indicator to begin any transformation is the presence of talented problem solvers. First, let's make sure we are on the same wavelength here. Just what is a problem? For this, I rely on the problem-solving methodology popularized by Charles Kepner and Benjamin Tregoe (KT Methodology) in their book, *The New Rational Manager* (Princeton Research Press, 1981). They define a problem as, "the difference between what is and what should be." Furthermore, in part, they break down what most of us call problems into three types of concerns. These three concerns are problems, decisions, and potential problems. It is great reading and I recommend it to all. And by all, I mean *all*, not just those interested in Lean.

Back to problems for a moment … Once a leader develops his plan, he has just created a whole series of problems. Just how has he done that? As soon as he creates goals, he now has created a new "should be." For example, if OEE (Overall Equipment Effectiveness) is 60 percent and the goal is to achieve 85 percent, the OEE "should be" 85 percent—*et voilà!*—the manager has created a problem for someone else.

Problem Solvers and Their Skills

Problem solvers who have requisite skills to solve problems are rare indeed. They must be able to:

1. Exhibit great observations skills and readily grasp a situation.
2. Turn it into a meaningful problem statement.
3. Know how to gather and sort through the data.
4. Analyze the data and information, including doing a statistical review.
5. Utilize root-cause problem analysis.
6. Create a list of possible solutions.
7. Sort through the options, comparing the options to the needs of the business, and weighing the risks of each solution.
8. Decide which one is the best solution.
9. Use project management skills to turn this solution into action plans.
10. Show the leadership to implement those plans, turning them into improved performance for the facility.

Point of Clarity   A key role of a good leader is to create problems where no problems previously existed.

It takes an awesome compliment of skills to be an accomplished problem solver. Most managers believe the scarcity of good problem solvers is solely a result of not having the necessary skills—by that, they mean the technical skills; and by implication, they believe that all of the requisite skills can be taught. These managers then believe that all that is required is to train these people in the necessary technical skills.

Our experience does not support this logic. While it is true that few people have the technical skill inventory listed below, there are some who have the technical skills yet are not effective problem solvers because they lack other requisite traits. So what are these other traits that are required? We will discuss tis later in this chapter in the section entitled, "Problem Solving, Your Personality and Group Problem Solving" but now let's jump into the heart of problem solving.

What's the First Problem-Solving Tool We Should Teach?

At the moment a lean transformation begins, everyone's job is to make improvements and this means changes. Many of the changes will be simple decisions to be made but some will require the root-cause removal inherent in problem solving. Regardless, several aspects of continuous improvement need to be addressed before we should take any action; for this we devised the Five Questions of Continuous Improvement. They are specifically designed to avoid the two major pitfalls I have seen that are a hindrance to good problem solving. Interestingly enough, one of those two pitfalls is not the difficulty in finding solutions. Rather the two pitfalls are poor problem definition and failing to use a disciplined, structured approach to solving the problem. All too often the problem resolutions approach used is "Ready, Fire, then Aim" and often without really being "Ready." We are too prone to act and then we are forced to redo the problem solving later in a more expensive fashion; rather than take the time to properly problem-solve now. The Five Questions of Continuous Improvement is a tool to focus and guide you through a proper thought process of continuous improvement. They are powerful questions, yet can and should be used by everyone. They are specifically designed to avoid both major pitfalls. The Five Questions are as follows:

1. What is the present state or condition?

2. What is the desired future condition?

3. What is preventing us from reaching the desired condition?

4. What is something we can do now to get closer to the desired condition?

5. What is our expectation when we do this?

    A. What will happen?

    B. How much of it will happen?

    C. When will it happen so we can "go see"?

> **M**anagement is prediction... "Theory of knowledge helps us to understand that management in any form is prediction. The simplest plan—how may I go home tonight—requires prediction that my automobile will start and run...."
>
> W. Edwards Deming. The New Economics for Industry, Government, Education 1993: 101, 102.

These Five Questions are seemingly simple; don't let that fool you. They too require training and skill to administer. Too often I hear this dialogue around the Five Questions. For example, someone might ask the quality manager the following ... with his answers in italics following the question.

1. What is the present state or condition? (QM) We have way too much scrap.

2. What is the desired future condition? (QM) We have zero scrap.

3. What is preventing us from reaching the desired condition? (QM) We are making defective product.

4. What is something we can do now to get closer to the desired condition? (QM) Quit making bad product.

5. What is our expectation when we do this? (QM) Everyone will be happy.

You can readily see this use of the Five Questions is going nowhere. Now let me modify this hypothetical dialogue a bit and add some "training through questioning" to the administration of the Five Questions. The annotation of (QM) will be the quality manager's comments and (PM) will be the plant manager's comments.

1. (PM) What is the present state or condition? (QM) We have way too much scrap. (PM) Agreed but we cannot solve all the problems. What is a specific product code or defect type we might make progress on? (QM) Well mislabeling has been a recurring problem and is No. 1 on our pareto for frequency. (PM) Good enough, mislabeling it is.

2. (PM) What is the desired future condition? (QM) We should have zero scrap. (PM) Clearly that is our Ideal State, but is that reasonable now? What might be a reasonable target condition? (QM) Actually we are at 5200 ppm and if we could get to less than 1000 ppm, we would meet the customer standard and get them off our backs. (PM) Ok, less than 1000 ppm will be our new target.

3. (PM) What is preventing us from reaching the desired condition of 1000 ppm? (QM) Our labelers are old and we don't have a maintenance program for them, the forklift drivers are to print the labels and to save time, they print several at a time ignoring the standard; several work stations share labelers and misprogram them, I guess I could go on. (PM) As you can see we may have many root causes which means we may have several problems. Can you break these down into separate problems? This step we call problem stratification. (QM) Sure, it looks like we need better definition. (PM) Ok let's call them: problem 3a, labelers need maintenance; problem 3b, multiple labels printed; and problem 3c, labelers are misprogrammed.

4. (PM) Now regarding each of these three problems, what is something we can do now to get closer to the desired condition? What might be a good countermeasure? "For 4a … for 4b and for 4c?" From here we may need to do some more data stratification or quantification. However, once we do have enough information to proceed, the PM might say … .

5. (PM), "Now what is our expectation for countermeasure 4a? What do expect to happen? How much do we expect it to reduce mislabeling? When do we expect that to occur and when can we "go see."

For the sake of brevity, I stopped with problem 3a. However, the PM would then guide the QM through questions 4 and 5 for problem 3b and 3c as well. If the thought process is thorough, the expectations expressed in the answers to questions 5a, 5b, and 5c should close the gap from 5200 ppm to the desired state of less than 1000 ppm and "voilà" we are well on our way to "problem solving our way to the Ideal State." This is one very clear example of lean thinking.

In this case, the plant manager is leading by teaching and I am sure progress will be made. But this highlights the most common pitfalls in problem solving, inadequate problem definition, and poor discipline and structure in solving the problem(s). Yet proper administration of the Five Questions is correcting that nicely.

From this example I wish to highlight four major points. First, this is leadership in its finest form. It is lean leadership (See Chap. 5) and it is teaching by questioning. Second, as part of the Five Questions, the concept of hypothesis testing is embedded in the questions. It is how we learn, how we teach, and how we solve problems. This is the heart of the scientific method. And third, the use of the Five Questions may or may not be problem solving. If the countermeasures involve root-cause removal, then you are problem solving. It could also be a simple decision or it could be standardization. This is why they are named the Five Questions of Continuous Improvement, not the Five Question of Problem Solving, nonetheless we improve by all three methods. Fourth, there is frequently a lot of

> **P**oint of Clarity The key to using the Five Questions is to remember a basic rule of problem solving. You make progress by solving one problem at a time … and in no other way.

> **R**ecall that the Five Questions are a focusing tool. By coaching the QM in this way, we have focused on something we can solve.

> **P**oint of Clarity A hypothesis is a statement of the form… "if we make the following specific changes we expect to achieve this specific outcome."

work that needs to be done between the steps as outlined above. Although this example was straightforward, often there is a lot of work to define question 2. For example, if there is no standard, some work may need to be done to define one. Likewise, regarding question 3, although very often the obstacles are known, that is not always the case. Once we focus properly, you may need to do some more data gathering or a Kepner–Tregoe specification to further define this concern. What is usually pretty straightforward is the answer to question 4, the countermeasures that comprise the first half of the hypothesis. What I have found is that of all the aspects of continuous improvement, often the easiest step is that of coming up with countermeasures. More often than not there are several you know will work and likely you will need to employ some model of decision-making to sort through the many options. For this I strongly recommend the decision-making techniques outlined in the Kepner–Tregoe methodology ("The New Rational Manager"). And finally the answers to question 5, the second half of the hypothesis test are normally pretty straightforward as well. After you have done all this work, normally the key question is really "when" can we do all this; which in turn is a matter of good project management, a skill in its own right.

Continuous Improvements, Problem Solving, and Standards

In a lean plant, concerns can be broken down into three categories or types. We will have concerns when we have:

- No standard, a Type 1 concern,
- A standard that is not met, a Type 2 concern, and
- A standard that is not ideal, a Type 3 concern.

The most difficult "problem" to solve, actually it is not a problem at all; rather it is state of "being lost." That is the "problem" of not having a standard, a Type 1 concern. Simply enough, with no standard there is no "desired future state." If you know the present state, but the desired future state is not yet defined, there is no "difference between what is and what should be," there is no gap to bridge. Hence there is no problem to solve. At this point, it is incumbent upon management to create a standard. Although this is often difficult, it is the task of management and management alone to create this standard (see Chap. 16, Planning and Goals and Chap. 5, Leadership). Likewise for Type 3 problems, it is incumbent upon management to create the new target if the existing process needs to be improved and moved closer to the Ideal State; this is simply a part of continuous improvement.

> "If you do not know where you are going, any road will get you there."
> **Lewis Carroll**

The most typical problem is Type 2, and these problems need to be solved by everyone. Such problems include the typical customer complaint, the production demand that is not met, the quality standard that is not achieved, and the delivery date not met. In addition, Type 2 problems include internal problems such as OEE not achieved or cycle-time degradation. Often these problems lend themselves nicely to the Five Questions methodology. The beauty of these problems is since there is already a standard in place, question No. 2 is answered. Problems of Type 1 are problematic as first, a standard must be created. This means that there must be some level of direct management involvement in resolving these problems, similar to Type 3 problems.

Early in the lean transformation, it must be made clear that everyone is responsible to solve problems. It is the challenge of management to engage everyone in problem-solving activities. As the transformation develops, it can be determined how this is approached. For example, some have used small group activities such as *quality circles* with great success. At a minimum, all employees should be taught the "Five Questions of Continuous Improvement."

It is also very crucial to have a small cadre of very talented problem solvers. Even in a facility of 500 people, only three or four are usually required. Many problems, especially in the early days of implementation, are easily solved by a wide range of personnel, including group leaders, production supervisors, technicians, and engineers. However, some problems will crop up that require more technical skills than the typical group leader, production supervisor, or technician will have. In addition, some of these problems require significant dedicated time to do the data gathering and analysis. Many production workers, even if they have the skill to solve these problems, do not have a block of available time so they can do the necessary data acquisition, reduction, and analysis. These three or four talented problem solvers should be versed in plant operation, as well as a wide range of problem-solving techniques.

Standardization and Problem Solving...a Double Burden

There is yet another issue with problem solving that is not well understood by many—that is, the process of standardization is just another name for the process of problem solving (for more details, this concept is explained more fully in the appendix at the end of this chapter). So those who are good at problem solving are also good at standardization. The opposite is also true: Those who are weak at problem solving will be weak at standardization. In this book, the emphasis on standardization and the emphasis on reduction of variation of all forms is repeated over and over. It is a hallmark of lean manufacturing to have standardized processes, and there is no substitute for it. So our problem solvers are now doing double duty. First, they are fixing problems—and now you learn that the same skills are required to execute the techniques of standardization.

Other Key Problem-Solving Skills

Beyond the Five Questions of Continuous Improvement, the first and most important technique is logical problem solving. I know of no technique superior to the Kepner–Tregoe (KT) methodology. These skills can be acquired by attending a training session taught by KT—this is really the very best training investment you can make. If for some illogical reason this is not in the cards, pick up their book, *The New Rational Manager* (Princeton Research Press, 1981) and teach yourself. KT also has a slightly less intensive program named "Analytical Trouble Shooting," which is also excellent. Several years ago you could have taken a class at the Ford Training Center entitled, "TOPS: An Acronym for Team-Oriented Problem Solving." In this class, you learned the KT methodology as part of completing Ford's 8D. However, when the Automotive Industry Action Group (AIAG) was formed and methods in the auto industry became standardized, the TOPS program disappeared. This was a large oversight and should be corrected. In my experience, the KT methodology has been proven to have the broadest applications and the highest success rate of problem solution. It does have one weakness, though: a scarcity of statistical techniques to assist in the quantification of variation and in statistical decision making.

This is where the Six Sigma statistical problem-solving techniques are so powerful. Actually, even though we say Six Sigma is a problem-solving methodology, it lacks many of the powerful logical tools inherent in the KT methodology. Instead, it relies on the simple DMAIC methodology (Define, Measure, Analyze, Improve, and Control). The Six Sigma tools are extremely powerful when making statistical decisions and in understanding, mathematically, the risks involved. These Six Sigma tools include statistical and decision-making skills such as multivariate analysis, hypothesis testing for both variation and averages, correlation and regression plus SPC (Statistical Process Control), MSA (Measurement System Analysis), DOE (Designs of Experiments), and Response Surface Analysis. Even if you do not have any Six Sigma blackbelts, someone should be well versed in DOE. It is not very time consuming to acquire the basics of DOE, and the applications abound in most manufacturing plants. Those trained in Six Sigma skills are much like the "Deming Statisticians" which are often talked about. For a good description of that group, refer to *Out of the Crisis* (MIT, CAES, 1982), specifically Chap. 16, by W. Edwards Deming.

A much overlooked set of skills are those required for group facilitation. These skills are not only helpful with groups, they are helpful any time a problem solver must interact with another person on any problem, such as getting information from the line workers. So these skills are very powerful in making a problem solver more efficient and effective. There are various places to acquire this training, and there is even a touch of it in Six Sigma training. However, I recommend you send those people who need these skills to the training as is provided by Oreil Incorporated (formerly Joiner Associates). Alternatively, buy *The Team Handbook* (Joiner Associates, 1988) by Peter Scholtes and teach yourself. And in this case I recommend you do an out-of-print search and get the original version written by Peter himself. Since his passing the book has been modified and I find the first edition to be a superior document and quite frankly a great deal. You can often find it in good shape for $0.01 plus shipping. So in one fell swoop you get a great book and a great deal…tough to beat that.

Finally, if you are in the discrete parts industry, especially electronics, I have found the so-called Shainin tools to be valuable. They have been published in a book by Keki Bhote entitled *World Class Quality* (AMA, 1988). If you want to pick up this book, do an out-of-print search and try to find the first edition if possible. It's superior to later editions and like *The Team Handbook*, usually a whole lot less expensive.

Which Trainings to Do First? The Following Three Are a Must!

- Kepner–Tregoe problem solving (clearly the #1 choice)
- Statistical problem solving (such as Six Sigma or at least SPC and DOE)
- Group facilitation training

So for this small group of dedicated problem solvers, all should be accomplished with the KT methods, all should be proficient at SPC and DOE. It would be good if one or more was an accomplished Six Sigma blackbelt, and if one or more had strong group facilitation skills and were skilled and experienced enough to do a facilitated "spin-around" on a broad variety of problems as described in the accompanying sidebar.

Brainstorming Rules, the Facilitated Spin-Around

The facilitated spin-around is a very powerful way to gather information and resolve several types of problems. For many data-driven, on-the-line production problems, this is not a good technique. First, it is normally not needed; second, it simply takes too long. It is good for issues with "soft data," such as opinions and for decisions that are emotionally charged, such as "should we require all employees to wear a uniform?". It is a technique that will reach very good decisions, but more importantly, because of the process itself, the members will have a strong commitment to implementing the group decisions. The facilitated spin-around is often required to reach consensus. It is critical that your organization be able to use these techniques and have facilitators capable of leading these groups. The technique goes like this:

1. The facilitator introduces:
   A. Himself
   B. The topic very briefly
   C. The objectives of the group
   D. The agenda
   E. The planned time frame for this meeting

2. If meeting ground rules need to be discussed (such as cell phone use, bathroom breaks, etc.), this is done at the beginning.

3. The facilitator reminds them of the brainstorming rules, which are as follows:
   A. One item per person per turn.
   B. Each person in turn (hence "spin-around" the room).
   C. The documenting and posting of each item is done on flipcharts. (I recommend you use flipcharts. They are much more personal and much more effective than using a computer projected onto a wall.)
   D. When an item is stated, no value judgments by others are allowed (at this stage). No agreement and no disagreement are permitted. All items are taken at face value.
   E. Discussion can proceed on any item, but only "to the point of understanding." Once it is understood, discussion ceases.
   F. No other talking or work of any nature is allowed; no side discussions are permitted. If there is a question, it must be directed to the group. This requires a great deal of both attention and patience, not to mention respect for the people and the process.
   G. No piggy-backing or modifying of any item is permitted without the author's agreement.
   H. Pass if you do not have an idea on this turn.
   I. Spin-around the room until all items are exhausted and documented.

4. The session then proceeds when the facilitator documents the first item and the spin-around continues.

The preceding procedure requires good facilitation, which is a topic in itself, but is unfortunately outside the scope of this book. I recommend you refer to *The Team Handbook* by Peter Scholtes (Joiner Associates, 1988) for further information.

In addition, this small cadre must be available whenever the plant is running. Generally, this entire group is not readily available, but it needs to be managed in such a way that at least one of these problem solvers is readily available.

> **P**oint of Clarity We will have a JIT plant, and thus we will need JIT problem solving.

When to Use Brainstorming??

Probably the most often used tool I see in SPC is the fishbone diagram (often called the Ishikawa diagram after its founder or a Cause-Effect analysis) and 90 percent of the time it is the wrong problem-solving tool. Why is that?? I find the fishbone is used most often because it "can" be used rather than when it "should" be used. For problem identification and root-cause analysis, it should be used—**only after all hard data sources have been exhausted**. Problem solving in a lean environment is to be data driven. Fishbone diagrams are filled with opinions, not data and in any problem solving; data, as a source of information, will trump opinions nearly all the time. So exhaust all your available hard data before you rely on opinions to guide your problem-solving effort.

Problem Solving, Your Personality and Group Problem Solving

First, we must define some methodology to solve problems. Various ones exist, but they all take the form of:

1. Some observation/evaluation occurs and it triggers the thought that "we have a problem." Someone then defines the problem.

2. Observations about the problem area must be made.

3. Evaluations must find the cause of the problem.

4. Solutions must be imagined, created, and compared to values.

5. Decisions are made.

6. The decisions are turned into action plans.

Second, let me say a word about individual personality development. C. G. Jung developed a theory of personality that is widely accepted today. In short, he said there were two major aspects of how we incorporate and handle information (Fig. 7-1).

1. The first aspect was the "observing scale." This is the way in which individuals accept information. It is done by "sensing it," by touch, feel, smell, and so on. We call these people "sensates." Or this information can come in via the process of "intuition." This is characterized by such things as "my gut feeling" or "I just sensed that…" Although to most people intuition is a lesser form of accepting information, its value cannot be underestimated, especially in problem solving. The twin sister to intuition is imagination, which you will find is incredibly important in problem solving, as is intuition itself.

2. The second aspect in Jung's personality model is the "evaluating scale." This evaluation gives meaning to the observations obtained. It has two extremes. One is "thinking" and the other is called "feeling"—although empathy may be

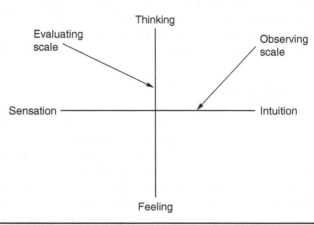

Personality Types

FIGURE 7-1   Jung's scales for personality types.

a better English word. Pure thinkers make evaluations based on logic and cold hard data. While those with a strong "feeling" function will make decisions based on what we call human values, such as community, human-worth, and quality-of-life to name a few. As you might expect, most businesses are filled with "thinking-sensate" personality types. If we plotted them on the scales in Fig. 7-1, the thinking-sensates would be in the upper left-hand quadrant.

The dynamics of personality development are such that at a very early age, people start to use a certain type of observation technique and evaluation format. Then, when they find one that works for them, they work to refine it so life is better for them. This process of personality development is almost entirely unconscious, but it is still easily characterized, even at a very early age. It is a dynamic of personality development, that as one end of the observing pole tends to work, for you, then the other end of the scale becomes subordinated. Hence, people tend to become sensates at the expense of the skill of intuition, and one type of information gathering dominates their personality. Sensing becomes the conscious way to accept information, and intuition is driven into the unconscious. It will remain there until something occurs to cause it to surface. Most people will develop a dominant way to accept information and to evaluate. This then becomes their personality type. For example, one could be a "thinking-sensate" or a "feeling-sensate," to name just two. Most people do not develop—at least until they are older—more than two of these dimensions.

Now here is the rub. Once a person develops a personality style, they use it and refine it. In fact, they can become very rigid and structured. In some aspects of life, this rigidity of personality will serve them well, but in time the very nature of life will expose them to different challenges, and this rigidity will harm their ability to resolve some of life's issues. For example, early in life a young man can be a very rigid "thinking-sensate." Then this fellow gets married and finds that this cold, hard, rational thinking

> **"A**t 20 years of age the will reigns, at 30 the wit, at 40 the judgment... . **"**
>
> Benjamin Franklin

does not work so well with his wife and kids and he needs to adjust. This dynamic is one reason why there are few really good problem solvers in our business who are also young. More on that later.

So how does this fit with problem solving? Unlike many aspects of work, while solving problems a person must have both strong sensing skills to make objective observations, but they also must have intuitive skills (maybe a better term would be "imagination skills") to foresee what could *possibly* happen, even if it is not currently happening. This is a crucial aspect of the observing process that must occur in problem solving. If a person lacks this intuitive skill (imagining), he can only envision what is occurring at that moment in time, and so if the problem just happens to manifest itself in the present context, he will see it—otherwise, he will be hampered by some level of blindness. At the level of quality demanded by most companies, it is very rare that we actually see the problem when it occurs; hence those who do not have this intuitive skill are hampered as problem solvers. This intuitive skill is not only needed at the observing stage but is certainly needed in the "creating possible solutions" aspect of problem resolution.

In short, what happens is this. When needed, most businesses hire individuals with a strong "thinking-sensate" personality type. Since they are strong sensates, they are usually deficient in intuition—this is natural, predictable, and largely unavoidable. Unfortunately, when it comes time to solve a serious problem and the intuitive skills are needed, the workforce is, alas, weak in this area. So the company, with the best of intentions tries to teach problem solving. Much of the methodology can be taught, but the intuitive skills can't really be acquired in the classroom. So in the end, real problem-solving skill development proceeds slowly. Since most companies think problem solving can be taught, they are disappointed and usually shrug their shoulders. In fact, the truth is that some problem-solving skills can be taught in the classroom, but for those "thinking-sensates" the intuitive skills of observation and solution conception are largely taught by life. Sometimes that means just enduring more of life's experiences to broaden your personality—or in other words, it's a part of getting older, and hopefully wiser.

Hence, my point is that few people have the total complement of skills to be really good problem solvers. Some of the skills can be taught in the classroom, but some skills come to work with the person's personality and are largely unteachable.

This polarization of skills within an individual has given rise to the concept of group problem solving. Given a group as small as five or six individuals, frequently we will have all four of the personality traits present in one or more of the people. What then happens is that those who are strong "sensates" are very active in the problem definition stage, but may be less involved in the highly intuitive stage of imagining possible solutions. On the other hand, a person with a strong intuitive tendency could assist in the step of creating possible solutions yet be less involved in the problem definition stage. With a small group, it's very possible to get all four of the poles adequately represented. Now, if the group also has the technical knowledge, it is likely a good problem-solving effort will be achieved.

I have taught groups and done controlled experiments with groups and find them to be very effective in generating superior solutions to many problems. But as with so much of life, if you get something good, it often comes with some baggage. And the group, when used for problem solving, does have some baggage—a situation that is threefold, at least.

- First, to be successful the group must be well facilitated; hence, yet another skill is required, that of facilitation, a scarcity in its own right.

- Second, although the process is thorough, it is not fast. Groups need to form and develop and mature. Then there is the whole issue of meetings. Solutions tend to be weeks or even months away. This is anything but JIT problem solving.
- Third, often the issues in lean implementation have to do with very technical issues, so the group may not have the requisite technical knowledge. Where intimate technical knowledge is a requirement, one well-informed person working alone is always more efficient and usually far more effective.

We have a great deal of information on group problem solving, group management, and group dynamics, and what we find is that these skills are not the most important ones to work on during the early stages of a lean transformation. Typical problems are usually more Lean specific, and the answers are needed quickly. These characteristics of a problem generally mean they are better solved by an individual.

On balance, it is much better to develop a small group of problem solvers and turn them loose to solve the problems individually. If you have three or four persons, it is easy to teach them technical skills, as well as statistical skills. Keep your eyes open for the select few who have a developed enough personality to be excellent problem solvers, and you will likely have a strong solid group.

An Example of the "Five Whys" Is Technique:

Jidoka is a weakness in lean implementations.

1. **Why** is *jidoka* a weakness in lean implementations?
   - Because quality problem solving, an integral part of *jidoka*, is hard to do.

2. **Why** is quality problem solving hard to do? Because
   - We avoid these problems rather than embrace them.
   - We do not have a sound continuous improvement philosophy.
   - We are not skilled at problem solving.
   - We do not have time to work on problem solving.
   - Our data are not good enough for solving quality problems.

3. **Why** are our data not good enough to solve quality problems?
   - Because the data are total reject data only, not stratified by type of defects.

4. **Why** are the data total reject data only and not stratified by type of defects?
   - Because we only need total reject data to answer the questions we are asked by management.

5. **Why** are the total reject data adequate to answer the questions from our management?
   - Because the questions we get from our management focus on production and yield and not quality.

6. **Why** does our management ask questions about yield and production and not quality?
   - Because management is more interested in yield and production than they are in quality (solution statement).

A Couple of Points about the "Five Whys"

- First, the "Five Whys" technique is seldom a straight-line linear process. For example, you will note that the second "why" has five possible answers and we only addressed one of them. We have two options when we have multiple causes.
  - We could answer why to each of these, which would give us a very branched problem solution. Quite frankly, all five branches might converge on the same conclusion.
  - However, the normal advice is to quantify the issues and follow the branch with the largest impact.
- Second, as is the case here, it is arguable if you have reached a root cause, which is the objective of problem solving. For example, why is management more interested in production than quality? Maybe it's because of their bonus structure, or just maybe it's the best thing for the company today. Nevertheless, with problem solving of human systems, often the root cause is not really found, and sometimes it isn't necessary. What is necessary is to find an "actionable cause **that reasonable people agree** should be changed."
- Third, seldom is the Five Whys, actually five; it may be six as it is here or any other number. It is however many "whys" it takes to get to a root cause which is a solution that reasonable people would agree will resolve all the symptoms of the problem.

Let's check this possible solution statement with the "therefore" technique. If "management is more interested in yield and production than they are in quality" is a good answer to our problem, then the "therefore" technique will logically connect the "Whys" in reverse order, starting with the solution statement.

An Example of the "therefore technique," Which Is a Check of the "Five Whys."

Management is more interested in yield and production than quality (the solution statement).

1. **Therefore**, the questions we get from management focus on production and yield, not quality.
2. **Therefore**, we only need total reject data to answer these questions.
3. **Therefore**, the data are total reject data only and not stratified by defects.
4. **Therefore**, the data are not good enough for solving quality problems.
5. **Therefore**, quality problem solving, an integral part of *jidoka*, is hard to do.
6. **Therefore**, *jidoka* is a weakness in lean implementations.

And it works, which tends to confirm the validity of our solution statement.

What Are Our Problem-Solving Principles?

We use problem solving to drive our facility toward the Ideal State; this is the Means to Lean. The Ideal State is both real and conceptual. Even if it is currently or practically unattainable, we will still use it as our desired long-term destination. Furthermore, we will always problem solve "into the unknown." Hence all problem solving involves experimentation and some "failures" are certain. Rapid Response PDCA is the continuous improvement tool.

> **P**oint of Clarity   Problem solving entails the removal of root causes of undesirable traits.

Everyone in the Organization Is a Problem Solver

It is very simple, everyone from the CEO to the floor worker is expected, actually required, to be a problem solver. Hence they will be taught various, job specific, methodologies of problem solving. Not only will there be teaching and training but the organization will also be designed so problem solving is supported at all levels of the organization. Fast forward to Chap. 24 and check to confirm that "problem solving by all" is in the foundation of our House of Lean. Yes, problem solving by all is a "foundational issue" to our entire transformation.

All Problem Solving Is Done Using the Scientific Method
(See Appendix 1 near the end of this chapter)

- First, we make some observations and we clearly identify a problem or opportunity and decide a change is in order.
- Second, we will state a clear hypothesis, which is of the form of, "if we make the following specific changes we expect to achieve this specific outcome." And of course, the specific changes must be defined and quantified as with the outcomes.
- Third, we will design and complete an experiment to test our hypothesis.
- Fourth, we will analyze the data.
- Fifth, we will decide what we learned and take appropriate action.

(Note how the scientific method is also manifest in the Five Questions of Continuous Improvement.)

There Are "Eight Issues of Change and Uncertainty" We Will Consider, Always, in Problem Solving

1. Normally objectives are loosely defined, frequently change, and many times are incompatible.
2. The environment we deal with contains unknown variables and irreconcilable uncertainties.
3. We can never, "have all the facts": many are unknown; most are unclearly defined; and some will change during the problem-solving process.

4. Problems come from interactive systems that are seldom linear in nature. A three-dimensional spider web is a better metaphor for their form.

5. Any change we make, even in taking measurements; and of course attempting to control will cause the system to change. All problem solving is dynamic with a moving present state and a moving terminal state.

6. At times the response we get will not be a result of our actions or intents but rather the perception of our motives. These responses are varied, frequently difficult to predict and sometimes serendipitous, other times disruptive.

7. For any given stimulus we apply we will get many responses. In advance, many of these responses are unknown and unknowable.

8. We will not let these unknowns and uncertainties cripple our problem-solving efforts. In the face of this reality we will employ Rapid Response PDCA, repeatedly.

Our Problem-Solving Mantra...Small, Fast, Lots ....

In a large part, because of these issues of change and uncertainty—our problem-solving mantra is "think small, think fast and think lots" of cycles through the rapid response PDCA process.

> **M**anagement Is Prediction
> "Theory of knowledge helps us to understand that management in any form is prediction. The simplest plan—how may I go home tonight—requires prediction that my automobile will start and run... ."
>
> W. Edwards Deming. The New Economics for Industry, Government, Education, 1993:101, 102

A Word about Hypothesis Testing

Keep in mind a hypothesis is a guess. It is not some wild, off-the-cuff guess. It is always based on observations and has theory as its antecedent; but in the end it is an informed, scientific guess. It is a guess in the form of, "if we do these specific things, we will get this specific result." Hypothesis testing is also about prediction. The entire topic of hypothesis testing, especially statistical evaluations, is beyond the scope of this book. But keep in mind from a coaching standpoint, the Five Questions will guide you nicely and the "coaching questions" regarding any hypothesis are twofold. They are as follows:

1. What specific things will you do? (Question 4)
2. What specific outcomes do you expect? (Question 5)

On occasion, such as when you do a DOE (designs of experiment), there is a very specific protocol for a hypothesis test. However, the vast majority of the hypotheses we make and test will not require such rigor.

Chapter Summary

The third leading indicator to successful cultural change is the presence of good problem solving in the organization. Since the "Means to Lean" is to problem-solve your way to the Ideal State, it is impossible to become Lean without good problem solving and good problem solvers. The first and most important skill is the ability to use the Five Questions of Continuous Improvement throughout the organization. There are other key methodologies, including the Kepner–Tregoe logical skills, the Six Sigma statistical toolbox, SPC, DOE, the Shainin Tools and the "Five Whys." In addition we learned that standardization is really problem solving by another name. Also, we studied the implications of personality on problem solving and why group problem solving often works very well. The Eight Issues of Change and Uncertainty dictate why our problem-solving mantra is "think small, think fast and think lots" of cycles through the Rapid Response PDCA process. Finally, the concept of predictability, problem solving, hypothesis testing, and management were shown to be interwoven in an intricate web.

Appendix—Problem Solving and Standardization: How Are They Similar?

Concerns and Problems

Many things bother us. We have many concerns but not all concerns are problems. A problem is defined by the Kepner–Tregoe methodology as a "concern" that meets three criteria:

1. There must be a difference between what "is" and what "should be."
2. It must be a present issue.
3. It must have a root cause that is unknown, and the root cause must need to be eliminated.

If our concern meets all three criteria, it is a problem and needs problem-solving logic to resolve it. Problem solving is the finding and removal of the root cause of the problem.

Let's say warranty returns, by our standards, should be less than 50 ppm but are currently 220 ppm. This is characteristic number 1. It is a present issue so it meets the second criterion, and if we do not know the cause, it meets the third criterion. Now we have a problem to solve. That seems clear enough.

Let's look at something a bit more mundane. For example, each time you take your 10-year-old car in for servicing, the mechanic finds $200 worth of necessary extra work. This should not be, you are thinking. So, though it should not be, it *is*, and it is happening right now. But can I find and eliminate the root cause of the problem that is manifest by the extra costs? Probably not. So in this case, without some possible corrective action on a root cause, we do not have a problem. But what we do have is a concern in the form of a decision to make—that is, do you buy a new car or continue to pay for the repairs?

On the other hand, what if—instead of having high maintenance costs right now— you are worried about them happening in the future? In this case, even though it is a concern, it is not a present concern, so you do not have a problem. However, what you do have is a threat or a possible future problem.

Each of these concerns—threats, decisions, and problems—require a different logical approach for their resolution. To study the various methodologies here is beyond the scope of this text, but with a quick trip to the bookstore you can buy *The New Rational*

Manager (Princeton Research Press, 1981) by Kepner and Tregoe, and you can learn all of these powerful techniques. I highly recommend it.

Concerns?

- Threats
- Decisions
- Problems

Each requires its own methodology to resolve.

So that briefly describes one key part of problem solving, but what does this have to do with standardization?

Standardization

Standardization is an attempt to get all the parties who perform some activity using the same skills and actions. It is an attempt to eliminate the variation that exists in any activity.

Just how does this interact with problem solving? Maybe a real live example can help clarify this question.

For example, at the Jayaroot plant in Juarez, Mexico, SPC (Statistical Process Control) was implemented to better manage the process and to monitor the quality of the finished products. This plant made electric lighting ballasts. The final process step, prior to packaging was to perform a 30-minute infant-mortality burn-in. Over 1000 ballasts were produced per hour and the normal defective rate was about 1.1 percent. The data was recorded hourly on a P (Percent Defective) control chart (Fig. 7-2), and the process was statistically stable.

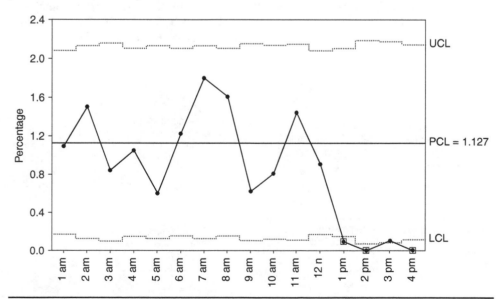

FIGURE 7-2   P chart for defectives.

This process was statistically stable and had normal variation about the mean of 1.1 percent defects. Then at 1 p.m., it went out of control. Note that the 1 p.m. data is below the lower control limit (LCL). I noticed this and asked what had been done about the change? Francisco, the supervisor said, "We did nothing because, although the process went out of control, the process improved; our defects have decreased to practically nothing."

The science of SPC is designed so that when a process goes out of control (i.e., a plotted point on the control chart lies above the upper control limit or below the lower control limit), we know with 99.73 percent certainty that the process has changed. It is then a requirement that the cause of this change must be found and understood. These causes have a unique name in the science of SPC. They are called special causes of variation. Then, if the special cause created an undesirable change (in this example, that would be for the defects to increase and hence the plotted point would be above the upper control limit), you then have a problem to solve. However, if the special cause creates a desirable change, as in this case, the proper action is to standardize (Fig. 7-3).

In this case, there are several possible solutions to this problem. However, the logical one, consistent with the company's objectives at that time was to go to a single source of circuit boards and purchase only boards from supplier C. A little more on that later.

Why did Francisco treat the change incorrectly? When a process goes out of control, it means the process has changed. Well, entropy being what it is, the change is usually bad, in the form of deterioration. In the preceding several months, since they had started with SPC, Francisco had spent so much time chasing root causes of problems in which the change had occurred in the "undesirable" direction that he understandably had overlooked this "avalanche of diamonds" that was given to him. In this case, the change signaled that some desirable change had occurred.

So Francisco and I investigated the process to find the root cause of the change... and what we found was amazing.

This line had three different suppliers of circuit boards, which we shall call suppliers A, B, and C. The process change that was seen on the control chart coincided to the minute (taking into account the lead time) when a new batch of circuit boards was introduced to the line. These boards were manufactured by supplier C. Upon further

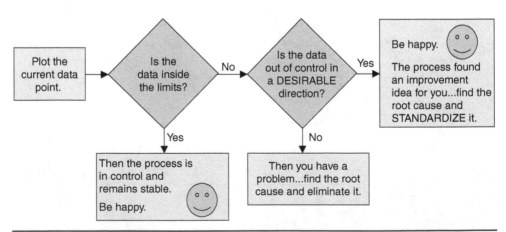

FIGURE 7-3 The logic of SPC.

investigation we found that all three suppliers met all quality standards but the hole placements on supplier C's boards had a Cpk (Process Capability Index) of 2.2, while the Cpks for supplier A and B were 1.37 and 1.43, respectively. We found that, although all three suppliers met the minimum standard of Cpk more than 1.33, supplier C was a superior quality supplier. Since two suppliers barely met the spec and this created over 1 percent scrap, it is only logical to conclude that the specification was too wide and should be reduced; however, that was not our problem—at least not at that specific moment.

We had found the root cause of this "desirable effect": The change to supplier C had caused the reduction of defects. This reduction of defects was not a problem since it was not an undesirable effect. Rather, it was just a change that was both unexpected and serendipitous. So, in a concerted effort to reduce the variation, our plan was to standardize on supplier C.

How Is Standardization Similar to Problem Solving?

In problem solving it is required to find the root causes of these "undesirable" forms of variation and then systematically eliminate them. In standardization it is necessary to find the root causes of these "desirable" forms of variation when they are found, and rather than eliminate them we wish to institutionalize them. Consequently, the skills required for good problem solving overlap a great deal percent with the skills of standardization.

A Little More for Your Amusement and Amazement

I told you earlier that there would be a little more… and so now I'll tell you "the rest of the story," as it's said, which goes like this.

Francisco, having been buoyed by the knowledge that we could reduce defects a full 1 percent simply by single-sourcing our circuit boards to supplier C, was energized. You see, the company was attempting to implement a Deming-type management system, and Point No. 4, of Dr. Deming's 14 Obligations of Management has to do with supplier development and going to single sources of supply. Francisco was energized beyond belief, and the first thing he did was calculate the possible gains. They exceeded $1,000,000/year for the quality losses alone. Next, he investigated this supplier a little. He found that supplier C was relatively new to Jayaroot Co. but was also the low-cost supplier for this board. Wow! Even more potential gains. Higher than a kite and filled with new-found energy, Francisco contacted the purchasing department and found out the following:

- Their objective was to go to a single source of supply.
- Although supplier C was the low-cost supplier, they had already been eliminated. In the future, boards would be supplied by A and B only. We were told that the decision had been made at central purchasing.

Well, not too discouraged, Francisco contacted the manager of the central purchasing department. He confirmed what Francisco had already been told. Upon questioning this manager, Francisco found the decision had been made, and no amount of data or logic would deter purchasing. Supplier C was out; A and B were in. Francisco argued so long and hard he almost lost his job over it. But the low-cost superior-quality supplier was eliminated and we needed to find another way to eliminate these defects.

I never actually found out why this decision was made this way, but the general manager for the plant just shook his head and said it did not surprise him. He went on to explain that they had similar problems in the past, nothing quite so clear as this, but in the past there was little they were able to do—just as in this particular case. Subsequently, I learned that this facility was strictly a cost center and all raw materials sourcing, costs, and metrics were managed at the home office in Chicago.

This example, more than most examples, points out the futility and destructive nature of centralized control taken to an extreme. We had transformed the manufacturing facility to operate with pull systems using cells operating at *takt*, and it was very Lean by many measures. From a manufacturing standpoint, it was a large success. However, it was not surprising that two years later, I was on the front steps saying goodbye to many fine workers as this facility was closed. You see, although it was a manufacturing success, it was a business failure.

The notice in the paper said they were moving the facilities to China to take advantage of the low-cost labor market there. This product sold for about $17, and had less than $0.12 direct labor in the cost to produce each unit; the home office burden was nearly $2.00.

CHAPTER 8

Cultural Change Leading Indicator No. 4, Whole-Facility Engagement

Aim of this chapter ... to explain in simple detail what engagement really is, how it differs from motivation and how management is the key to engagement of all employees.

What It Looks Like

You know what engaged employees look like? They are active; they are trying to get the task done; they are clearly focused on the objectives at hand; when they reach roadblocks, they show high initiative and immediately take action or seek help. They don't fret, they don't wait, and they don't make excuses ... . They are not only "moving and shaking," but engaged employees are actively getting the right work done in the right way to meet the right goals. In leanspeak they are hitting cycle time, following standard work producing good product each and every cycle. If this cannot be done, they are immediately "pulling the cord" to escalate the problem, just in time.

The Five Elements of Engagement

When I teach engagement to a new management team, I teach that engagement has five measurable behavioral traits. An engaged worker, floor worker, or manager alike:

1. Knows what to do to accomplish the task at hand. That is, they understand and can execute the major steps to their job.

2. Knows how to do the job. That is, they have the skills, techniques, and talent to produce a quality product safely and efficiently.

3. Has the resources to do the job. They have not just the physical tools but the training, support, and leadership as well. The resources also include such items as a plant-wide support system of JIT (just-in-time) problem-solving, JIT planning, JIT management support, and JIT maintenance, to name a few.

4. Wants to do the job (more on this later).

5. Wants to do the job better (also more on this later).

As management teams attempt to "make" employees engaged, they almost always give a lot of attention to the last two attributes. That is, they believe: If only we had enough "employee want-to," then we would have engaged employees.

They could not be more wrong!!

Engagement Is Not the Same as Motivation

The vast, vast majority of workers who enter your facility have tons of "want to,"—at least at day one. Most of the workers I have encountered—be they factory-floor workers, engineers hired right out of college, line supervisors hired from the outside, or recently appointed or promoted managers—act the same. They want to do a good job, and, furthermore, they instantly have ideas about how to improve their new work. Hence they are already motivated… at least they have this at day one. (See Chap. 6 on Motivation.)

> **P**oint of Clarity Workers come to work motivated at day one, it is management's role to keep them motivated.

So items 4 and 5 are not the real problem to getting employees energized and engaged. No, the problem lies in items 1, 2, and 3. These are the sole province of management to provide. If employee engagement, at all levels, looks a lot like responsible management (see Lean Killer No. 10), don't confuse them. Engagement IS a very close blood relative of responsible management, but is not responsible management. Rather it is one of the direct results of responsible management.

So in a phrase, getting worker engagement is nothing more than supplying good management.

Continued Engagement Requires Fuel

There are two major areas, as managers and supervisors we need to address to sustain the engagement. First, your facility will be under a siege of continuous change. As such you will need to make sure you have mechanisms in place to assure that you keep your worker up to date with the "know what to do"; "know how to do"; and have the "resources" to execute their job duties. Supplying "know what," "know how," and the "resources" is more than any one single thing, but it is largely a maintenance program for your training delivery system. Second, you will need to become very skilled in the use of the five intrinsic motivators to sustain the motivation they once, and hopefully, still have.

> **"P**ut everybody in the company to work to accomplish the transformation. The transformation is everybody's job. **"**
> Deming Point No. 14.

We Are Not Yet Done—Recall It Is "Whole-Facility" Engagement

When I speak to management teams, they are always interested in attaining high levels of engagement. And by this they mean engagement of the workers on the floor, the ones actually adding value to the product. Quite frankly, in my experience I have found this is easily done. Most workforces readily agree to do more, they only need the management to make it clear that is what they really want and then off they go.

That said, there is a larger problem—the engagement of the management group itself.

Take a look back at the 10 Lean Killers in Chap. 2, each is some matter of management that is not handled well. Not one of the 10 Lean Killers is controlled by the hourly workforce. Hence there is some aspect of "not knowing what to do" or "not knowing how to do it" or some "resource" deficiency that creates each of these Lean Killers. This is just another manifestation of what Dr. Deming taught us when he said that 94 percent of the problems are system created and not deficiencies in the worker.

Like I said earlier, "…the bottleneck to getting worker engagement is not the worker, but the manager." And later I said, "…there is a larger problem, management engagement." Or, as I like to say, "It's all about management. The rest is just details." This is a hard sell to many management teams. Mostly because of incorrect paradigms of both motivation and engagement not to mention that the management team normally does not wish to change themselves.

Fortunately it is recognized by some folks. A friend of mine, a recently retired Colonel from the Army told me one of his instructors in Military Science told him: "The bottleneck is always at the top of the bottle."

Let me give you an example. In a recent lean implementation we were teaching the use of SPC (statistical process control) and in particular its application in problem solving. We started with the facility management. Their *sensei* gave them a basic class and each manager was assigned a problem to research and solve using data driven problem solving. Soon the management team was using SPC charts and other statistical techniques to solve a wide range of problems. Then the managers, not the *sensei*, did some training with the mid-level managers and supervisors; a really great sign. Very quickly the tools of SPC were common jargon and data based, statistical decisions were being made with a huge and rapid resultant gain in plant yield. However, since they were just getting started in their use of statistical problem solving, they were also making a variety of errors. They asked me to review where they were and what they needed next to take the problem solving "to the next level." Make no mistake, they were energized, they were motivated and they were ready for more.

About 10 days later, I met them at the plant and prior to implementing any new training, we reflected (learn-do-reflect) on the last several months of activities. Following a brief discussion, we had each of the managers and supervisors make a short presentation on their specific problems; what they found and specifically what help they next needed. First to go was the plant manager, Roberto. He explained his problem, but because of some technical problem he could not yet handle and, he was not able to proceed. He had been at a standstill for several days and could go no further. I sensed a "teaching opportunity" and asked this question of the group, "Is Roberto still engaged in his problem?". First, a hush came over the room; after all this was the highest level person from the plant and my question certainly had a critical overtone. Next, one brave soul, Ralph, the maintenance manager, wanting to support his leader, said he thought Roberto was engaged. To which I replied, "Ok then exactly what is he doing?" To which Ralph replied, "He is waiting for help from you." And seizing the opportunity I said, "since when is 'waiting' a sign of engagement?" Immediately an energetic, discussion ensued and later I summarized, "Make no mistake, in this case Roberto is no longer engaged, he is no longer 'moving and shaking' to make his problem disappear … he is waiting. He is disengaged. But let's not be too harsh on Roberto, rather, let's learn from this. Roberto is 'waiting' because we have not yet developed a JIT training system to support him. I am certain he 'wants to' solve this problem. But today he is 'waiting' because he and his upper management have not created a system to support his JIT need for training."

The case described on the preceding page is a common occurrence in a lean implementation. Roberto, not only was leading this change effort, but he had to learn himself. And soon enough, the new skills he learned were no longer adequate. He had grown and developed but had reached a point where he was missing an element of engagement; he did not "know what to do" next. And in this case, Roberto did the right thing, he asked for help. He was no longer engaged, but it was not a disengagement of his choosing, it seldom is. He needed some support, and supplying that support is almost always a function of Roberto's management.

Very often the paradigm of engagement gets confused and equated to motivation. Although they are related as explained earlier, they are different phenomenon. It is incumbent upon the management team to understand both concepts and lead the changes and manage the business in such a way as to foster both.

> **"** ... **W**e are in a new economic age. Western management must awaken to the challenge, must learn new responsibilities and take on Leadership for Change. **"**
>
> Deming Point No. 2.
> Sherkenback

I have yet to see even a single lean implementation which had full management engagement that was anything less than very successful. Unfortunately this is not the norm.

Just how does the entire management team get engaged? That's not too complicated. First, start by taking a serious look at the 10 Lean Killers in Chap. 2. Then learn and practice the Six Skills of Lean Leadership.

Chapter Summary

Engagement is the fourth leading indicator of cultural change. Employee engagement is all about management doing a good job to supply them with the answers to what they need to do, how they need to do it, and then making sure they have the resources to do the work. Finally a limiting aspect of worker engagement is insufficient management engagement.

CHAPTER 9

Cultural Change Leading Indicator No. 5, a Learning/ Teaching/Experimenting Environment

Aim of this chapter … to explain many of the aspects of this critical cultural change leading indicator that is almost always minimized and frequently even overlooked. We will discuss how we learn and especially how learning is more about "pulling" out much of what is already there in the workforce and less about "pushing" in more stuff. We will explain JIT training. Also we will discuss how the factory floor is a laboratory and manufacturing is a series of experiments. We will reiterate the concept of PDCA (Plan, Do, Check, Act) and contrast it to many of its imposters. Finally, we will discuss the relationship of the teacher, the student, and the system … when learning does not occur.

> **"W**e get too soon old and—too late smart."
> Carolina Andreoletti Tam, my maternal grandmother

Survival via Continuous Improvement—via Learning and Teaching

At the heart of a lean transformation, is the concept of continuous improvement. And I have come to appreciate that at the root of this concept of continuous improvement, literally its foundation, is the concept of continuous learning. And real learning, the type that can be applied in a productive sense, is acquired through experimentation and living this reality—and in no other way. Consequently I have come to appreciate that at the heart of a culture of continuous improvement is the subject of learning; along with both its antecedent and its logical result, which is teaching.

> **P**oint of Clarity Transforming an organization to a lean culture is a life long "learn by doing" experience.

I have come to this conclusion through careful observation. I know of no comprehensive studies that have correlated the learning/ teaching environment to success but I have worked with many facilities in my 20 years in industry and many more organizations in my 25 years of consulting. I have seen a number prosper and unfortunately I have been on the doorstep saying "good bye" to some fine workers when two were closed. I have my own database where I evaluate many characteristics of these companies and learning/teaching environment is one such cultural category. What I have found is this: Those who made a successful transformation to a lean culture, without exception, had a deep commitment to teaching as a management tool and a deep commitment to learning for all. Furthermore, they made no distinction to the class of employees who must learn. All employees, management included were engaged in learning and those facilities that prospered the greatest and the fastest were characterized by a top management that was both very introspective and learning oriented. Furthermore, each of these successful transformations had a management team that was lean competent, often very competent. Normally, these management teams did not start out with this level of competency, but through "learning by doing," they become lean competent. On this basis, I assert that creating a learning/teaching/experimenting culture is a leading indicator to success for a lean transformation.

As you recall from our discussion of cultures in Chap. 4, the second key cultural characteristic is that a culture must be flexible (in addition to strong and appropriate) because the internal—but also and more importantly the external business environment—is changing. New suppliers are entering the market, new manufacturing, engineering, and management techniques are being developed daily, and in addition all companies are striving for continuous improvements in what they already do. The key ability to be flexible and respond to all these external changes lies solely in a company's ability to change. And highly intertwined with this ability to change is the ability to learn and modify behavior. In fact, the single largest leading indicator to having a flexible culture is the ability to teach, learn, and experiment. To teach, to learn, and to experiment—to what end?

So we can learn our way to higher levels of competency.

In the long run competency only will protect us. Patents, innovation, and even geography can protect us for some time—but they are too transient. Patents can run out and be morphed by modest technical changes; with employee levels of company loyalty shrinking and managers and engineers changing companies at the drop of a salary increase, innovation is way too transportable to protect any company; and the Internet and other dynamics have made geographic protection a complete false sense of economic security. Only competency and the tools to achieve ever-increasing levels of competency will allow us to become a stronger money-making machine, with higher levels of security for all and propel us to be the supplier of choice to our customers. Long-term, continued, ever-increasing competency is the direct result of having a learning, teaching, experimenting work environment. This is both the key cultural change leading indicator to create a flexible culture and the key cultural change leading indicator for sustained success.

> **P**oint of clarity   The purpose of formal education is to enhance your ability to learn from experience.

We Learn by Doing

Earlier, in Chap. 2 (Lean Killers 1–10), I made the point of how much there is to learn in a lean transformation. Furthermore, the lean mantra regarding learning is "you learn by doing" and the skills needed for lean transformation are not only intellectual in nature, they are behavioral as well. This concept is not well understood and unfortunately not well practiced. Much like learning to play basketball, the skills to transform a company to a Lean Manufacturing system must be practiced. Grasping the intellectual understanding alone is insufficient for the transformation and insufficient to lead the transformation—even for the C-Suite—especially for the C-Suite.

Recall that exhibiting the qualities of lean leadership requires that all leaders be both life long learners and life long teachers. Also keep in mind that the key factor in culture creation and culture change is driven by the few key people at the top ... that is, the management, the very top leadership. Hence if they are learning and teaching as a daily activity, you will be well on your way toward creating a culture with learning and teaching as core elements.

While reading about Lean, studying Lean, or visiting other facilities to learn how some others do it may be good ways to enlarge your understanding, they are insufficient methods to learn "how" to transform your company. So just how do we learn?

We experiment—that's how we learn; and that's how we learn to transform our facility.

> **P**oint of Clarity   Executing a lean transformation is not a spectator sport, nor is it an intellectual exercise alone ... *everyone* must be practicing.

We Learn by Doing Experiments and Hypothesis Testing

Toyota, among others, is known for its teaching-learning environment. It is manifest by the management emphasis they place on both. What is not commonly understood is that all those inside the entire Toyota Production System (TPS) *have been taught* that their entire production system is a series of ongoing experiments and hypothesis tests. Keep in mind a hypothesis is a guess. Not some wild, off-the-cuff guess. It is based on observations and has theory as its antecedent; but in the end it is an informed, scientific guess. It is a guess in the form of, "if we do these specific things, we will get this specific result."

The Factory Floor ... an Ongoing Experiment

For example, let's say you have a production-by-hour (PBH) visual display on the floor and you have an hourly goal to achieve. What you are really saying is, "if we supply our operators with these raw materials, these tools, give them this training (these specific things), we should produce this volume of defect-free product (this specific result)." Then by producing the product, that is, exercising the "specific things we supplied" and entering the actual production data, measuring "this specific result," each hour and comparing this result to the goals, the PBH then becomes a check on that hypothesis. And in so completing this cycle, we have now learned if the "specific things" done achieved the "specific results" as predicted.

It is incredibly interesting to note that in 1939, long before anyone heard of the TPS, and long before most of it was even created, Walter Shewhart, speaking of the mass production process in a plant, wrote:

> "... specification, production and inspection, correspond respectively to making a hypothesis, carrying out an experiment and testing the hypothesis and the three steps correspond to the three steps in a dynamic scientific process of acquiring knowledge."
> "Statistical Method from the Viewpoint of Quality Control" (Dover Publications, 1984)

The uniqueness of thinking of all production as a hypothesis, an experiment, and a test is not new to American industry. But like much of what Deming and Shewhart taught, it got lost here. We forgot it, we failed to practice it, and we lost it. And now as we study and learn from Japanese systems such as the TPS, we are relearning the things we cast aside 75 years ago. My grandmother was right, "we get too soon old and too late smart."

And at the heart of all this is what Shewhart called a *dynamic scientific process of acquiring knowledge*. As I study manufacturing systems like those at Toyota, Honda, Nissan, and Mitsubishi to name a few, what sets them apart on the surface is their process stability and their cultural strength. However, below the surface when we ask "what is the foundation they build on?," the answer is that they cultivate and nurture an environment of continuous learning.

Every experience is a learning experience. They think in terms of hypothesis testing so if the production proceeds as planned, they do not ignore it and take it for granted. Rather, they **consciously** conclude that "the specific things we designed into the plan, achieved the specific result we desired." If they did not get the specific result they hoped for, or they need to improve on these results, then they conclude there is something they must change. And accordingly they **consciously** create a new hypothesis, perform a new test, and change some behaviors.

Not only do they use the scientific method exhaustively, they rigorously practice *hansei* which is a Japanese term for reflection, learning, and self-improvement. It starts with a rigorous, very rigorous review. In the example in the insert, this *hansei* was a "reflection" on a project I was a part of.

Why I don't Like, but Still Can Appreciate *Hansei*

I have only been part of one rigorous *hansei* experience. And it was not pleasant—not at all. I was unaware of what *hansei* was, however, my host Eiji told me it was a reflection on how we had done on our project. Since we met all the design objectives, we were done early and finished under budget, I thought it would be a breeze. I was looking forward to it. Dumb me. What happened that morning, in my opinion was brutal. Every single step of the project was scrutinized and we were asked dozens of questions like, "did you think of this?" and "why did you not do this?." Every single thing we did was placed under an after-the-fact high-powered microscope and picked apart incessantly and mercilessly. Questions were relentless, often stupid and superficial, asked by managers who had only a rough understanding of the task at hand. Sometimes the questions were very personal. In my opinion, it felt like a brutal skinning done by a bunch of Monday morning quarterbacks, using dull knives no less. When we left, I asked Eiji how he felt. He was more than OK with it, he said he had learned much. We discussed it over lunch and he told me, "Wilson-san, our project was good, very good, but not perfect; there are things we can learn. This is just how we do things." I understood but still did not like it. As I reflect many years later, maybe, just maybe, it explains part of the success of Japan…after all, "this is just how they do things." If you want to find some good examples of project reflections, go to the Internet and search for Army After-Action Reviews (AAR); they have some wonderful information.

Not Only Is the Factory Floor on Ongoing Experiment but so Is the Production Schedule, for Example

The entire production schedule is a hypothesis... "if we do these specific things we will get this specific result." And the list of "specific things we need to do is complicated indeed." We then produce the part which is the experimental portion of the hypothesis, yet another complicated activity.

> **"R**esults of an experiment should be presented in a way to contribute most readily to the development of the knowing process. **"**
>
> Shewhart, 1939:105

Finally we complete the order and ship the parts. Either meeting or not meeting the shipping date is the test of our hypothesis. If we produce a "non-expedited, full lot, are on time," we proved our hypothesis, because "we got the specific result of shipping on time." If on the other hand, we shipped late, used extra resources, or shorted the shipment, we then would have disproved our hypothesis. This means that one or more of the "specific things" we did, was not accomplished as we had planned it. Our test disproved our hypothesis and there is one or more things wrong with the list of "specific things" we had to accomplish. Either way we learn. If we prove our hypothesis, that is, shipped on time we learn we have a predictable system and probably should keep our standard. If we disproved our hypothesis (shipped late), we have to make improvements. Generally we learn a great deal more when we fail to prove our hypothesis as now we have found a weakness in our system and we can fix it. Once we realize—**and accept this fact**—this we can learn and improve from every schedule we make.

Furthermore, every inspection station is a "test" of the hypothesis and its experiment. It is the check of the "plan" and "do" portions of PDCA. And it is the point at which we need to decide, "what is my next action?" and "do I ship or do I reject?".

PDCA and the Scientific Method...

PDCA, the plan, do, check, and act cycle, is based on the scientific method, yet goes beyond the scientific method. The plan, do, and check portion is largely the scientific method and from this process we can learn a great deal indeed. In a lean transformation we teach PDCA and it goes beyond the experiment and the learning from the experiment; it takes you to the continuous improvement cycle via the "Act" step. And by "Act" we mean **thoughtful action**, which may include consciously and responsibly deciding to not act at all. As you implement the "Act" portion, you now have converted spot learning into continuous learning and the cycle just goes on and on and on and the learning continues and continues and continues.

PDCA—Acting on the Results and the Means

In the industrial setting, the actions following plan, do, and check (PDC) should be of two types. There should be action focused on the "results." For example, if we achieved our specific objectives, do we now standardize or do we set another plan to improve further? In addition, following the check, we should determine if actions regarding the "means" of our PDC were the best we could use. The means are the skills, techniques, and tactics we used to achieve the results. For example, did we plan well? If not, what should, could we do next time? Did we have the right group of people involved? How do we do

> **"I** shall assume that knowledge begins and ends in experimental data but it does not end in the data in which it begins."
>
> Walter Shewhart

> **"T**he only person who is educated is the person who has learned how to learn and change."
>
> Carl Rogers

this better next time? The action steps should guide us to the next "thoughtful action." It is not only learning about the "results," it is learning about the "means," that is, *the process that created the results.*

PDCA and How NOT to Learn

By the way it's really is PDCA not PDHM, PDCI, PDCRAS, nor is it PDCT.

PDHM Is Plan, Do, Hope, and Move on

You see this often in the field of human relations for things like personnel development. Someone puts together a developmental plan for the growth and development of some young engineer and then they turn them loose to execute it. The follow-up is usually to "hope" it works and "move on." Common characteristics that foster this imposter to the PDCA cycle are when goals are long term by nature or in areas in which the management is not interested enough or patient enough to delve into the details. And since the project, for example, is a long time away they often are not interested because it is not one of the short-term things they commonly focus on. It requires real long range thinking rather than the fire fighting they are so used to and adept at. Another common example would be new program launches. They put together a complicated launch process, turn it over to some junior executive to execute and then "hope" it works and "move on." Frequently we see PDHM with management dictums that may follow some incident such as an equipment failure or operating incident. It will be decided, for example, that the machine needs its oil checked periodically and then no one creates a mechanism to make it happen. The earmarks of this cycle include lousy if any goals; completely inadequate or nonexistent follow-up systems; even though many examples are directly related to the value stream. That's PDHM. Certainly no one learns from this.

PDCI Is Far More Common than PDCA

It stands for plan, do, check, and ignore. Anytime a variable in your value stream that has an instrument gets no attention, you likely have PDCI. I see this occur in spades at the management level when some company has a large number of KPIs (key performance indicators). The plant manager, for example, may need to report on these 56 KPIs but there is little he can do about all of them, consequently he selects a group he can work on via PDCA. The rest, likely the majority, then go through the PDCI cycle. The key issue here is poor focus and sometimes inadequate resources. Again, very little learning occurs with PDCI.

PDCRAS Is Increasingly More Common

It stands for plan, do, check, record, and then store. This is pretty self-explanatory but includes the vast majority of data an operator may write down as a result of his daily activities. It also includes a wealth of numbers that are now available due to the proliferation of computer-generated data. Most of this is collected not because it "should be," rather it is collected because it "can be." Almost always that is a lousy reason. I find the main culprit here is that the facility really does not understand and use the Three Rules of Data Management that were taught by Ishikawa:

1. Know the purpose of the data
2. Collect it efficiently
3. Take action based on the data

I find in most companies today that they are "drowning in data, yet starving for information." Clearly an issue of focus. In this method, sometimes you learn where to go get the data if you might need it in the future … but not much else.

PDCT, to Me, Is the Saddest of All the Forms

It is plan, do, check, and tinker. Tinkering is the term for adjusting a process when no adjustment is necessary. It is a direct result of not being able to distinguish common-cause variation from special-cause variation (W. Edwards Deming, *Out of Crisis*, MIT Center for Advanced Engineering Studies, 1986). The problem is that the worker is trying to do his best. He is attentive and taking action. Unfortunately the action he is taking is making the process deteriorate rather than improve. This conscientious worker is unwittingly adding variation to a process, thereby increasing costs and hindering productivity. He is not trying to do this, he is trying his best to do exactly the opposite. The fault is not his, the fault lies in management who have not told him how to properly adjust his process. They, more often than not, do not know how to do this themselves. Nonetheless, today this is the biggest source of variation in the industrial world … overadjustment or tinkering. To stop this, management must learn and teach the science of statistical process control. Then everyone will understand the basic concepts of variation and learn how to manage them.

A Mantra for Initiative Are:

- Start where you are
- Use what you have
- Do what you can

JIT Training

Throughout this book you have heard and will hear me discouraging the extensive use of classroom training. I find most businesses favor classroom training because they view it to be financially and time efficient to get a whole bunch of people "up to speed." Quite frankly I find most of this training to be pure waste … not all, just most. First, this is almost always a manifestations of a culture that has very short-term perspective and is focused solely on financials as the key metrics to drive the business. They will think of efficiency first and ignore effectiveness. When it comes to training, efficiency means nothing if the training is not effective. Second, training is behavioral related and scare little can be taught in a classroom setting. It can be introduced, basics often can be discussed but the changing of the work habits, the behavioral aspect, to be effective must be taught on the factory floor using the real equipment in the real environment. Sometime, if you have good simulators—and I don't mean computer models—I mean physical simulators, you can effectively complete a portion of the training in the classroom. Third, for learning to be effective, the student must want to absorb the materials. Also, the student must immediately apply the techniques he has learned or they will soon enough be forgotten. This is seldom the case for the majority of the students in this classroom training environment. This further diminishes any learning that did occur and makes the training even less effective. Finally, there is nothing wrong with doing a little classroom training to explain the basic concepts of Lean and getting people's feedback is a good idea. It should be a 1-hour introduction with I hour for Q&A. It should not be 8 hours or 3 days of "Death by PowerPoint."

Quite frankly a work environment, supported by the guidance and support of lean leadership, is the best teacher.

Since "Death by PowerPoint" is currently an industrial epidemic that is being proposed by corporate management as a means to introduce any new topic, this puts people in a bind. They then ask, "If Death by PowerPoint is no longer an option, just how do we get started?".

Simply enough, you get started! Recall "the Mantra for Initiative." Quit talking and go grab your *sensei*, and consistent with True North (see Chaps. 16 and 17), pick a problem, any problem. Start with the Five Questions of Continuous Improvement. As soon as you run into an obstacle that hinders problem solving, seek assistance from your *sensei*. Be ready for questions! And then when it has been determined you will need some cool lean tool, he can teach it. He can supply JIT training. In this way, you have "pulled" in the specific tools you will need to be successful. Furthermore, since you want this tool to solve your problem and can apply it right now, you will very likely learn it. Following this, you then go back to the Five Questions and continue on your journey to becoming Lean. In this manner, you are problem solving your way to the Ideal State … you are truly practicing the "Means to Lean."

> It is more obvious in acquiring motor skills than in acquiring intellectual knowledge—but no more true—that learning is an active process rather than a passive one. The necessary effort will be expended only if there is a "felt" need on the part of the learner.
>
> McGregor, Douglas (2005)
> *The Human Side of Enterprise*,
> McGraw-Hill, Kindle Edition

Teaching by Questioning

From the time I was in junior high tutoring my friends in math, to being a lab instructor in college, to being a manager in an oil refinery, to my consulting practice today, I cannot recall any protracted time period in which I was not instructing or teaching in some form. Certainly all my work life, either in manufacturing or consulting, I have been teaching. And outside of work I have been teaching. I coached soccer for 32 years. In addition I frequently volunteered at our church to teach the Bible or other adult learning programs. I have been taught and I have taught … all my life.

And I was always thought of as a good teacher. In junior high, after tutoring some students, I was asked to do a second class. On several occasions I attempted to retire from soccer coaching but someone would hear I was available and ask if I would teach.

Twice, the very day I ended one soccer coaching assignment, I was asked to coach my next team. Soccer is a love of mine and I had a very tough time saying no, but people lined up to get me. My church sought me out as an instructor numerous times. I thought I was a good teacher and others did as well.

Then, over 20 years ago, two things happened that made me question both my ability to train and the value of classroom training. First, I was asked to return to a company I had consulted with a few years earlier. They wanted additional training on SPC (statistical process control). Upon returning, I found that many were students who had earlier taken my 36-hour class. They had done well in the classroom and with their projects and I remembered them. They remembered me and gave the other students the praises of my teaching ability and how easy it would be to learn this new material from me. Of course I was flattered. But as I began to teach them a second time, I found they had not retained much at all, rather it was mostly new to them, just a couple of years later.

Well, it took a while but I finally realized the weakness in my teaching. You see I was so caught up in the opinions of others who believed that I was a good teacher, I believed it. I really was okay by the standards of the American educational system. I could teach what that system viewed as good … unfortunately locking up a bunch of people in a room, giving them a book with some ready-made problems and then asking them to puke back this information hardly constitutes teaching … in my opinion—now. But worse yet, it never constitutes training. You see we must distinguish between learning and training. Learning is the intellectual acquisition of knowledge, while training is behavior modification. After people are trained, they can do new things. Teaching is the delivery of both. And in my SPC class, I thought I had trained these students … unfortunately I had not.

That very week I began to question just how good a teacher I was. I was so distraught that I changed my whole consulting practice. At that time it was about 80 percent teaching and 20 percent on-the-floor problem solving. I intentionally and quickly reversed these numbers and limited my teaching to 20 percent of my billable hours. I simply was not happy teaching—if no one was learning.

Second, and serendipitously shortly thereafter, I had a discussion with Joseph Juran. Every year in El Paso we held the International Quality Forum. In 1993 he came to El Paso as the keynote speaker. That year we had around 700 attendees and I was one of his hosts, so I got to spend a great deal of time with him. I was also the chair of the education subcommittee; consequently we had numerous conversations about training and education. He, like many good educators drew a clear distinction between training and education and I recall clearly he said, "most of what the industrial world calls training is really just superficial education and a full 90 percent of it is wasted." He went on to say that he thought that many were well intended, just not close enough to the work to really see the lack of benefits. Others were supplying the education as a response, trying to superficially, inexpensively and expeditiously placate some audit deficiency. Neither yielded the desired results of improved behaviors.

Let me share with you things I have learned that changed completely the concept of good teaching in my mind.

First, I have not had a client yet who did not have plenty of intellectual talent and curiosity on staff to do almost all that is needed for a lean transformation or implement

SPC or institute a program of value engineering. Whatever it is they wish to do, I have not found intellectual muscle to ever be a weakness in any client.

Yet they, both the workers and their management, almost all believe the opposite. I find this to be both amazing and disempowering.

It has been so commonplace, I have expressed it as three of the key beliefs of Quality Consultants … specifically ….

- The vast majority of skills needed to vault your enterprise into becoming a lean enterprise are already inherent within your organization.

- The key to success is largely an issue of unleashing these skills.

- One major and very critical tool needed to unleash these skills, and hence to become a lean enterprise, is to learn and employ the Six Skills of Lean Leadership at all levels of the organization.

Second is a part of teachings of psychology today, especially found in the writings of C. G. Jung. He popularized the concept of the unconscious and how powerful it is. In simple terms he believed we have a vast reservoir of thoughts, recollections, facts, and information which we have been exposed to. Some has been repressed, some has been forgotten, and even some we may not have realized we assimilated. None of this is in our conscious mind, hence we are not aware of it. We don't know we know these things. These data are stored away in our unconscious and not available for immediate use. However, these thoughts and facts could be brought into our conscious mind by some trigger. One of my favorite authors, M. Scott Peck wrote:

> When beginning to work with a new patient I will frequently draw a large circle. Then at the circumference I will draw a small niche. Pointing to the inside of the niche, I say, "That represents your conscious mind. All the rest of the circle, 95 percent or more, represents your unconscious. If you work long enough and hard enough to understand yourself, you will come to discover that this vast part of your mind, of which you now have little awareness, contains riches beyond imagination."
>
> Peck M. Scott (2012-03-13). *The Road Less Traveled: A New Psychology of Love, Traditional Values and Spiritual Growth* (p. 243). Touchstone. Kindle Edition.

Consequently, with this being a basis, it may be more advantageous in problem solving and in learning to work with what is already there … the 95 percent rather than try to augment the 5 percent by pouring in more "stuff." The numbers are certainly in your favor.

Third is the word educate, a synonym for teach. Educate comes from the Latin word, *educere*, which literally means to "lead forth" or educe, to draw out. The implication is that the information and knowledge very well may be there and the function of the education process is to draw it out.

Since these folks have a good database of conscious knowledge to begin with, then over and above that, they have a largely untapped reservoir of information stored away in their unconscious. Maybe we should first try to "draw out" this information before we try to "teach" them some more things? Just how can we coax this information from them and pull it out, without pushing in a bunch of stuff, which very likely is not even relevant?

Questioning—That's How

After all Socrates knew that. It seems like we forgot some of the real lessons of the past. We forgot the teaching of Shewhart and Deming but even before that we forgot about Socrates … go figure out.

I have found that since most of the knowledge is already present, most of the learning needed can be drawn out. Questioning will do this. There are two other aspects of "teaching by questioning" that promote long-term retention. First, it requires you meet the student where they are. Hence, they are the client, it is focused on them. They are important and they take it more personally, because it is. Second, as soon as you ask the first question you can assess their current level of understanding. So the system builds on what they already know. You can make a strong connection to their prior data and they will assimilate the new material much better. It will have greater context and make more sense to them, and they will retain the information for a much longer time.

Alternatively you can push some information at them based on some preconceived notion of where you believe they should or might be. From personal experience I can tell you this does not work nearly as well.

There clearly is a place for learning in the classroom but it is grossly overused. If you are attempting a lean transformation and you want to engender a high degree of engagement, then teaching by questioning should be your primary teaching mode.

There is a great deal of information on questioning and leading by questioning in Chap. 5, including the Five Questions of Continuous Improvement, how to ask open-ended questions, questioning follow-up techniques, and persistence in questioning. I easily could have put that information here, but it is such an integral part of lean leadership, I left it there.

Reflection as a Training and Growth Tool

As you begin to transform your culture, you will find that reflection is a strong everyday tool you will use. Not all activities need the formality and structure of an After-Action Report or *hansei*. But all should be reflected upon. For example, if you are having a meeting, start with reflection on "what has happened of relevance since we last met?". When you finish a meeting, reflect on "what should we stop doing, what should we keep doing and what should we do better?". In all activities "learn-do-reflect" is the lean mantra.

Growth and Failure

I can tell you with certainty that the greatest learning experiences in my life have been when I failed. Almost anyone over 50 will tell you the same thing. Jerry Yeagley, the great soccer coach from Indiana University, once told me "The coach's two best friends are the bench and a loss."

> If ... "post mortems" comprise no more than a search for a culprit in order to place the blame, they will provide learning of one kind. If, on the other hand, it is recognized that mistakes are an inevitable occurrence in the trial-and-error process of acquiring problem-solving skills, they can be the source of other and more valuable learning.
>
> McGregor, Douglas (2005)
> *The Human Side of Enterprise*
> McGraw-Hill, Kindle Edition

He went on to explain that success, if not balanced with humility, breeds complacency and even arrogance. And in a student, those are not traits conducive to learning. Failure gets everyone's attention.

Well, that means that you have to make failure possible by challenging your people, but you also must make failing acceptable, at some level, or they will not accept the challenge. In many businesses, the reality of failing is highly and overly criticized. This creates two compounding and very negative effects. First, when people encounter failure, and they will, they hide it and no one learns. This then gets compounded by the fact that people then become risk averse and won't accept the slightest of challenges. By being risk averse, they try to avoid failures; when they occur, they are even more likely to be hidden—a double negative whammy to learning. Furthermore, please, note this intolerance for the reality of failing, is an integral part of Lean Killers Nos. 9 and 10 in Chap. 2.

A Final Word or Two....

I have used and studied the TWI (Training within Industries) methodology. I have found it to be the very best method to deliver training to the factory floor, and beyond. It was developed during World War II and, in its original form is still the very best

> "Let's face it, we're all imperfect and we're going to fall short on occasion. But we must learn from failure and that will enable us to avoid repeating our mistakes. Through adversity, we learn, grow stronger, and become better people."
>
> John Wooden

training delivery method I have seen. I have a copy of the original documents thanks to Mark Warren and his efforts to get this from the National Archives and elsewhere. It is impactful material and if you are undertaking a lean transformation, it is a must read.

However, they make a basic statement repeatedly which I find misleading, if not incorrect. They say simply, "if the student did not learn, the teacher did not teach." The reality is that if the student did not learn, then teaching did not occur. I guess that is logically correct. The implication, made clear in the materials, is that the teacher was deficient in some measure. By placing the burden on the teacher in this manner, I find a problem. I, like many teachers, can give you many circumstances where, given the same training some people learned and others did not.

On the other hand, there is an old proverb that says, "when the student is ready the teacher will appear." This adage then places the burden on the student to be ready. Likewise I have seen many circumstances were eager and ready students were given instruction but not really taught.

While both may be partially true, neither is helpful to improve the situation. Unfortunately both are bad paradigms and focus attention on the people.

What I teach is "the teacher will teach and the student will learn when the system is ready."

You see, as Dr. Deming taught us, at least 94 percent of all problems are system related as we have discussed earlier. I am sure anyone can point up an instance or two where the people being trained or the trainers were not up to the task at all. This is the minority report; the majority of the failings are due to an inadequate teaching/learning system. System problems include such items as teaching today in the classroom and expecting students to apply them in the distant future; not teaching the managers and then the managers cannot support as needed; not teaching the managers and hence they cannot ask questions relative to the new teachings; teaching skills that do not align with the current problems to solve and the list goes on.

Each company must develop a culture of teaching and learning through experimentation, so we can then learn their way to greater competency and become better, faster, and cheaper as we support our customers, stockholders, and employees. And it is not a coincidence that this teaching/learning skill set are two of the Six Skills of Lean Leadership.

Chapter Summary

Possibly, the most overlooked cultural change leading indicator is the aspects of learning, teaching, and experimenting. In fact, when this is achieved, and only when this is achieved, will you have the needed mechanism to have a flexible culture. In most transformations this leading indicator is frequently minimized and sometimes even overlooked. In addition, a much overlooked aspect of your workforce is their current skill inventory. Learning is more about "pulling" out much of what is already there in the workforce and less about "pushing" in more stuff. Finally, all of this learning, teaching, and experimenting is centered on the factory floor, which through the use of hypothesis testing is a large laboratory and changing and sustaining the manufacturing system is done through a series of experiments. Also we debunked many of the imposters of the PDCA cycle, including PDHM, PDCI, PDCRAS, and PDCT.

PART III

The Solution.......
How to Implement Lean Manufacturing

Part Three addresses the solution to lean implementation. Technical topics of Lean are discussed, and the formulas to lean implementation, as well as numerous case studies to amplify the text are included. Chapters 10 to 13 address Lean Manufacturing and how it relates to the Toyota Production System, the critical topic of variation and its relationship to inventory, a simplified explanation of Lean Manufacturing, and the significance of lead time in the lean system. Chapters 14 and 15 address the five strategies for becoming and sustaining Lean respectively. With this foundation, Chap. 16, the very heart of the book, tells us how to implement the lean transformation at both the corporate and the value stream levels. In Chaps. 17 to 19 we address three technical topics. In order they are planning and goals, constraint management, and cellular manufacturing. These are followed by several clarifying cases in Chaps. 20 to 23. Finally in Chap. 24 we have not only the House of Lean but the system-wide evaluations used throughout the text.

CHAPTER 10

Lean Manufacturing and the Toyota Production System

Aim of this chapter … to explore what Lean Manufacturing is and what it is not. In so doing, we will define Lean Manufacturing from several different perspectives and compare it to other philosophies, including how Lean Manufacturing and the Toyota Production System (TPS) compare. The creator of the TPS, Taiichi Ohno, wrote a great deal about the TPS, and we will explore his thoughts on its uniqueness. Finally, we will discuss the limitations of Lean Manufacturing and its applicability outside the "Lean Stereotype."

The Popular Definition of Lean

The popular definition of Lean Manufacturing and the TPS usually consists of the following:

- It is a comprehensive set of techniques that, when combined and matured, will allow you to reduce and then eliminate the seven wastes. This system not only will make your company leaner, but subsequently more flexible and more responsive by reducing waste.

- Wikipedia says "Lean is the set of 'tools' that assist in the identification and steady elimination of waste (*muda*), the improvement of quality, and production time and cost reduction. The Japanese terms from Toyota are quite strongly represented in 'Lean.' To solve the problem of waste, Lean Manufacturing has several 'tools' at its disposal. These include continuous process improvement (*kaizen*), the '5 Whys' and mistake-proofing (*poka-yoke*). In this way it can be seen as taking a very similar approach to other improvement methodologies."

What Is Lean?

The TPS is often used interchangeably with the terms Lean Manufacturing and lean production. Regarding the technical issues of TPS and Lean, I will frequently use these terms interchangeably. It is called Lean because, in the end, the process can run:

- Using less material
- Requiring less investment

- Using less inventory
- Consuming less space
- Using less people

Even more importantly, a lean process, be it the TPS or another, is characterized by a flow and **predictability** that severely reduces the uncertainties and chaos of typical manufacturing plants. It is not only financially and physically Leaner, it is **emotionally much leaner** than non-lean facilities. People work with a greater confidence, with greater ease, and with greater peace than the typical chaotic, reactionary—change-the-plan-hourly-and-then-still-work-overtime-and-then-still-expedite-it-all manufacturing facility.

Becoming Lean Is Not Just about Results

The typical businessman is looking for results when he plans to make a lean transformation. Properly executed the results will follow and they will satisfy all three constituencies: stockholders, employees, and customers. The results will be that your facility will become a better money-making machine—that has improved security for all—and will be the supplier of choice for your customers. What often gets minimized or even ignored, but is of equal importance, is that a lean transformation is not just a result; it is simultaneously a "means." The "Means to Lean is to problem-solve your way to the Ideal State." It is as much about the process that creates the results as the results themselves.

A large criticism I have with many businesses is that they are driven by results almost exclusively. And normally these businesses have a very strong emphasis on short-term gains (Chap. 2, Lean Killer No. 6). I also read and hear a lot about how bad management by objectives (MBO) has become and many are touting a new technique called MBM, management by means. Frankly, I belong exclusively to neither school yet see the benefits of both. First, although means are important, by itself MBM will not really work since to assess the method, the "check" step requires you measure against some metric, so results are inherent in the MBM process. Likewise, MBO taken to an extreme is also a destructive management practice. Lean, properly executed, provides the proper balance. The use of *Hoshin–Kanri* planning provides the result-oriented approach and focusing on the process improvements provides the balance to include the "means." For example, Toyota Motor Corporation published "The Toyota Way, 2001" (Toyota Motor Corporation, 2001) to explain to their expanding base of manufacturers and suppliers, what the Toyota principles are. Based on their foundation of Continuous Improvement and Respect for People, they detailed their five key principles. One was "Kaizen, to improve business operations and processes continuously, always driving for innovation and evolution." Another was "Challenge, we form a long-term vision, meeting challenges with courage and creativity to realize our dreams." Their *Kaizen* principle speaks to both means and results and their challenge principle certainly speaks to a long-term vision. So you see, inherent in "The Toyota Way, 2001" is a balance of results with means and likewise, a lean initiative will yield that balance.

What Does Lean Look Like?

I will frequently get asked, "just what does a lean facility look like?". Well that is not so easy to answer but I specifically look for three things.

First, I Look for Transparency

This is the concept that the information on the floor will explain itself. Can I tell at a glance if the cell is producing good product and producing as per plan? For this there are production-by-hour boards, *heijunka* boards, and *kanban* along with segregation locations for defective product to name a few. There are also floor markings for inventory locations and *andons* to signify equipment status. In addition I check to see that these data are relevant and being used.

Second, I Check for Patterns and Measures of Consistency

For example, if there are four cells to manufacture one product, do all the cells look and perform the same? Is the pattern of production consistent? Is the pattern of shipping consistent? Is the work, as dictated by the *heijunka* board, being followed? These are all measures of both consistency and predictability. These patterns and measures of consistency can often be uncovered by asking the following questions about the process. Is the process:

- Self-ordering?
- Self-explaining?
- Self-regulating?
- Self-improving?
- Where what is supposed to happen, does happen, every time, on time?

As I then walk around the facility to begin my assessment, I will stop at one cell and ask a few questions. I seek out the team lead and the discussion might go like. "Hello Harold, I am the new lean advisor your company has hired and I am trying to assess the status of your lean maturity, could you help me a bit?" At which time Harold almost always says, "sure." I then go on, "Harold, how do you know you have the correct amount of raw materials on hand?" Harold quickly replies, "Well that used to be a huge problem with people running around chaotically trying to avoid stock outs but we would invariably shut down this cell three or four times per shift due to materials issues alone. No more. You see, the floor is marked with the tape. Yellow is raw material container locations, blue is finished goods and you'll note the forklift lanes are marked in red. These are plant-wide standards. Each raw material location is designed to hold about 1.5 hours of production and our materials handler is on a 60-minute cycle. Nonetheless, we occasionally get to the last container. But if look at the floor markings under the bottom container (he points to a square box on the floor) you will see it says that that container holds 32 minutes of production parts. A quick look at the clock and we can tell if the materials handler will arrive in the next 32 minutes. If not, we turn on the *andon* for an emergency restock. Works great! In the last 2 weeks we have had only one stock out, yesterday. Turns out our materials handlers' wife had a baby, and there was forklift safety training scheduled that day as well. We had to really scramble. The super recognized it at the team meeting, held over one guy but we still missed one part in my cell. As team lead I submitted a corrective action report and I think we can solve this with a little cross training."

In this example, you can envision the patterns that exist and also see that the cell is meeting virtually all of the five criteria.

Third, I Check to See if Everyone Can Answer the Question, "Are We Winning?"

To do this I would expect to see Plant Wide Metrics posted in a conspicuous place so all workers can see how well the entire team is performing. Likewise there needs to be value stream and cell metrics, which are clearly related to the plant metrics, appropriately located to answer the, "are we winning?" question at that level. In addition to the "results" of a lean process, there must be some visibility of the "Means to Lean," which is "problem solving to reach the Ideal State." There should be some record of problems being solved, problems solved, or the results of problems having been solved, especially at the cell level. Frequently this is covered if the cell metrics have charts showing not only instantaneous performance but performance over time with trend lines. These are simple to produce yet the impact is huge. First, if the team is winning in the long run, any short-term problem in any given day is not so demotivating. Also it is always refreshing to look back and see how much progress has been made. Keep in mind that a "sense of accomplishment" is one of the five intrinsic motivators.

Lean and the Toyota Production System

To further explore the depth of what a Lean Manufacturing system really is, we will look deeply at the TPS. Not because the TPS is the best lean system around, although it may be. Rather, we will look at the TPS because it is the best documented system and it has proven itself over a very long time. It has not only proven itself but stands as the example of "Lean done extremely well."

What Did Ohno Say about the Toyota Production System?

If we really want to understand what the TPS is, we certainly must listen to what its creator has to say. In any discussion about the TPS, Ohno will be my arbiter. A great deal has been written about Lean and the TPS, but some of it misses the point. If there is any question about what Lean or the TPS is, we will use Ohno's thoughts as the final word on the topic. His most notable writing on the subject is his book, *The Toyota Production System, Beyond Large-Scale Production*. In it, Ohno makes three key statements, which when taken together define his TPS.

- "The basis of the Toyota Production system is the absolute elimination of waste." (p. 4)
- "Cost reduction is the goal." (p. 8)
- "After World War II, our main concern was how to produce high-quality goods. After 1955, however, the question became how to make the exact quantity needed." (p. 33)

Taken together, we could then write a definition such as, "the TPS is a production system which is a quantity control system, based on a foundation of quality, whose goal is cost reductions, and the means to reduce cost is the absolute elimination of waste."

The TPS and Lean Manufacturing Defined

I find that all these definitions miss part of the essence of Lean Manufacturing. I do not think Ohno defined it more carefully because oftentimes we do not feel the need to define those things that are very near and very obvious to us. I believe others may not define it better because they simply miss the point, while others may understand it but not be able to articulate it. My definition of the TPS is a manufacturing system that:

1. Has a focus on **quantity** control to reduce cost by eliminating waste
2. Is built on a strong foundation of process and product **quality**
3. Is fully integrated
4. Is continually evolving
5. Is perpetuated by a strong, flexible, and appropriate culture that is managed *consciously*, *continuously*, and *consistently*

I refer to this as the "fivefold definition" of the TPS. It is a much better characterization of the TPS, but unfortunately to those within the TPS this is old news, while to those outside the TPS this definition may not be very meaningful. So, as we go further into this chapter, I will try to put these five aspects of the TPS in context.

Who Developed the TPS?

The accepted architect of the TPS is Taiichi Ohno, the chief engineer of Toyota for many years. Others contributed greatly, including Shigeo Shingo and the members of the Toyoda family, but Ohno gets most of the credit for its creation, development, and implementation. Ohno may also get most of the credit because he wrote the most about it, or maybe just did the most about it. Neither is important. What *is* important is that the TPS specifically, and Lean Manufacturing in general, is a tremendous contribution to society and manufacturing in particular, and we owe a huge debt of gratitude to those individuals who created it and caused it to mature even further.

The Two Pillars of the TPS

Ohno describes the TPS as consisting of many techniques that are designed to reduce the cost of manufacturing. His method of reducing cost is to remove waste. This waste elimination system, the TPS, is built on two pillars.

Just in Time

The first pillar is *just in time* (JIT). This is the technique of supplying exactly the right quantity, at exactly the right time, and at exactly the correct location. *It is quantity control*. It literally is at the technical heart of the TPS. Most people envision this pillar as inventory control, and this is a part of it. However, JIT is much more than a simple inventory control system. What's surprising to a large number of practitioners is that at the heart of quantity control—at the heart of JIT—is a deep and abiding understanding and control of variation.

> **"A**fter 1955, however, the question became how to make the exact quantity needed...**"**
> Taiichi Ohno

Jidoka

The second pillar is *jidoka*. This is a series of cultural and technical issues regarding the use of machines and manpower together, utilizing people for the unique tasks they are able to perform and allowing the machines to self-regulate the quality. Technically, *jidoka* uses

> **P**oint of Clarity The TPS is a QUANTITY control system.

tactics such as *poka-yoke*, (methods of fool proofing the process) *andons* (visual displays such as lights to indicate process status especially process abnormalities), and 100 percent inspection by machines. It is the concept that no bad parts are allowed to progress down the production line and it is supported by a culture focused on real-time problem solving . This not only is needed to protect the customer and reduce scrap costs, it is a continuous improvement tool and is a key element in making *kanban* work. It is a violation of *kanban* rules to allow bad parts to be transported.

The Original Western View of the Japanese Techniques

In the early writings on Japanese techniques (the term "Lean" had not even been coined), there was a very simple explanation of the basics. It is worth reviewing some of the more salient books to get a simpler and clearer description of how the Japanese made this new manufacturing system work. I have found that too many "experts" have clouded a basically simple system to the point that many folks have lost sight of the basics.

Not so with Schonberger and Hall. These early writers ("Japanese Manufacturing Techniquees" Free Press, 1982, Richard Schonberger, "Zero Inventories," Business One Irwin, 1983, Robert "Doc" Hall) painted a very simple picture. First, there were two aspects, JIT and total quality management (TQM). The whole system started with reducing the set up times so that pieces could be made and delivered in one-piece flow to the next station just in time. If the product unit was defective, production was stopped and corrections made so that total quality was achieved. With very little inventory in the system, problems of quality and flow were immediately obvious so problem solving could take place. Workers on the floor were trained in problem solving and the entire system was based on finding and fixing problems. This caused resultant increases in both quality and productivity. This then highlights some of the most basic concepts of Lean that have gotten lost in the "fog" created by the "experts."

1. Lean is all about reducing lead time.
2. To reduce lead time, reduce inventories.
3. When inventories are reduced, problems become more obvious.
4. All the lean tools (5S, *Heijunka*, *Andons*, etc.) are means to make problems obvious.
5. Problem solving is "the Means to Lean" and it is everyone's task.
6. Problem solving reduces the variation.
7. As the variation is reduced, the lead time is reduced and "voilà" we are back to step 1.

What Is Really Different about the TPS?

Technical Issues

So just what is different about the TPS? What makes it so revolutionary? The answer to that is not so simple. First, let's look at the technical aspects, particularly some of the industrial engineering aspects. These technical skills and tactics are the foundation for the quality and quantity control aspects (items 1 and 2 from our fivefold definition of the TPS). Some very old engineering techniques are used within the TPS. In addition, some old techniques with new twists, as well as some totally new techniques, are also included. Later, we will address the deeper issues of the integration (item 3), the evolution of the system (item 4), and the cultural differences (item 5). But as I said, let's first discuss some of these technical issues. We will compare several aspects of a manufacturing system as either Normal Mass Production Model (MassProd) or as Lean. And where the TPS stands out among lean facilities, we will highlight that as well. These manufacturing aspects include the following:

- The makeup of the typical production cell/line and how quality is handled
- Handling multiple models of a product
- The use of "pull" versus "push" technology
- The issue of changeover times
- How parts and subassemblies are transported in the plant
- How finished product demand and supply variations are handled
- How quality is managed
- How cycle-time variations are managed
- How line availability is managed

In the next subsection, we shall review them in order.

The Typical MassProd Production Cell/Line

The makeup of the typical MassProd production cell/line is normally a flow line. The typical flow line has work stations lined up in a row. It has inventory in front of and after each station. These work stations operate more like production islands rather than a connected process. The production of each station is maximized to improve equipment utilization, they call this "optimization" and product is pushed to the next station. This station-by-station "optimization" and "push production" are extremely key issues. The production rate is determined by the plan and all too frequently each island has its own plan. Normally product is produced, one model at a time, in large batches. There is an attempt to finish each batch before the appointed delivery time. The typical quality system is inspection based, is normally done by humans, and rework is not only common but is necessary and designed into the process flow. This differs greatly from the lean solution, which is typically a *cell, with a pull system, balanced at takt, flowing one piece at a time, with a system of jidoka.*

We will explain each of these lean concepts more clearly later, but the key differences are that all process steps are placed close together—in a cell, for example—so one-piece flow is possible. There is no buildup of inventory between processing steps. This close connection and lack of inventory allow the product to "flow." Furthermore, all work stations are balanced, in that each station has the same cycle time. This allows synchronization of manufacturing. Furthermore, the balanced cycle time is designed to be at *takt*, which is the demand rate of the customer. Product is made at the same rate at which the customer wishes to withdraw it, and production only occurs when the customer removes the product. This is the "pull" concept, which is different from "push" production. Quality is managed by the *jidoka* concept, which not only finds and removes any defects, it initiates immediate root-cause corrective actions. Using the preceding reference techniques, the lean process itself, rather than any of its individual steps, is optimized, forgoing the optimization of local work stations. Many lean facilities have very good *jidoka* systems, but the TPS is second to none in the application of *jidoka* principles, largely due to the way Toyota has managed, and continues to manage, their culture.

Handling Multiple Models of a Product

The handling of multiple models of a product is a uniqueness of Lean. In the typical MassProd where we have multiple models, they are normally produced in large batches even if they use the same production facilities. Generally, the reason for the batch operation is because when switching models, changeovers are required. The lean solution employs the concept of model mix leveling. Say, for example, we have three models of our product. They are A, B, and C, all produced on the same process facilities. The model mix is 50 percent A, and 25 percent of each B and C. We then would produce these three models simultaneously in our cell, producing in this order ABACABACA-BAC. Of course, to do this Single Minute Exchange of Dies (SMED), or quick changeovers, or more likely its refinement, OTS (one-touch setups) must be very mature in this cell. Surprisingly enough to those not acquainted with the TPS, this is not as difficult as it sounds. In addition to accomplishing the leveling and knowing which model to produce, a *heijunka* board is employed to schedule the leveling of the various models. The *heijunka* board will accept the production *kanban*, which are removed from the product when the product is withdrawn. These *kanban* normally are circulated directly to the pacemaker step and the replenishment signal, the *kanban* itself goes directly to the production cell, normally bypassing planning altogether.

> **P**oint of Clarity   Lean Manufacturing is a batch destruction technique.

Pull Production Technology

Pull production is a confusing concept to many, but it is a critical aspect which makes Lean work. It is one of the key tools used to avoid overproduction, both local overproduction at a specific work station and overproduction at finished goods. Pull production is the concept that the production process is not initiated by orders or schedules. What triggers production is customer consumption. To implement pull, the business must make the huge cultural change of responding to what the customer **does**, rather than what the customer **wants**. It is the "take one, make one" concept and this production mode is called "replenishment." It is virtually impossible to have a pure pull system in most manufacturing environments, but all pull system have three characteristics.

- First, no production is initiated until actual consumption has occurred.
- Second, since the production process sometimes has mismatched rates or machines which require changeovers, storehouses are often needed to maintain flow. Hence, the production process must feed the storehouse, at a rate greater than the consumption of the customer. However, to maintain control, all storehouses must have a maximum upper limit on inventory. In fact the entire production process must have an upper limit on inventory to make pull a reality.

- And third, the production is triggered only when the next downstream customer in the production process comes to pick up the product. This aspect has gotten lots of press from the GM supervisor who said, "Yeah, I understand pull, cuz in a pull system we ain't sendin nuttin nowhere, somebody comes to git it."

MassProd systems use "push" production systems. They are the opposite of pull systems. They rely on schedules and forecasts and try to "push" the product to the next work station. They do not have controls on inventory so it is possible to have work in process (WIP) explosions with highly variant lead times and inventory problems galore. But the real killer problem that is solved by pull production is variation caused by scheduling changes, often initiated by customer order changes.

The downside of a pure pull system is that it must be a "make to stock" system. You must have inventory to make it work, and recall inventory is one of the named seven wastes. Sometimes, holding this inventory is not practical so the inventory buffer is handled with a time buffer or approximated with a FIFO (first in first out) lane.

Finally I only know of two pull systems that have proven themselves in industry. These two are *kanban* and CONWIP.

Changeover Times

The issue of changeover times is generally an item that is largely ignored in MassProd. Long changeover times are often just simply accepted. Consequently, you have two options when employing model changeovers. First, you can shut down the entire line when the machinery undergoes a changeover. Of course, this causes a loss of production. In this case, all equipment needs to be oversized to account for the downtime. This then requires a greater initial investment. Alternatively, what is most often done is to oversize only the equipment that needs the changeover, to account for the downtime, and design inventory buffers in front of and behind the machines so the rest of the line can continue producing during the changeover. This helps keep the investment in the rest of the line down, at the expense of work-in-process (WIP) inventory. In addition, the conventional wisdom is that if the changeovers are done less frequently, the time to do the changeover, which is a form of downtime, can be distributed over a larger volume of parts, thereby reducing the per-part cost. This conventional wisdom leads to long runs and extremely large inventories before and after the machinery requiring the changeover. The lean solution is a large paradigm shift. That is, in Lean we do not accept that the changeovers will take a lot of time. The technique employed is SMED or quick changeovers. With SMED technology, the need to oversize the machinery is dramatically reduced. In addition, the need to hold inventory is also reduced. SMED technology was developed and refined by Toyota. Shigeo Shingo not only designed much of it, he published several books on the subject and is considered the architect of SMED technology. SMED is a staple in the lean toolbox.

Transportation of Goods

How parts, finished goods, and subassemblies are transported in the plant in the MassProd seldom looks like a lean facility. Nearly all materials are "pushed" and huge volumes of WIP build up between processing steps, taking up space, and greatly inflating both inventory and operating costs. More and more, even in a MassProd facility, raw materials are being handled like in a lean facility—Lean in the sense that *kanban* cards are used for raw materials replenishment. However, there the similarities end. In the

MassProd system, often *kanban* cards are used but the entire *kanban* system is not employed. Rather, *kanban* rules are seldom followed; the most egregious errors being:

- The failure to reduce the number of *kanban* to achieve process improvement
- The willingness to transport defective products

These two rules are always followed in a lean plant, along with the other four rules of *kanban* (see Chap. 11 for a discussion of *kanban*). However, in MassProd, seldom is *kanban* used for subassemblies and for finished goods. Rather the demand is scheduled and the volume is pushed through the system until it shows up at the storehouse. Huge volumes of inventory accumulate in the process, and the lead time is both long and unknown. The lean solution is again a huge paradigm shift. In Lean, a pull system is used as described above, and inventory is volume, time, and location controlled. In addition, *kanban* was an invention of the TPS and they use it not only for raw materials, finished goods, and subassemblies, but for such items as tools as well.

Handling Product Demand and Supply Variations

How finished product demand and supply variations are handled in a MassProd facility is a function of the scheduling. As demands change or production rates vary, the scheduler tries to manipulate the planning program to respond. This is a terribly ineffective solution that leads to not only a lot of overtime, a lot of expediting, a large number of late shipments but very large inventories as well. In addition, it leads to very high levels of uncertainty and stress. Adjusting to these demand and supply variations in this manner is not only ineffective, it is frequently impossible. The planning model is designed to be updated weekly or sometimes daily, for example, but the production volume is changing hourly or by the minute, thus there is no way the planning model can be responsive enough to the actual production line needs.

In the lean solution, the effort to respond to demand and production variations is accounted for by two factors. First, the variation in the processes is considered a huge problem and so is attacked from that standpoint. Stable processes are a foundational issue in Lean and are taken care of early in the transformation. As a result, variations are dramatically reduced through good process management. Second, the variations that remain are managed by segregating the inventory into logical categories and responding to inventory withdrawals with problem solving when appropriate. These inventory segregations are:

- Cycle stocks (normal pickup volumes)
- Safety stocks (to account for internal variations in production)
- Buffer stocks (to account for external variations in demand)

So, for example, when the customer comes to pick up product in a quantity greater than the scheduled demand, some of the buffer stock is removed. Once removed, buffer stock removal rules require that immediate corrective action be implemented (likewise, corrective action is required for the removal of safety stocks). In addition, by using a *kanban* system, once the product is picked up by the customer, the *kanban* cards are circulated and show up at the *heijunka* board at the production cell. The cards for buffer and safety stocks are a different color, which signals there has been a change that requires some countermeasure by the production cell, such as scheduling some overtime to replenish the inventory. If the *kanban* cards are picked up each hour, then in two hours or less, the signal that something unusual has happened is sent to the pacemaker process. The cell can then implement some countermeasure for replenishment. If that information goes to the typical MRP (Manufacturing Resource Planning) system, a countermeasure may be a full week away. The lean solution is both more responsive and relies less on centralized planning functions for daily operations. It should be noted that for production floor scheduling, the classic models of MRP, MRPII, SAP, ERP, BRP, SCM, and any other generic models are simply inadequate. Many efforts have been made to make them responsive enough, but they have internal inadequacies when it comes to shop floor planning. In most cases, the best system is *kanban* for triggering production. The inadequacies of these scheduling tools such as MRP, is a topic outside the scope of this book but your experience has already told you that your scheduling model will never make you Lean. However, if you wish to read about this, it is covered well in *Factory Physics*, (McGraw-Hill, 2008) by Hopp and Spearman.

Managing Quality in MassProd

How quality is managed in MassProd is changing somewhat. A few years ago bad quality was only a volume and financial issue. If production rates could not be met due to quality drop out, it was a crisis, but if rates could be met, then the motivation to reduce defects was one of finances alone, using short-term economics and immediate effects only. Since the rework of product was commonplace, this system caused little improvement in quality.

However, more recently, driven by customer demands for a higher-quality product, all manufacturing facilities in the supply chain have been affected. Now almost all firms still in business, at least claim to have a continuous quality improvement philosophy. However, old habits change slowly and most do a better job of talking about quality than actually delivering it. Typical quality systems rely heavily on inspection done by humans and have a high dependence on attribute data. In the end, they generally just sort out the defective product. Frequently the only place good inspection is done is at final inspection and test.

These systems are not designed to be used for continuous improvement and the defect data are employed mainly for yield and maybe quality cost data. Sometimes problem solving is completely absent; most often it is superficial at best. These facilities are not staffed for problem solving, and so quality problems not only persist, but frequently even the ones that were "solved," rise again. Finally, the attitude toward problems, in the typical MassProd facility is one of trying to avoid dealing with them.

Compare all these efforts to supply quality to the lean facility whose primary job, like the MassProd model, is to protect the customer. However, quality management runs much deeper. First and foremost, rather than just inspect and sort the product, the process is scrutinized deeply so it will improve and the need for both process and final inspection will be dramatically reduced. The amount of emphasis placed on process control in the lean facility is staggering when compared to a typical non-lean plant. The result is that the lean plant develops much more robust processes with more stable cycle times and improved process and product quality. This isn't rocket science, simply good old-fashioned hard work with a deep-seated belief in process management.

In the lean plant, the primary purpose of inspection data, which is largely variables data rather than attribute data, is for problem solving. Real problem solving is deep and practiced by all. *Jidoka* is used in all steps of the manufacturing process, not just at final test. The underlying principle of *jidoka* is that no bad parts are allowed to progress in the production cycle, even if it means shutting down the production line and ceasing production until the root cause of the problem is found and removed. Using this technique, it is clear that quality is equal to production. The second underlying principle of *jidoka* is that it is a continuous improvement process. *Poka-yoke* are used extensively and quality problems become immediately obvious through the use of *andons* and other forms of operational transparency. The uniqueness of Lean, in the area of quality, lies in four areas. Note that the first three of these distinctions are cultural, rather than technical:

- First, a quality problem is not just a reject, it is failure of the system, which is owned by all.

- Second, the quality problem is not bad news, rather it is often good news, signaling a weakness that can now be understood and corrected, thus leading to a more robust system, rather than be ignored and forgotten only to reappear again.

- Third, everyone participates in the technical solutions to problem solving.

- Fourth, and finally, the system uses a system of tools such as *poka-yokes* to attain 100 percent inspection.

How Cycle-Time Variations Are Managed

In MassProd, cycle-time variations are not considered a problem. They are seldom quantified; hence they are largely unknown and ignored. Average cycle times are understood, but to maintain average rates, large volumes of inventory are held between stations. As long as average cycle time is maintained, the variations do not affect the overall production, but only at the cost of huge inventory volumes (for an example of this, fast-forward to Chap. 24 and perform the Dice Experiment). Frequently, without understanding the damaging effects of cycle-time variations, efforts will be made to reduce inventories. This is almost always met with a significant drop in overall production rates.

It is extremely common in MassProd facilities that even when individual stations perform at design rates, on average, the overall lines do not perform at the design rate. This is caused by the interaction of variation and dependent events that occur on the line. (This is completely explained by the Dice Experiment in Chap. 24.) In the lean facility there is always a significant effort to optimize the cycle times, including significant efforts to reduce the variation in the cycle times. In the lean facility, both the average and the variation of the cycle times are known, understood, and managed.

Frequently, there will be specific tools at the line to signal if cycle time is being met or not. Common among these tools are the *heijunka* board. Where *heijunka* boards are not used, it is common to see counters and/or clocks to assist the workers and advise them if the process has slowed for some reason. These tools are not supervisory in nature, they are supplied for diagnostic purposes so process deterioration can be found and corrected. In addition, the concept of transparency of lean systems has the element known as standard work (SW). SW is a set of tools, one of which is a flow chart with cycle times for each process step, so a supervisor, engineer, or manager can evaluate how well the process is performing and assist in process improvement.

How Line Availability Is Managed

In MassProd, line and machine availability, like cycle time, is seldom measured, thus it is typically not known or managed. If production should fall behind schedule, the typical response is usually overtime and expediting. In the lean facility, availability is known, understood, and managed. The key issues in availability problems usually center on two issues. The first issue is materials, either defective material not allowing the process to perform, or stock outs. The answer to both can be *kanban* or other pull system tools. The second problem that adversely affects availability is usually equipment downtime. Tools like *andons* exist to signal problems, but the key tool used in the improvement of availability is total productive maintenance (TPM).

Summary of Technical Issues

The following essentially summarizes the majority of the technical issues in the TPS:

- Production cells flowing using pull production systems
- Balanced, so synchronized flow is achieved
- Producing at *takt* rate
- Using *kanban* to reduce inventory
- Rate and product mix leveled to minimize inventory
- Using cycle/buffer/safety stocks to handle internal and external rate fluctuations (while keeping cell production stable)

What we find by reviewing this list of tools and techniques in the lean facility is a clear lack of uniqueness. The use of cells and line balancing to minimize labor costs are engineering techniques that have been around for a long time and clearly preceded the TPS. The use of strict inventory controls, operating at *takt*, SMED and *poka-yoke* techniques are not really new, but are applied with a rigor and a vigor in Lean not seen in MassProd. So they are different only in the way they are applied, but they certainly are not revolutionary. So technically there is a difference in the application, but compared to the last three aspects of the five-part definition of the TPS definition, the technical and operational differences are relatively minor.

The Integrity of the TPS

The third major difference is the integrity of the system. Most people think honesty is a synonym for integrity. It is not. Integrity has come to mean honesty because so many people misunderstand the word "integrity" and misuse it. This is not all that uncommon. Words will often change meaning due to misuse. Integrity, integration, and the word integer all have the common root in the Latin word, *integrare*, which means to make whole. The word integrity means "state or quality of being complete; undivided; unbroken" (from my *Webster's New Collegiate Dictionary*). The TPS is a highly integrated manufacturing system and is distinct from most manufacturing systems. It is integrated internally as well as externally; horizontally as well as vertically.

How is it internally integrated? What are the characteristics it has that lead us to call it integrated? First, it operates as one system from beginning to end, which is the supply to the customer. It is a whole system, with the customer being the pacesetter of the process by consuming the product. A series of pull signals, starting with customer consumption, cause the system to operate in unison: a take-one-make-one system. Every technology is challenged to make the system "flow." Every technology is challenged to speed up the flow of the product through the system. Every technology is challenged to reduce the distance the product travels and the time it takes to deliver the product to the customer. Every effort is made to shrink the system in both time and space; this makes it more compact and smaller. The opposite of an integrated system would be a disintegrated system. A system that is scattered and a series of islands connected in some fashion. The TPS is designed to eliminate these islands of production whether they are batches or large "monument" machines requiring very long setup times.

Second, and most importantly, the TPS is integrated because there is a deep understanding of the concept of systems and that the system is the entity that requires optimization. They fully understand that "the system optimum is not necessarily the sum of the local optima."

Most manufacturing systems strive for local optimum conditions that are often at odds with the overall optimum. For example, they will strive to have high machine or manpower efficiencies, even if that means overproduction and finished goods sitting in the warehouse without sales to pay for them. The goals system and the accounting system in most plants are geared to drive toward these local optima. If managers proceed without an understanding of the local impacts on the system performance—that is, proceeding without an understanding of the integrity of the system—it is easy to push the entire system to a nonoptimum location: or simply put, to waste money. When a manager is striving to optimize the performance of his division amid the changing demands of the customer—a process that includes delivery, personnel, and raw materials problems—it is often difficult, if not impossible, to maintain focus on what is best for the system.

So internally the TPS is integrated as it operates as one entity—in a synchronous fashion—and always strives to optimize the system to supply maximum value to the customer.

However, the TPS was designed to be externally integrated as well. It is externally connected to both the customer and the supply chain. It is connected to the customer because the customer wants value and that is the foundation upon which the TPS is built. Most production systems are built on the concept of providing volume at low cost, which does not connect them as directly to the customer.

The TPS is connected to the supplier chain as if the supply chain was an extension of the plant itself. Ohno started incorporating what he calls the "cooperating firms" in TPS concepts before the oil crisis of 1973. In his book, *The Toyota Production System, Beyond Large Scale Production* (Productivity Press, 1988), he stated:

At this time, as early as 1955, the primary, if not sole, concern of all customers was the laid-in cost of an item, and yet Toyota was busy teaching their suppliers the TPS and techniques such as *kanban*. Today, it is commonplace for a supplier to seek

> **"T**hen after the oil crisis we started teaching outside firms how to produce goods using the *kanban* system. Prior to that, the Toyota Group guided cooperating firms ... in the Toyota system.**"**
>
> T. Ohno

advice and support from their customers. Prior to 1974, this was revolutionary. Nonetheless, it had the effect of better connecting suppliers to Toyota, which is integration.

Many companies have tried to mimic this concept of integrating suppliers. What most do is talk about a cooperative long-term relationship based on trust and mutual support. However, in the end, they just hammer the supplier to produce a lower-cost product without working with them on how to produce it less expensively. A few minutes after the discussion of trust and long-term relationships is finished, the topic of costs comes up. Altogether too often this is a discussion that ends with the open or veiled threat of "if you can't cut the costs, we'll be forced to find someone else who can." So much for mutual trust and that long-term relationship.

Also, having worked with many customers and suppliers alike, one thing I find very common is the case where customers will demand things from their suppliers that they themselves are not capable of doing. Just because they are not competent does not mean they cannot be demanding. This is not so with Toyota. Ohno refers to this as the "My plant first principle." Without exception, I have found that for any production technique they demand, they are also capable of assisting the supplier if not outright teaching them. I find this interesting that customers will require skills and standards from their suppliers, which they themselves do not have. I have often wondered how they can then evaluate if their suppliers are complying, much less be able to assist them.

Toyota has an integrated production system for many reasons, but at its deepest core are these concepts.

- They fully understand what the customer wants: *value.*
- They know how to provide value, using a system-optimizing production system, and forgoing local optima.
- They have readily available information that tells them how the system is producing.
- They are willing and able to respond to the system if it is not optimized.

A Philosophy of Continuous Improvement

Fourth among the differences is the reality that things can always be done better, faster, cheaper, and with less waste. In addition, Toyota has shown a great awareness of the world of manufacturing with their ability to be introspective and questioning. They clearly understand the concept of continuous improvement and recognize that their system is never fully optimized. If the TPS is to improve, it must change, it must evolve.

At the foundation of continuous improvement is the education process. Not only are workers and supervisors trained, but suppliers are also able to receive excellent training. They recognize that the minute you stop trying to learn is the minute you stop improving (See Chap. 9, on A Learning, Teaching and Experimenting Culture).

A Culture That Is Managed

Finally, and certainly the most important of all differences, is the culture. It is a culture of consciousness. They are aware of what is going on. It may sound odd, but most manufacturing plant managers are only modestly aware of how their plant is functioning.

They are hampered by little time on the floor, poor information systems, unclear goals and objectives, ever-changing philosophies, and myriad operational problems, including low line availability, poor quality, and delivery problems. Altogether, these issues then create a culture of chaos and firefighting. It is no wonder they are not on top of the action. The culture of Toyota is one of awareness, and this is no more obvious than in their characterization of such things as inventory and operational problems.

Ohno describes inventory as not only consuming space, raw materials, and time, but he explains that it hides the real problems of a system so they cannot be found and corrected. It is an oddity of the TPS that they seek out problems. The vast majority of managers yearn for the day when they will have no problems. Not so with Toyota. Awareness of problems, awareness of value supply, awareness of operational performance—pick your topic. Awareness is a cultural characteristic of the TPS.

There is a great deal more about the culture of Toyota (and cultures in general) in Chap. 4, but it too only scratches the surface of this topic. There is one cultural aspect, however, that deserves a little coverage here: *continuity*. W. Edwards Deming, in his book *Out of the Crisis*, (MIT, CAES; 1982) speaks about "constancy of purpose." Toyota is the poster child for this. They have exhibited the same principles for over 70 years. The principles apply to them and to all those who work with Toyota, especially their supply chain. The principles have been maintained through numerous management changes and through crisis after crisis, including those crises that have threatened the very existence of the company. Through all this, their principles have not changed. This type of continuity is almost unheard of in industry, and it is the key reason why the culture has been both so strong and flexible and why it has endured so well.

The Behavioral Definition

Reduction of waste and the specific definition of waste is almost a unique contribution of the TPS. Ohno defined waste in ways that no others had really thought. He described seven types of waste:

1. Transportation
2. Waiting
3. Overproduction
4. Defective parts
5. Inventory
6. Movement
7. Excess processing

Others have tried to remove these wastes, but the TPS has carefully defined them and made the continual reduction of waste an effort that is almost religious in its fervor. This is the definition of the TPS at the behavioral level, at the action level. However, to really understand the TPS, one must go deeper into what Ohno has written.

The Business Definition

In his book, *The Toyota Production System, Beyond Large-Scale Production* (Productivity Press, 1988), when asked what Toyota was doing, Ohno is quoted as saying:

Beyond this discussion, there is little that would lead you to believe this is a business model. There are no discussions of markets and market share, no discussion of stocks and earnings per share, no discussions of paybacks and return on investment; only this discussion on the principle of cash flow, or more pointedly, how to use manufacturing to improve cash flow. He has no other financial discussions, not even the calculations to justify manufacturing projects. The TPS is clearly not a business system; it is a production system.

> **"All we are doing is looking at the time line, from the moment the customer gives us an order to the point when we collect the cash."**
>
> T. Ohno

Several Revolutionary Concepts in the TPS

To appreciate the TPS and its genius, it is worthwhile to view some of its more revolutionary concepts.

- Supplying value to the customer
- Reducing lead times
- Focusing on the absolute elimination of waste; especially the waste of inventory

None of these concepts are new, but some aspects of the TPS are attacked with such a religious fervor that they seem almost a uniqueness of the TPS. Many businesses, long before the TPS became popular, would attack one or more of these issues, but none has packaged it in such an integrated way. Nor do they attack it with such single-mindedness as is seen in the TPS.

The Supply of Value to the Customer

This is a revolutionary concept as used in the TPS.

From the early years of mass production, the metrics of manufacturing focused on two parameters. Cost and production rate were the two items of major importance to manufacturing firms. Sometime later, in the 1960s or even the 1970s, quality became a major issue. The typical plant manager was always working to meet the production schedule to reduce the costs of production and make sure the product met the quality standards.

Of course, today, these three factors are always on the mind of any businessman. However, Ohno began to think in other terms. He basically said, "I know what my plant needs from my perspective, but what does my plant need from my customer's perspective?" His answer became his key metric, it was **value**. He described it as those things the customer was willing to pay for—which is what he called value.

Value-Added Work

To be defined as "value-added" work, the activity must meet these criteria:

- It is something that adds to the form, fit, or function of the product.
- It must be something that the customer is willing to pay for.

Suddenly, much of what a typical plant does is now questioned if this concept is truly employed. Of course, producing defective units is not a value-added activity, but consider packaging and transportation, for example. We go through serious design efforts to develop good packaging so we can transport the parts of an automobile, for example, to the final assembly plant. We utilize design tools such as Failure Mode Effect Analyses (FMEA) in the design of the packaging to assure we have no losses during transportation, and that the end packaging is suitable for our customer. Now Ohno calls not only the packaging we so carefully designed waste, but the transportation costs as well. The customer does not care that the steel came from Brazil. He does not care that the steel was packaged and transported to Mexico, where it was stamped into a wiper blade holder and then packaged and sent to Detroit to be assembled into a wiper blade assembly so it could be packaged and transported to the automobile assembly plant in Kentucky, where it could be installed on a car, which was then prepped and transported to Seattle, Washington for sale to some customer.

The customer does not care that this wiper blade traveled 25,000 miles and went through four packagings and four unpackagings, hundreds of handlings, and four tiers of suppliers with all the associated costs, before it was even attached to his car. His primary concern is that he gets good value for his expenditure.

This understanding and application of value is truly revolutionary.

Another way to look at this concept is by comparing it to what is called the Golden Rule: "Do unto others as you would have them do unto you." This is great advice and quite frankly a real stretch for me and others to aspire to. It requires that you think carefully about a situation, and think about exactly what you would like them to do to you. The question is, "How would you like them to act upon you?"—and then proceed to do the same with them. This level of introspection and detachment is extremely difficult, yet when followed it will lead to a higher level of awareness and a higher moral awareness as well. It is then hoped that this awareness will lead one to a more appropriate action on their part.

But I maintain that there is a flaw in this logic. It requires that you act on them *as you would like to be acted upon*. Well, what about their wishes? Sometimes these things can be reduced to simplicity for understanding. Apply this to something simple, such as buying a present for someone. If you follow this maxim, you will end up getting them what *you* want. Well, *maybe* that would be a good gift, but should we not get people what *they* want? I think so. And that is what makes getting gifts so difficult at times: It requires a high level of empathy—a quantity that is in increasingly shorter supply in our narcissistic world. Therefore, I believe the Golden Rule should instead be, "Do unto others as they would wish to have done unto themselves." Then they get what *they* want, not what *you* want.

This is what Ohno did. He put himself in the shoes of the customer and looked at value. Well, you might argue that this is what most typical plants do. On this point, I doubt this is the case, however. Plants look first to survive and second to prosper financially. Concepts to the contrary make for good discussions, but in the end if the place makes money, it stays in business. If it does not, it disappears. Not too complicated. So the typical plant, while looking at production rates, cost, and quality—and seemingly looking to the customer—are actually only looking internally, in order to survive and prosper.

Don't get me wrong, there is nothing wrong with surviving and prospering. After all, the customer counts on and needs the product; the people need the jobs created by the plant; and in the end the plant is providing several societal needs. Ohno simply took a huge leap beyond the normal thinking at a plant and in the end really tied his manufacturing system to the customer, letting the customer decide how he should redesign his system. He truly was connected to the customer, and that connection helped provide what the customer really wanted: *value*.

Reduction of Lead Times

In his book, *The Toyota Production System, Beyond Large-Scale Production* (Productivity Press, 1988), Ohno commented on what Toyota was doing, writing that:

The point here is that the TPS is clearly a system whose function is the reduction of lead times. In Ohno's writings, he does not really stress this point beyond what has been quoted. Instead, he focuses on the means to achieve reduced lead times, which are, of course, waste reductions.

The means of reducing lead time is through waste reduction, but the benefits of reduced lead time go well beyond the obvious savings regarding the waste that was eliminated. The beauty of reducing lead times can be seen in a variety of activities in the typical business. Take, for example, when a specification needs to be changed. This always raises the question of product obsolescence and the scrapping of those obsolete units. However, the bigger business question is, "Who pays for the obsolescence?" As you reduce lead times, the impact of this obsolescence is reduced. This fact is not missed by the typical production manager.

> **"A**ll we are doing is looking at the time line from the moment the customer gives us an order to the point when we collect the cash.... *And we are reducing that time line by removing the non-value-added wastes.***"**
>
> T. Ohno

> **P**oint of Clarity   A key metric in the TPS is lead time; the key goal is lead-time reduction.

However, lead time reduction goes way beyond that and can be seen in two parameters that all managers want, but few know how to obtain. A plant with shorter lead times is both more *responsive* and more *flexible*. It is more responsive in terms of being able to change when the customer's schedule changes. It does not matter if the change is volume or model mix or both. With reduced lead times, a plant is better positioned for both types of changes.

There is yet another benefit of reduced lead time that is not discussed at all in the TPS literature: future business. It is not part of the TPS literature because the TPS is designed for a business that has a secured customer and some sense of future commitments—hence, a relatively stable demand. The concept of *takt*, for example, states that there is a commitment of product demand by the customer. This is not true in all businesses. Take a typical job shop in which each job is unique—possibly a cabinet maker or your standard air conditioner supplier. It is not easy to calculate *takt* for them, yet unbeknownst to most of the mass production world, lead time is THE key metric for them. Having a short lead time not only improves their quality responsiveness and cash flow, it dramatically increases their possibility of getting future work. If the salesman can quote short lead times and deliver, he will get a lot of business. He will literally steal business from long-lead-time suppliers even if he does not have the lowest cost. Interestingly enough, in my experience I always find that those with the shortest lead times are also the suppliers with the lowest cost and highest quality.

Point of Clarity The way to become more responsive and more flexible in a manufacturing business is by reducing lead times.

In fact in our rapidly changing world, although short lead times do not guarantee success for this type of business, long lead times will almost certainly guarantee failure. So in a nutshell, short lead times for them are equal to future business. (Chapter 13 has more information about lead times.)

Initially, most engineers and manufacturing managers do not see the power of lead-time reductions. It is no underestimation to say that at the very heart of a plant's flexibility and responsiveness is the topic of lead times but Ohno sought to reduce lead times so he could get paid sooner.

However, once lead-time reductions are achieved, a number of other equally powerful manufacturing qualities are unleashed. First, the plant becomes both more responsive and more flexible. These are both obvious and powerful manufacturing skills to have. However, the fact that you have a shorter lead time has a significant impact on variation reduction, most notably the variation in the production schedule. Think about it, how accurate is your production schedule today or one day out? Generally, the schedule is pretty good. What about one week out or even one month out? As you go to one week out, or one month out, you can count on some demand variation. But what about six months out? How much do you think the production demand will change? Thus, it's simple: In order to reduce this demand variation, reduce the lead time.

Through the Absolute Elimination of Waste

In his book, Ohno states "The TPS, with its two pillars, advocating the absolute elimination of waste, was born in Japan out of necessity." Think about that: the "absolute elimination of waste." Not the reduction of waste, but its *absolute elimination*.

Ohno categorized wastes into the following seven principle types.

1. **Overproduction.** This is the most egregious of all the wastes since it not only is a waste itself but aggravates the other six wastes. For example, the overproduced volume must be transported, stored, inspected, and probably has some defective material as well. Overproduction is not only the production of product you cannot sell, it is also making the product too early. An interesting note about overproduction is that, in my experience, I have found that nearly all of the overproduction is planned overproduction. It is planned, and often for a variety of good-sounding reasons. However, upon scrutiny, I find that nearly all planned overproduction should be eliminated. For example, to ensure they have sufficient finished goods, many companies plan for extra production and purchase extra raw materials because they will have quality fallout during the process. This planning process is really just guesswork and adds considerably to the variation in the process. Even worse, many companies work hard to fine-tune this planning process so as to minimize the waste of planned overproduction. *Thus, we have the already scarce supply of technical manpower working to remove the planned-overproduction, which is caused really by the planning process, which saw a need because there is a quality problem which affects production quantities.* So why not attack the quality problem and get rid of all this waste, including the waste of the lost technical manpower? Sounds simple, but it is often overlooked.

2. **Waiting.** This is simply workers not working for whatever reason. It could be short-term waiting, such as what occurs in an unbalanced line (see the story of the Bravo Line in Chap. 21), or longer waits, such as for stock outs or machinery failure.

> **P**oint of Clarity   Don't work at getting good at something which should not be done at all!

3. **Transportation.** This is the waste of moving parts around. It occurs between processing steps, between processing lines, and happens when product is shipped to the customer.

4. **Overprocessing.** This is the waste of processing a product beyond what the customer wants. Engineers who make specifications that are beyond the needs of the customer often create this waste in the design stage. Choosing poor processing equipment or inefficient processing equipment increase this waste also.

5. **Movement.** This is the unnecessary movement of people—such as operators and mechanics walking around, looking for tools or materials. All too often, this is frequently overlooked as a waste. After all, the people are active; they are moving;

> **P**oint of Clarity   The TPS is a batch destruction technique.

they look busy. The criterion is not whether they are moving, it is: Are they adding value or not? I can't think of any example of people movement that is value added. Work design and workstation design is a key factor here.

6. **Inventory.** This is the classic waste. All inventories are waste unless the inventory translates directly into sales. It makes no difference whether the inventory is raw materials, WIP, or finished goods. It is waste if it does not directly protect sales.

7. **Making defective parts.** This waste is usually called scrap. But the phrase Ohno uses, "making defective parts" is classic Ohno. Most people use the term "scrap," so they view the defective part as waste. Ohno moves far beyond this. He not only categorizes the part as scrap, but the effort and materials to make it. Ohno was a natural process thinker. In this case, he not only lamented the loss of a production unit but the fact that people spent valuable time, effort, and energy to make the unit—all of which was lost, not just the production unit.

The TPS Is Not a Complete Manufacturing System

The TPS is not a complete manufacturing system. In fact, it is only a part of a manufacturing system. To better understand what part of a manufacturing system it is, or rather what it is *not*, we need to return to Ohno's book for a moment. While discussing flow as the basic condition, he writes:

These two sentences are so simple that their significance is missed by almost everyone. However, the implication to these two sentences, especially to those wishing to undertake a TPS

> **"A**fter World War II, our main concern was how to produce high quality goods and we helped the cooperating firms in this area. After 1955, however, the question became how to make the exact quantity needed.**"**
>
> T. Ohno

transformation, must be thoroughly understood. For example, let me paraphrase it a bit:

> From the end of WWII until 1955, we had focused our attention on improving the quality of our goods. By 1955, we thoroughly knew how to provide quality to our customers. We could discuss the key quality concerns with them, we could determine how to supply quality and we could provide it to a very high level. We used a long list of tools to achieve these communications skills, but the most important two were the simple customer quality questionnaire—which we statistically analyzed, of course—and we also became very proficient at Quality Function Deployment (QFD). Long ago, we ceased using inspection, especially visual inspections by humans, as a means of achieving quality. Instead, we moved to process control as a means to make the process more robust. To do this, we first became proficient in data gathering and analysis using such techniques as those that Ishikawa outlines in his writings. Now the vast majority of our data is used for process improvement rather than product evaluation. In addition, we became very proficient in a wide range of statistical techniques so we could analyze and make better decisions with our data. The four fundamental statistical techniques of measurement system analysis (MSA), statistical process control (SPC), designs of experiments (DOE), plus correlation and regression (C&R) are widely understood at even the supervisory level in our plants. We also made all levels of personnel responsible for root cause problem resolution, which means we trained them in various levels of problem solving—the "Five Whys" being the cornerstone technique. Another significant effort was the transition in our quality, process, and product data. Initially, the majority of our data was attribute data on the product. We moved from a high percentage of attribute quality characteristics of the product to variables characteristics of the process. We recognized early on that high levels of quality could not be achieved if we used attribute data, so this meant we needed to correlate the attribute defects to process parameters, and so we became skilled at this very early in our quality efforts. We had been very committed to providing quality products and had been working very hard on quality. With the help of Deming and a unified effort pushed by JUSE, we could supply excellent quality and our costs and losses associated with quality were very low. We had what most Westerners would call a very mature manufacturing system that could consistently produce high-quality goods and deliver them on time for a reasonable price. Quality was no longer a production problem. Now we needed to look at the losses caused by producing the wrong quantities—especially the wrong quantities produced and delivered to the wrong places at the wrong times.

If I were Ohno, that is what I would have written, because that is where they were as a manufacturing company. So what Ohno built, the TPS, had a foundation of quality, but his TPS is not a quality system. Yes, it had *jidoka*, but we will learn that *jidoka* is there to support JIT. In addition, and as an example, Ohno makes nearly no mention of Cp and Cpk which are the two accepted measures of process capability or "process goodness." Nearly every book you read on manufacturing and process quality reduces the concept of quality to Cp and Cpk for all measurement data, yet Ohno hardly mentions it. One has to ask why? Well, it is because of what he said—they simply could supply high-level quality, and quality improvements were not what they needed to focus on to reach a higher level of manufacturing excellence. Their focus, as he says, was on quantity. Make sure you spell that: q-u-a-n-t-i-t-y.

> **P**oint of Clarity   The TPS was built on a foundation of quality and the focus is to control quantities. (See House of Lean in Chap. 24.)

The implication of this information, to some company that wishes to embark on the journey into Lean, is usually quite sobering. Ohno says they just spent seven years—seven very focused years—learning how to deliver *quality*, after which they embarked on the journey of *quantity* control. I can say with assurance that most companies who entertain the option of mimicking the TPS do not have the sound foundation that Ohno describes in his writings. After all, that is what made them Toyota and the TPS only took them farther along.

Herein lies an interesting aside. It is part of the genius of Ohno and others like him that they do not care if you try to copy them. They know that you can't unless you have undertaken and built the foundation that they had in place when they started their individual journey of quantity control.

Most think they can bypass this step and are always disappointed to find that there are no shortcuts. If you want the benefits of the TPS, the foundational issues must be addressed. It does not mean you should not embark on the journey. I am not saying that. In short, the foundational issues must be addressed, or your effort will be in vain. However, it is possible, with good guidance, to attack the foundational issues as well as the quantity control issues, simultaneously.

To Not Understand This Concept Is Dangerous

This concept alone, specifically that the TPS is built upon a strong foundation and that one huge element of that foundation is high levels of delivered quality, is the reason most companies fail while trying to implement a lean transformation. I do not mean their efforts yielded less than they had hoped for, I mean downright failure.

For example, I got a call from a potential client who was trying to mimic the TPS. They called me when their production rates flagged and on-time delivery had dropped below survival levels. They described their efforts as a JIT Implementation. Over the phone, it took me about two minutes to diagnose their problems. They had tried to install a JIT system without the help of an expert. They had plunged headlong into an inventory reduction effort to improve lead times but were burdened by two major flaws. First, they had only a superficial understanding of the TPS. Thus, they had, in effect, placed a high-powered rifle in the hands of a child. Second, even if they understood the basics of JIT, their system was not able to undergo inventory reductions without attacking the underlying and necessary foundational issues, principally the reduction of process variation (See Chap. 11, Inventory and Variation). Consequently, the JIT system exacerbated, rather than solved, their problems. My advice to them was to hire an expert, immediately, to help them. They said they did not have the resources to do that. My next best recommendation was to undo what they had done and return to the method they previously had, so they would at least survive. I am not sure exactly what they did, but I later learned that the business had closed—and with it over 200 people lost their jobs.

Still others implement the TPS system and fail to achieve what the system is capable of delivering. Quite simply, there are no shortcuts, and to that end, shortcuts of understanding are the most devastating (See Lean Killer No. 1 in Chap. 2). So if you want to embark on any transformation, make sure you have both a thorough understanding of the transformation, as well as a commitment to implementing it. (See the Five Tests of Management Commitment in Chap. 24.)

A Critical and Comparative Analysis of Various Philosophies

A number of philosophies exist—some of them still popular—that are an effort to improve business. Most focus on manufacturing and are generally a response to the success achieved by the Japanese in the automobile industry. In no particular order, they are:

- Theory of Constraints (TOC)
- Deming's Management Technique
- Total Quality Management (TQM)
- Crosby Approach to Zero Defects
- Six Sigma and its hybrid, Lean Sigma

I will touch on each briefly since there are many books and articles available to those who wish to research this further.

The Theory of Constraints

The Theory of Constraints (TOC) is a concept developed by Eliyahu Goldratt while he was trying to create a planning program to make chicken coops for a friend. The TOC addresses three major concepts. First, it covers process bottlenecks, the logic of problem solving, and contains a touch of business theory that nicely simplifies the topic of money in a manufacturing business. His system is strong on inventory reductions, reduced lead time, and reduced batch sizes, all needed to accelerate cash flow—much as Ohno discusses. There the similarities end, however. His theory is very weak on quality and many other aspects of waste. I have found that learning and applying the TOC is often a solid place to start for many businesses before they embark on a journey into Lean Manufacturing. On the other hand, if you have a pure make-to-order system, with multiple routings and highly variant machine cycle times, many tools in the lean tool kit, seem very difficult to apply for those just learning Lean. Some of the tools and techniques of the TOC become more effective. Since almost no business is a pure make-to-stock system, it is a good idea to have an understanding of the TOC as you embark on your lean journey.

Deming's Management Technique

Deming's Management Technique, along with his 14 Obligations of Management and 7 Deadly Sins, is pure gold. The wisdom contained therein is simply wonderful, but the problem is that few have found a way to turn it into a solid management or business practice. His writings contain a number of solid thoughts and principles, but it is not clear that they are woven into an overall philosophy—at least not one that many can apply. I have found plenty of companies who have embraced many of his teachings, but only a few who have been able to turn it into a clear business or manufacturing system. Some have tried to do so under the name of TQM. Most writings about TQM picture it as a comprehensive philosophy that supports the principle of continuous improvement in a business. In the design of the TPS, and in Lean, it is easy to see the extensive and profound influence of Deming, his teachings, and his 14 Obligations of Management.

Crosby's Approach to Zero Defects

Crosby's approach to zero defects is an idea that had a great deal of traction in the 1980s, and many companies made improvements based on the concept of quality cost reductions. At that time, companies could survive with quality levels that were measured by percents of defects, and rework was a way of life. Today, quality levels have improved dramatically and are measured in the parts per million (ppm). No one who is really serious about quality embraces this philosophy today. First, it is based on a fallacy that zero defects are achievable. Second, the key metric used to drive the effort is "quality costs." With quality costs, there are at least two major problems. The first issue is that many of the major quality costs are not quantifiable. For example, what is the true cost of a customer return? What is the cost of a field warranty failure? What is the cost of the loss of future business? However, even worse than that, the American system of cost accounting does not capture quality costs very well. Our financial accounting systems are designed to take care of two business needs: the profit and loss statement and the issue of taxes. It is not designed to capture things like the cost of poor quality.

Consequently, a zero defects program finds itself unable to account for the costs, and then the engineers whose job is to reduce these quality costs (as well as justify their own existence) find themselves buried in the cost system trying to extract the savings from their efforts. They then become experts in the cost system, and so quality issues suffer.

Six Sigma

The concept of Six Sigma has achieved a great deal of publicity and, quite frankly, a great deal of success lately. Most of the publicity is directly or indirectly related to the earlier efforts put forth by GE, which was made very public by their now-retired CEO, Jack Welch. Six Sigma owes its roots to Motorola and the efforts of Mikel Harry. I first read about it as a design concept and tolerancing mechanism popularized in the publication *Six Sigma Producibility Analysis and Process Characterization*, put out by Motorola University. The purpose of this technique was to start at the design phase and try to produce a "Six Sigma" quality product. This was defined as one with less than 3.4 defects per million opportunities. And the focus was largely on variation reduction. Later, the concept of Six Sigma was broadened to become a problem-solving technique, and the Six Sigma curriculum has become standardized with Six Sigma Blackbelt and Greenbelt certifications now available. At this time the concept was broadened from one of variation reduction to working on three critical aspects of manufacturing: cost, delivery, and quality with an emphasis on lead-time reduction (delivery). The three categories of Six Sigma projects were those classified as:

1. Cost … critical to cost or CTC
2. Delivery … critical to delivery or CTD
3. Quality … critical to quality or CTQ

Since the early days at Motorola, the Six Sigma concept has grown and has various degrees of success. Welch in his writings claimed that GE sent $10 to the bottom line for each dollar they spent on Six Sigma efforts. Most large GE facilities had a complement of Blackbelts and Greenbelts and actually set up these internal consultants as cost centers. Today, to most people: *"Six Sigma is a project-based, problem solving initiative which uses basic as well as powerful statistical methods to solve business problems and drive money to the bottom line of the business"* (from www.qc-ep.com). In fact, more recently it has morphed into almost exclusively a cost reduction program and each project is evaluated against an ROI or other financial measure of success.

This being the case, Six Sigma is neither a manufacturing system nor a manufacturing philosophy at all; rather, it is a fine set of tools that can enhance problem solving in any sort of business, manufacturing, or otherwise. The Six Sigma problem-solving concept is a sound one and helps make the company a more powerful money-making machine.

Summarizing, Six Sigma was created by Motorola to achieve Six Sigma levels of quality focusing on variation reduction. Through success at GE and other companies, it morphed to become a cost-reduction program—every project needs a clear ROI. Lacking a strong culture-changing mechanism, it rarely leads to sustainable change. In its early stages, Dr. Deming would have vigorously supported it. Recall he said that management is prediction. Well, with reduced variation you can you get better predictions hence better management. On the other hand, the way Six Sigma is managed today, I am certain he would not be a current supporter at all.

Six Sigma is not a manufacturing philosophy nor is it a culture-changing mechanism. Yet when it is contained within a lean transformation and managed for what it originally was...solid statistical problem solving focusing on variation reduction...it can be a real asset to a lean transformation.

So What Really Is the Defining Difference between Lean and the TPS?

When Jones and Womack published their landmark book, *The Machine That Changed the World* (Rawson Associates, Macmillan, 1990), they either created or popularized the term "Lean Manufacturing." They called it Lean because it generated products using:

- Less material
- Less investment
- Less inventory
- Less space
- (And) less people

The term Lean Manufacturing has since become synonymous with the TPS, but there are at least two differences. The first is a rather subtle difference and has more to do with the implementation of Lean, while the second is the fundamental difference between Lean and the TPS.

The first difference is subtle and is lost on many people. It has to do with the starting point of the journey into Lean. Recall that quantity control is the defining characteristic of Lean. When Ohno started in 1955, he had in place an extremely sound quality control system. His foundation of quality control was more than sound, it was very mature. In fact, the first application of Toyota's *jidoka* system predated the Toyota Motor Company. It was done in the Toyoda Spinning and Weaving Company in 1902. Today, few companies have this same solid and mature foundation of quality when they embark on a lean transformation. So they must simultaneously work themselves out of a serious quality problem while trying to implement quantity control measures. Consequently, to implement a lean transformation today, companies must embark on a renewed effort in quality control. Hence, lean efforts today have become synonymous with not only quantity control but also *quality control*, which was never an issue with Ohno.

The second difference is more obvious. Many businesses can become Lean by simply following the outline in this book. They will achieve large gains in profits, be able to reduce lead times, become more flexible and responsive, and generally become a better business. Quite frankly, this is not too complicated. What it takes is sound leadership with a decent plan; a motivating work environment focused on intrinsic motivators; a few problem solvers with a willingness to implement change; a management that promotes a learning/teaching/experimenting work culture and total engagement of the workforce. Add to this more than a touch of good old-fashioned hard work and couple those attributes with sufficient doses of both humility and introspection, and you have enough to make you Lean.

The difficulty is not getting there, but staying there. Here is where the lean facilities, which have sustained their effort, stand out—and, of course, the granddaddy and greatest of them all is the TPS. Toyota has not only been an innovator in improving manufacturing techniques (and that may be the understatement of the century), but they have sustained this excellence for over 50 years. They have done this by not only implementing lean techniques, but also by managing the culture in such a way as to sustain these gains through every kind of change and challenge imaginable. They manage their culture consciously, continuously, and consistently.

Ohno was a master at changing the culture and then creating the type of environment that would sustain those cultural changes. Herein is the main difference between Toyota and many other firms—some of which are very Lean. Toyota has been able to manage their culture in such a way that the gains are sustained. Sounds simple, but simple it is not. (We will touch on the subject of cultures in Chap. 4-9, but you will find that this is only a brief introduction to the topic of cultures.)

In simple terms, the TPS is a production system, focusing on quantity, and was built on a sound foundation of quality control. Lean is also a quantity control system but nearly always, in the lean application, the quality control system must also be developed. Second, the TPS is a manufacturing system that is driven and supported by the Toyota culture. Other lean firms, at least in the first several years of implementation, seldom have the strong, focused, mature culture like Toyota. However, with serious work, specifically on the culture, these lean firms can have a manufacturing system that approaches the TPS in excellence. That is why we can say that although the TPS is Lean, not all Lean Manufacturing is done to the standards of the TPS.

Do not let that be discouraging. Ohno and Toyoda embarked on the development of the TPS over 65 years ago and they built on what others had done before them, particularly those in the Toyoda Spinning and Weaving Company. They did not create the Toyota culture in just a few years. It took decades of hard work, decades of dedication, and literally decades of trial and error to create the culture they wanted. But the key is that they figured out what they needed to do, what Ohno calls "out of necessity" and then not only did it, but managed it with a long-term philosophy of growth and integration. You can do the same, out of your necessity, but only if you're willing to make the short-, medium-, and long-term commitments and sacrifices that Toyoda and Ohno made.

Where Lean Will Not Work ... or Not Work Quite So Well

Limitations and Lack of Understanding

Imitation is the sincerest form of flattery .... This is a natural truth, and it is no more apparent than in the number of ways in which Lean is trying to be imitated. Everywhere you look you find Lean this and Lean that. Do an Internet search on

Lean and it is amazing what you will find. There is lean management, lean education, and lean health care to name just a few. Then, some practitioners use Lean as a new lead-in for some double-barreled title to spiff up or differentiate some already mature field such as Lean Six Sigma and Lean Software Development to name a few. Lean is "in" and all the salesmen know it!

It is interesting to me that I find many people who just can't quite see how lean principles apply to their business when it is a natural fit for their situation. And yet I also see some people stretching the principles of Lean so far out of shape as to make them fit what they would like Lean to be. For example, I was assisting a small group trying to use lean principles to guide their improvement efforts in education. They had spent many hours developing the "7 Wastes of Education." I asked how they decided on seven. Their reply was "That's what the TPS has." They had spent a tremendous effort to catalogue the wastes and then force-fit them into seven categories. I found it interesting to say the least, and nonproductive to say the worst.

Some Questions

However, this brings up some questions that some of us should address:

- What are the limits to Lean?
- Where will it not work so well?
- Where will it simply not work at all?
- Does it apply to the production line only?
- Does it apply to the staff functions in the manufacturing plant as well?
- Does it apply equally well to all aspects of manufacturing, regardless of product or customer?
- Can it be applied to all businesses regardless of product or customer?
- Does it only work in manufacturing and not in the service sector?
- Are there applications for it in the nonprofit sector as well?

A Two-Part Discussion: The Enterprise Level and the Product Level

The TPS and, hence, Lean Manufacturing was designed based on a certain set of business conditions. It would seem to make sense that the uses of Lean Manufacturing might have applications beyond the production of automobiles. Likewise, it seems reasonable to assume that if you diverge significantly from these basic business conditions, the applicability of Lean Manufacturing concepts and tools might wane as well. So it would make sense that the application of lean concepts and tools may not have infinitely wide applications.

In fact, after some review it seems that at the enterprise level—or some may call it the business level—the enterprise must be driven by four basic concepts. Lacking any of these, the enterprise is poorly suited to use Lean as a primary business philosophy. These basic concepts are as follows:

- The enterprise must be in a competitive free-market environment. For those entities that are not struggling for profits and or survival, there is simply insufficient motivation to undergo the discomfort of the huge cultural changes it takes to implement a lean transformation.

- There must be a clear customer focus. The enterprise must know who the customers are, what they need, and what they want. The enterprise must continually work to supply their needs and work to be ever improving in both finding and meeting the needs of the customer. In leanspeak, these needs and wants of the customer are value.

- In supplying value to the customer, a key strategy must be the elimination of waste.

- The business must have a long-term focus, even at the expense of short-term gains.

Hence, entities that do not have a strong customer focus, that are not interested in survival and growth, and that are not willing to drive out waste over the long term would not be good candidates for Lean as a guiding business philosophy. In fact, in almost all cases such as this, Lean simply will not work.

Some examples of this are covered in the following sections.

Sports Teams

Professional sports teams are very bad candidates since they are focused almost solely on short-term gains. They may look only to the end of the season for the Super Bowl, for example, but if that is out of reach, it is not uncommon for them to make a huge change in the middle of the season. They will fire the coach and release veteran players with their large salaries, with the hope of becoming more competitive next year.

Recall the Florida Marlins who, immediately after winning the World Series, completely liquidated their high-paid roster and sunk to the bottom of the league the next year. I am surprised some ticket holders did not start a class-action lawsuit against them. Whether to liquidate a roster like that is legal or not, I do not know, but it certainly is not aimed at providing value to the customer. Nor is it the sign of a business striving to compete. It is the height of arrogance toward ticket holders.

Sports teams may be the perfect place that Lean is doomed to failure. They have no survival issues at all. In fact, they are a monopoly with practically a guaranteed income via television. Second, they have no interest in waste reduction at all. Indeed, they intentionally increase wastes as they can pass the costs on with impunity. Third, there is no sense of long-term stability. In fact, their mantra is "What have you done for me recently?" Finally, all their protestations to the contrary, they no longer consider the individual fan to be their customer. With the price of tickets so high, the vast majority of tickets are bought by businesses. In addition, the majority of the income is from television so the networks are their customers, in reality.

Charities

What about charities? They have no profit motive and consequently there is insufficient motivation to make Lean work. In fact, I have worked with charities that have the end-of-the-budget-year problem of not spending all their grant money. So fearing they will get less next year, they find ways to spend the money. Rather than reduce waste, they frequently create waste.

Not-for-Profits

What about other not-for-profits? A lot is said about Lean in the government—the entity that is supposed to be serving you and me. I think they have lost sight of who their customers really should be. To apply Lean as a guiding philosophy in the top levels of government management, I see no hope whatsoever. The top few are interested in survival, but the survival issue is not the survival of the business (government), rather it is their individual job survival that is of importance to them. Their primary focus is on the self-serving survival issue of reelection. Reelection efforts are fueled by money, which is received through such things as PACs, which are largely supported by businesses. The "customer" of the high-ranking government official is more likely to be a PAC or large donor than Joe citizen. The PACs and large businesses are not interested in Grandma Jones getting her Social Security check. They are interested in their own self-serving purposes. So in the application of Lean as a business philosophy, this is a complete misfit.

I do see *some* hope for the application of many lean tools but not from the top persons in government. Lean tools are exactly what is needed at the level of government with which you and I interact. For example, at the Social Security Office, or the Department of Motor Vehicles, it has tremendous applications. At the "service provider" level, far removed from the top-level politicians, all these agencies use processes that could easily benefit from applications of the tools of Lean Manufacturing. Since the lean tools are so powerful at waste reduction, some clever politician who wants to make a name for himself has a powerful tool at his disposal. If he applied it at the right time in the right way, he could use it as a way to get reelected and promote Lean in government along the way. Waste reduction is so sorely needed in the U.S. government. Quite frankly, it is a lean opportunity well past its time. The key will be how to apply it and by whom?

The Health-Care Industry

Some possibilities in using lean tools exist for health-care companies, but not as many as I would like to see. In the small doctor's office, with one or two doctors and a small staff, Lean should work fairly well. These offices are usually customer sensitive, interested in making money, and focused on the long haul.

However, in the hospital, serious issues are present that prevent lean applications. The first problem is … "Exactly who is the customer?". The hospital will tell you it is the patient. But that is only partially true—and is more untrue than not. So why is the patient not really the customer? Well, think of what makes a customer a customer. Generally, to be a customer, you:

- Are courted or otherwise sought out by the provider
- Use whatever they are selling
- Pay for it
- Can complain about a problem and get action

Well, in the hospital situation, seldom do I hear about someone shopping around like you would for a present or a new car. If the typical patient needs an MRI or knee surgery, it is rare that they shop about to find a low-cost provider. Rather, most people just go where their doctor sends them, unless their insurance will not pay for that location, then they instead go where their insurance will cover the cost. So guess who the hospital's customer is by this measure? Here, they are courting the doctors and insurance companies, so by this measuring stick, *you* are not the customer. You use the service, but who pays for it? Well, here comes the insurance company. Guess who qualifies as the customer in that instance? In fact, I have never been to a health-care facility where they didn't first check my insurance before they checked me—emergency room service included. So, just who is the customer by this measure? Well, I guess you see my point. The large providers are mostly disconnected from the patient as being the customer. If they claim the patient is the customer, they are confused, at the very least. Hence, there is little hope here to apply Lean as a business philosophy until some dramatic changes occur.

But Lean Can Apply...Just Less Broadly

In the cases just listed, the primary driving forces of the business are so distant from the driving forces behind Lean that I cannot envision lean principles becoming the guiding philosophy of the business or entity.

However, do not lose hope. In every case I have mentioned here, there are a series of processes, within the business, where lean process tools still have some limited application. In many processes internal to these entities, the customer is much better defined and quality characteristics can be determined, measured, managed, and improved.

For example, we see the best as well as the worst applications of lean principles in professional sports. While the removal of waste is not important in most sports, watch the execution of a football team during their 2-minute drill. That is a very special application of driving out the wastes in the "process to score a touchdown." Another outstanding example is a NASCAR team while it is undergoing a pit stop. Where else can you see four tires changed, a car fueled up, the windshield washed, and the driver get a drink in just seconds? Here, every wasted motion is eliminated in the drive to create the shortest pit stop possible. Or, just because a professional football team is not a lean enterprise that doesn't mean you can't get a good-quality hot dog, with a minimum wait at the concession stand when you attend the game, even if it costs you a 500 percent premium. Likewise, just because your hospital is more interested in your insurance company than they are in you does not mean they can't serve their patients a good meal, on time, or get you to your MRI on time.

Remember that "wherever there is a value stream, lean principles will apply." Even if the top levels of the business of government are driven by large businesses and PACs, that doesn't mean you can't get your driver's license renewed using a lean process. In government, as in all the other examples given, lean principles may work at some level—these applications are just limited.

What I am saying, and will repeat to avoid confusion, is that those entities that are not customer-focused, without a survival motive and without a concerted effort to reduce waste and provide value to the customer—those entities that do not embrace a long-term philosophy of growth and service can't become lean enterprises, and cannot become candidates for Lean as a guiding business philosophy until they change. They, however, can still utilize some of the process management tools of Lean for some of their internal, particularly lower-level, processes.

However, if the business does meet these criteria, is competitive with a clear customer focus on supplying value by driving out waste, and is in it for the long term, does that mean Lean as a business philosophy will necessarily work for it?

The answer to that question is yes, but to a varying degree due to three basic conditions of the product. Those conditions, when combined, are known as the *Lean Stereotype*. Specifically, the more the business fits the Lean Stereotype, the more it will be able to utilize the lean tools in its battle to survive and become more profitable. The Lean Stereotype is the specific type of business for which the strategies, tactics, and skills of Lean Manufacturing were developed. Consequently, the more a business approaches this stereotype, the more the business will be able to directly apply the strategy, tactics, and skills of the House of Lean.

The Lean Stereotype Is a Business in:

- Manufacturing
- Discrete parts
- Stable product demand

Lean Applicability: Continuous Process Industries

The first and smallest negative effect on the applicability of Lean is the shift from discrete parts manufacturing to the continuous process industry. This would include industries such as petroleum and chemicals manufacturing, food processing, and pharmaceuticals to name a few. My background was in petroleum refining and we were able to make almost total applicability of the earlier mentioned principles, with one major exception: destroying the batch. We were able to see and work at waste reduction, we were able to produce to *takt*, create flow, and use pull systems. There was one large problem with refining: The unit you deal with is the batch. And the batch is sometimes, out of necessity, large. Other times, it was possible to reduce the batch dramatically with resultant improvements in lead time. As you might expect, this batch reduction could translate into improvement in those two wonderful business weapons: flexibility and responsiveness. In most cases, reduction of batch size could be easily done. However, because the history of the petroleum business is to make batches, that paradigm was very hard to change. The momentum was very much against batch size reduction. Another factor in these businesses that must be understood is the continuous process industries, such as refining and chemicals, where the capital investment per employee is much larger. For example, in refining, it is not uncommon to have $4,000,000 of capital investment per employee. By contrast, many of my current clients who are typically tier 1 automobile suppliers, have $3000 to $50,000 of capital investment per employee. This high capital investment will cause those businesses to view their wastes somewhat differently.

Lean Applicability, Unstable Demands

The next largest negative effect that makes the operational techniques of Lean less applicable is unstable customer demand. When you compare the three-year contractual demand that most tier 1 automobile suppliers are blessed with to the come-and-go demand that many job shops face, it is easy to visualize that this is a huge obstacle to implementation of lean principles. For example, there is no *takt* to calculate, so synchronizing with the customer is difficult, and since the life of a job is often very short, continuous improvement requires a completely different philosophy. However, it is still possible to create a pseudo-*takt* to use in synchronized supply with the customer and synchronized production flow. Often flow can still be balanced and, most importantly, the flow velocity can be accelerated by reducing the batch size. In the typical job shop, lot size reduction is a powerful tool (see the Story of Excalibur Manufacturing in Chap. 13 and the Story of the Bravo line in Chap. 21 for specific examples). To keep lead times reasonable, quick changeover (or in leanspeak, SMED) technology must be very strong in this type of business. In most job shop applications, to create a "flow" for each job is not too difficult, and all the basic principles of Lean apply.

The complexity is borne in the concept that lots of products, or jobs, exist. These jobs have different routings with variant cycle times and complex interactions of people and machinery. Yet examples of businesses that have made huge improvements by applying the principles of Lean abound. Two principles that seem to repeat time and again are the use of SMED technologies to reduce setup times, and the use of small batch sizes to reduce lead times. The efforts in job shops to reduce lead times pay triple dividends.

- First, with short first-piece lead times, rework is reduced dramatically. This is an extremely powerful quality weapon that should not be overlooked by these businesses (See Chap. 11, for a specific example).

- Second, short lot lead times translate into quicker deliveries with improved cash flow.

- Third, the ability to quote shorter lead times is a power weapon to acquire future business.

Lean Applicability: The Service Sector

The third and even larger negative effect that reduces the applicability of Lean is seen when the business is not manufacturing, but instead is typical service sector work such as that of a hotel, restaurant, or hospital. Service sector work has at least two very large problems.

- First, but less important, is the demand instability a typical service sector business must deal with. Seldom do they have the stability of the tier 1 automobile supplier. Sometimes doctor's offices and dentist's offices have a fairly stable demand rate, but they are the exception rather than the rule in the service sector. Compare them to the comings and goings of customers at the typical hotel or restaurant.

- Second, and more importantly, unlike manufacturing, the service sector has a paucity of specifications. The time it takes to check in at a hotel, get served at a restaurant, and get your oil changed have no real specification. Consequently, the metrics of the service sector are often up for grabs, and it becomes very difficult to measure the quality. Do not forget that the foundation of Lean Manufacturing is of good quality. It is hard to build on a foundation that barely exists.

So How Should We Proceed When We Have to Deal with These Effects?

All this is interesting, at best, and possibly contrary to progress, at worst. Those who ask about the applicability of Lean are often looking for the "correct formula" to remedy their ills. All too often, they are trying to find a simple, proven, ready-made solution to what are often complicated problems. It's sad to say, but there is no ready-made formula; there is no "silver bullet" to solve these ills.

By developing the TPS, Ohno obviously found his remedy, and using the same logic, you too will need to find your solution—and do this by using his logic, which may or may not mean you will end up using his tools of improvement.

> **P**oint of Clarity   Do not let what you cannot do, prevent you from accomplishing what you can do.

Ohno said the "TPS was developed out of necessity." My advice to you is this: Find your own necessity and then develop what you need to, for your unique circumstances. And keep in mind that although the TPS may not apply totally to your situation, I am equally sure that some of it will.

However, if you, for the moment, ignore the TPS itself and instead focus on the logic and method that Ohno used to create the TPS, you will find how to apply his method to your situation. In other words, you will find "your necessity" and then with some of the following:

- Good old-fashioned hard work
- Sound logic
- Good problem-solving skills
- A determination to get through resistance and failures
- Truly motivated introspection
- A touch of humility
- Sufficient courage

You will see that Ohno's method applies totally to your situation as well.

So Just What Is Lean?

Frequently we can describe a complicated concept with just a few words, we call these aphorisms. So what is the aphorism for Lean? I have thought long and hard about that and cannot catch the essence of this powerful concept in a simple phrase. Maybe that means I really do not know it well enough, or maybe it is just so complicated that no single aphorism can properly describe it; of this I am not sure. That said, you have heard and will hear me use a number of aphorisms throughout this book, let me summarize.

- At the cultural level ... "Lean is the creation of a culture of continuous improvement and a culture of respect for people."
- Considering the three constituencies of stockholders, employees, and customers, and having a balanced approach to meeting the needs of all three ... "implementing Lean is how we make 'your facility a better money making machine, a more secure workplace for all and the supplier of choice to your customers.'"
- At the process level, Lean is all about creating the "results" hence ... "Lean is 'better, faster, cheaper.'"
- While we are achieving these results we need to use the right "means" hence ... "the 'Means' to Lean is problem solving our way to the Ideal State."

- The Path to Lean is to "balance externally, balance internally, flow where you can, pull where you can't then standardize."

- But my favorite and the one you hear me say most often and especially as we are problem solving is … "Lean is the search for capacity and the creative use of that capacity." On the floor, problem solving, this is what we are creating … capacity. And think for a moment and ask yourself, "what problem do we have that capacity, in some form, will not solve?". Said another way, I often say that given unlimited resources, problem solving is a piece of cake.

When we find and eliminate a defect-creating mechanism, we have immediately created the ability to make more product right the first time. That is capacity. When we remove the waste of transportation and move work stations closer, we have created space. That too is capacity. When we eliminate unneeded work steps, we have created capacity in the form of labor which is now able to do more somewhere else. However, especially as it relates to labor usage, we must creatively use that capacity. And I mean just that … we must create something with this time we now have. If we reduce the workforce and lay off people, well the results from that are both destructive and predictable … that is not being creative, quite the opposite. So we must both create the capacity and creatively utilize it. I am not sure that is the best aphorism describing Lean, but it is the one I use most often.

Chapter Summary

The TPS is a quantity control system built on a solid foundation of quality, and was the manufacturing system perpetuated by the strong culture of Toyota. Herein are the two differences between Lean Manufacturing and the TPS: the strong Toyota culture and the solid quality foundation. Both are strong manufacturing philosophies designed to make your business a better money-making machine, a more secure work environment for your employees, and the supplier of choice for your customers, through the total elimination of waste, thereby supplying what the customer wants: *value*. It needs to be understood that Lean is primarily a manufacturing philosophy and is not a business philosophy. Finally, the tools of Lean were designed for—and work best in—what I call the Lean Stereotype, although you are only a little hard work and imagination away from applying these tools very broadly.

CHAPTER 11

Inventory and Variation

Aim of this chapter…to explore why many early JIT (just in time) efforts failed and introduce why we have inventory, why we need inventory, and the two key business reasons why we strive to reduce inventory. We'll explore the dynamics of inventory creation and its relationship to variation and dependent events, including making the sample calculation for the three types of inventory: buffer, cycle, and safety stocks. Finally, we will discuss the powerful tool designed by Ohno called *kanban*.

Background

In the 1970s, it became clear to a select few that the Japanese, most notably Toyota, had found a better way to manufacture cars, which caused a number of very interesting things to happen. First, and most notably, the majority of the automobile manufacturing world went into a huge case of denial. This was heard as "that will work in Japan, but not here" and a variety of other statements that could politely be said to have lacked insight.

However, some with a little more insight, curiosity, and humility asked, "Could there be something to this?" Well, from that small group came a series of efforts to try to capture parts of the Toyota Production System that were serving Toyota so well. The piece that seemed the most appealing was the JIT concept. It was rapidly popularized as an inventory reduction effort, which in fact is only a part of what it really is.

JIT practitioners came out of the woodwork and many companies went about implementing *kanban* and slashing inventories to reduce the high cost of producing and managing the inventory. Some went about using the slogan of "Zero Inventory" and slashed inventory with such fervor it was as if they were pursuing the Holy Grail of manufacturing. Inventory had become a bad word, much like "scrap."

Unfortunately, many of these efforts were grossly misguided. Their only focus was on inventory reduction. They reduced inventories as if it were an independent entity that had no relationship to anything else. JIT implementation efforts became nothing more than aggressively slashing inventories. Those that had this approach often caused irreparable damage. They found they needed to expedite nearly everything, needed to work large amounts of overtime, and then still frequently missed delivery dates. Others found the worst of all scenarios. They not only missed shipments but as they cut inventories they found that production rates flagged significantly. Due to these misguided efforts, many companies ceased to be competitive and some even went out of business.

Just Why Do I Have and Why Do I Need the Inventory?

The smart ones, once they got in trouble said, "Wow, maybe there's more to this inventory reduction than just slashing our inventory volumes. What's going on here?" The really smart ones asked—and answered—two basic questions. These were questions that were overlooked by the vast majority of the JIT implementers of the past and are frequently overlooked by the managers of the present. These questions are:

- What is the basic purpose of inventory? Meaning, if it's so bad, why do I have any at all?

- What is causing the need for the inventory? In other words, why can I not seem to operate without it?

What Is so Bad about Inventory?

So what's so bad about inventory? Simple: Inventory costs a lot of money. First, there are the raw materials and operating expense it costs to produce it. Next, we must handle it, which means we need more people and machines like forklifts. Then we find ourselves moving the material around, usually more than once before it gets to its desired location. This in turn requires space and transportation and neither are free. Next, we must keep track of it, which means even more people, computer programs, and reports galore—almost all of which are filled with errors. We then try to fix these errors. The way we try to fix the errors is to use ineffective band aids like cycle counts which then take more people again, more time, more computers, and worst of all more reports and more meetings. In addition, we must care for this inventory to make sure it does not get damaged. And finally, we must ship it before it becomes obsolete.

I have dealt with several firms who say their cost of inventory, *obsolescence excluded*, exceeds 25 percent per year of the product value. That is 2 percent per month, and if your company is operating on 12 percent earnings on sales—well, the impact is huge and you can see why many firms wanted to get rid of it. What else can be done to make such a huge bottom-line improvement? It's no wonder many firms jumped on this bandwagon of inventory reduction in the name of JIT.

All these liabilities of inventory are obvious bottom-line opportunities, and yet the greatest advantage of reduced inventory is not even mentioned here. In fact, it is often not even recognized. In just a minute, we will get to that crucial advantage which so few see and even fewer appreciate.

Question 1: "What Is the Basic Purpose of Inventory?"

What is the basic purpose of inventory? This is not a complicated question—indeed, it is a rather simple one actually, but it is asked by a scarce few, and answered by even fewer. First, let's make sure we are on the same sheet of music here. I am talking about the use of inventory in a classic for-profit business. The objective of those companies is generally to do just that: make a profit. In those businesses, the purpose of inventory is singular and simple: You should only hold the inventory you will need to protect your sales. Yes, sales; not production but sales.

I see no other reason to hold inventory. Any amount beyond this is an expense that is not justified, yet to hold less undermines your ability to supply your customer—and nothing will hurt profits like failing to sell. It is a very simple concept. Well, it is pretty simple, anyway, and yet is still missed by many.

We only hold the inventory we need to protect sales. There is a relationship between the amount of inventory and the volume of sales. So in a nutshell, if we know the sales volume and can understand the relationship, we could calculate the inventory we need to protect those sales. The relationship is a simple mathematical calculation, but just how is that calculation made?

> **P**oint of Clarity   The only economic purpose of inventory is to protect sales.

First, let's address finished goods inventory. This inventory calculation comprises three types of inventory, each with its own calculation. The three types of inventories and the basis for their calculations are:

1. Cycle stock —this is to cover the picked-up volume by the customer
2. Buffer stock—this is to cover for external variations, usually demand fluctuations
3. Safety stock—this is to cover internal variations, usually production issues

The total inventory is the sum of the three types of inventory, which are discussed below.

Cycle Stocks

This is the volume you need on hand to take care of the normal demand pickups by your customer.

For example, if your customer picks up each Wednesday, you will need their ordered volume ready for pickup then—and not before. Unfortunately, the information handling system, production, and the delivery system are not instantaneous, so we need some volume in cycle stock that is above the bare minimum the customer will pick up. So, to make sure you can achieve the customer's needs, we will calculate the cycle stock's volume to be the production rate multiplied by the replenishment time plus some arbitrary safety factor we will call Alpha (see the section "Finished Goods Inventory Calculations" later in this chapter). The stock replenishment time (see Fig. 11-1) is the sum of four variables:

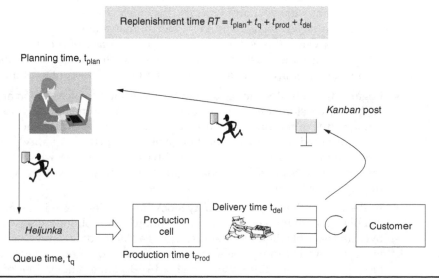

Replenishment time $RT = t_{plan} + t_q + t_{prod} + t_{del}$

Planning time, t_{plan}

Kanban post

Delivery time t_{del}

Heijunka

Production cell

Customer

Queue time, t_q

Production time t_{Prod}

FIGURE 11-1   Replenishment time.

Planning Time   This is the time that the order takes to be processed and sent to the production line.

Waiting Time   This is the time the order is waiting to be processed. This is sometimes referred to as queue time.

Production Time   This is the time it takes to produce the desired quantity.

Delivery Time   This is the time to get the lot from the production line to the storehouse.

Buffer Stocks
This is the incremental volume of inventory, above the cycle stock's inventory volume, which is held to account for **external variations** and is calculated based on historical data of the variation of these external causes.

Safety Stocks
This is the incremental volume of inventory currently held that is above both the cycle stock and buffer stocks. It is held to account for **internal variations** in supply to the storehouse.

Question 2: "Just What *Is* Causing the Need for Inventory?"
The need for each of the three types of inventory is caused by different factors. These factors are as follows:

- For the cycle stock, the need for the inventory is caused by the size of the picked-up shipment, which, for a constant demand product, is a function of how frequently the shipment is picked up. In addition, some inventory is needed to cover the time it takes to plan the shipment, make the shipment, and move the shipment within the plant. This is the replenishment time calculation.

- For both the buffer and safety stocks, the need for inventory is one of the world's best kept secrets. The need is caused by variation. When we have more variation in the system, we need more inventory. The buffer stock size is usually determined by two variables: changes in customer demand and variations in delivery conditions. Often, this is due to weather, or in the case of products that cross an international border, customs can be an issue. So, the sources of variation for this volume of buffer inventory are somewhat outside of the control of the plant.

- Regarding safety stock in particular, the large sources of variation are issues of supply to the storehouse. These sources of variation include such items as line outages due to machinery failure or stock outs. Poor cycle-time performance can cause production to fall short of goals, and of course quality problems can also be a major cause of variation. All three of these issues, which happen to be the three aspects of OEE (Overall Equipment Effectiveness), are largely under the control of the plant.

- In the case of the buffer and safety stock inventories, there is a simple way to calculate the volumes needed. If the variation of the volume swings is calculated over a reasonable time frame and stated as a standard deviation, your variation is now converted to numbers so we can have a common understanding of it. Now if you have a stable system and hold 2.33 standard deviations of inventory, you will have enough inventory to cover about 99 percent of these deviations, presuming your data are normally distributed. Since, in reality, there are many possible sources of variation, assuming the normal algorithm is reasonable. With weekly shipments, that would mean about one undersized or late shipment every 2 years.

What Creates the Need for Work In Process (WIP) Inventory?

The preceding discussion of inventory was focused on finished goods inventory, although the concepts are general and apply to WIP and raw materials inventories as well. In many plants, the problematic large inventory is not in finished goods but in WIP. What causes the need for WIP inventory?

Take a simple cell, for example. Let's say we have a six-station cell and all work stations have 60 seconds of work, which is also *takt*, and that there is no inventory between stations and we have a pure pull system with one-piece flow. When station 1 finishes a piece, so do stations 2 thru 6—and in unison, all six pieces of in-process work are simultaneously pulled from the previous work station every 60 seconds. This is perfect synchronization of process flow, the Ideal State.

But for the moment let's imagine that the cycle time for station 4, although it averages 60 seconds, varies from 50 to 70 seconds. When station 4 performs at 50 seconds, it finishes its process and then station 4 has a 10-second wait time before its product is pulled by station 5. There are 10 seconds of wait time, which is a waste for station 4. *But this is not a production rate problem.* The cell will still produce to *takt*. It is just that the operator at station 4 will sit around a while. On the other hand, when station 4 takes 70 seconds to produce its work, that subassembly is held up and station 5 is starved for work for 10 seconds. This delay passes through all the workstations of the cell in a wave, and that piece is produced on a 70-second cycle time.

So let's recap… If the station that varies—in this case, station 4—operates faster than *takt*, station 4 must wait for the subsequent station to pull the production. However, when station 4 just happens to operate slower than *takt*, station 4 will slow down the whole cell on that cycle and there will be no recovery. So even though the station may have a 60-second cycle time *on average*, any time the cycle time is above average, the production rate drops. This concept is known as the effect of variation and dependent events. (The dependency is that the "next step" depends on the "prior step" for supply.)

So the solution is, guess what?… You got it! *Add some inventory.* We will need to add inventory both before and after station 4, the one with the variation. We need the inventory in front of station 4 so when it produces faster than *takt*, say at 50 seconds, there is raw material available to keep it producing. We also need the inventory after station 4, so when it is operating slower than *takt*, say at 70 seconds, there is raw material to supply station 5. Then station 4 can have the variation AND maintain production at *takt* on average.

So the answer to the question is "The need for WIP inventory is caused by variation."

This destructive relationship of variation and dependent events interacting to cause a reduction in production rates is critically important. It is understood by a scarce few, yet you need to understand it if you wish to implement a lean transformation. At this point, I suggest you go directly to Chap. 24 and perform the dice experiment to begin to understand this relationship. It may be the best 60 minutes you can invest in your understanding of Lean.

> **P**oint of Clarity   Inventory is a necessary response to system variation so rate can be maintained.

So this inventory—whether it is buffer and safety stocks in finished goods or WIP between work stations on the line—is nothing more than a response to variation so rate can be maintained. Once you understand this, it is easy to see that the answer to inventory reduction is to first reduce the variation, and then the reduction in inventory can be made with no loss in production.

Although inventory is a waste, it is one of those wastes which we classify as "necessary under the current conditions." We wish to reduce it, but few businesses can survive without some inventory at some point in their process on the way to the customer. Remember that inventory is needed because of the variation that is present. Later in this chapter, you will learn that variation "is the inevitable differences...," hence it is unavoidable. So it is not possible to eliminate it totally, but we strive to do just that.

Just What Is This Key Advantage of Inventory Reduction That Was Alluded to Earlier?

To answer this question, you will need a lesson in physics, such as those in *Factory Physics* (McGraw-Hill, 2008), Hopp and Spearman's book. The particular law of factory physics to which I refer is known as Little's Law. In leanspeak, it states that the WIP in any system is equal to the throughput rate multiplied by the lead time.

Little's Law: WIP = TH × CT

Where:

$$WIP = \text{the work in process ... units of inventory between any two stations}$$
$$TH = \text{throughput rate ... units produced/unit of time}$$
$$CT = \text{cycle time ... which they define as the average time from when a job is released into a station to when it exits...} \textit{in leanspeak, this is called lead time.}$$

If we replace the term CT with Lead Time (LT) and then rearrange them, we get:

LT = WIP/TH ... which is Little's Law in leanspeak

As you can see, as WIP is reduced, lead time is reduced in direct proportion.

Consequently, we can double the throughput rate or cut the inventory in half. Both will reduce the lead time by 50 percent.

And as lead times shorten, flexibility in production will improve, as will the plant's responsiveness to changes—all this and cash flow will improve as well. *Lead time, more than any other metric, is the most descriptive measure of the health of a Lean Manufacturing system.* And improved lead times come about largely by reducing inventories.

Lead time will be further amplified in Chap.13 since it deserves a chapter of its own. In addition, it will be highlighted in several case studies throughout the text.

About Variation

Production process variation is everywhere. It affects every aspect of every step of your process and every specification of every part of your product. It is present in the materials, the manpower, the methods, the measurement, and the environment of all that we do to manufacture our products.

It is inevitable.

It is the enemy.

It is the enemy to not only good product quality, it is the enemy to rate, it is the enemy to rate stability, and consequently it is the enemy to operating costs—but most of all it is the enemy of bottom-line profits. Nothing is more basic to improving the manufacturing systems than the reduction of variation.

Variation in a production process is to be understood, sought out, and destroyed.

Variation is "the inevitable difference of the individual outputs of a system." I have taught that definition for years and am not sure of its origin. I believe I can thank either Walter Shewhart or Donald Wheeler for it, but I am not sure. But what is important is that it is a clear representation of variation. It is:

- Inevitable
- Applicable to all outputs—in fact, it is applicable to every characteristic of each output
- System generated—in other words, every part of everything that went into making the individual output varies

Sometimes, especially as it applies to attribute data, I use a definition I found taken from the writings of Walter Shewhart. Most people refer to this book as *The Western Electric Handbook*, but it's real name is *Statistical Quality Control Handbook*, (AT&T, 1956). It is the three-part definition of variation and it says:

- Everything varies.
- Individual items are not predictable.
- Groups of items, from a constant cause system, tend to be predictable.

It does not really matter which definition you use. In a process, they both converge and you find that variation is the enemy to both process stability and process capability. It is the enemy to the very foundation of a lean effort and must be understood and aggressively reduced at all times, and in all processes.

So, to summarize...all systems have variation; hence, all systems will need some inventory to maintain rate. However, inventory is a waste, but at some level it is a necessary waste, so you will want to scientifically and economically minimize it. To minimize the inventory, you need to reduce the variation—there is no other productive way. In short, inventory reduction is reduction of variation by another name.

Point of Clarity   To reduce inventory levels, reduce variation.

Buffers

Whenever there is variation, we need inventory to compensate for the variation if we wish to maintain the production rate. This is not quite a true statement. Specifically, when we have variation, we need *something* to compensate for this variation, to maintain rate. We talk about inventory as being a countermeasure for variation, but in a more general sense we need a buffer. A buffer is some resource we have in excess that is designed to account for the fact that production cannot be in perfect lock-step with consumption.

Buffers Come in Three Forms:

1. Inventory
2. Capacity
3. Time

There are three types of buffers: inventory, capacity, and time.

- Finished goods inventory is a buffer because we must accumulate finished goods between customer pickups. WIP inventory is a buffer. It is a natural response to variations in the production system, including scrap production, machine downtime, changeovers, and cycle-time variations, to name just a few.

- Excess rate capacity in a machine that requires changeovers is a buffer, a capacity buffer.

- When we do not have a good understanding of our lead times—which for the sake of argument vary from three to five days—we may enter a time buffer of six days into our planning program to make sure that when we release an order, it will be completed on time. In addition, a typical lean strategy is to run a plant for 2–10 hour shifts. This strategy coupled with some overtime, provides both a time buffer and a capacity buffer.

Kanban

Basics

Kanban means sign board. A *kanban* can be a variety of things, most commonly it is a card, but sometimes it is a cart, while other times it is just a marked space. In all cases, its purpose is to facilitate flow, bring about pull, and limit inventory. It is one of the key tools in the battle to reduce overproduction. *Kanban* provides two major services to the lean facility.

- It serves as the communication system.
- It is a continuous improvement tool.

Types of *Kanban*

Kanban provides two types of communication. In both cases, it gives the source, destination, part number, and quantity needed.

- Parts movement information, the transportation *kanban*—this is like a shopping list.

- Production ordering information, the production *kanban*—among other things, this is primarily a production work order.

Kanban Rules

The Six Rules of *kanban* management provide several unique functions. The rules and functions are listed in Table 11-1.

Kanban Calculations

Let's analyze a production *kanban* system. Recall that the *kanban* represents the entire inventory in the system. To assure delivery to the customer we

> **"T**he *kanban* method is the means by which the Toyota Production system moves smoothly."
>
> **Taiichi Ohno from Toyota Production System**

will use a management policy with our finished goods inventory. It will involve the use of three types of finished goods inventory. To assure we have stock on hand for the normal pickups by the customer we will carry cycle stock. In addition, in order to provide supply to the customer we will carry stocks to handle external demand variations and the internal supply variations of the finished goods. Hence we need a buffer and a safety stock volume, respectively. So our inventory management philosophy will

Rule No.	Ohno's Rule	Function	A Simpler Interpretation of Ohno's Rules*
1	Later process goes to earlier process and picks up the number of items indicated by the *kanban*	Creates pull, provides pick-up or transportation information. The replenishment concept is formed here	Take only what you need
2	Earlier processes produces items in a quantity and sequence indicated by the *kanban*	Provides production information and prevents overproduction	Only produce what was withdrawn
3	No items are made or transported without a *kanban*	Prevents overproduction and excessive transportation	No items are made or transported without a *kanban*
4	Always attach a *kanban* to the goods	Serves as a work order	Always use *kanban*
5	Defective products are not sent to the subsequent process	Prevents defective parts from advancing; identifies defective process	Quality only
6	Reducing the number of *kanban* increases their sensitivity	Inventory reduction reduces waste and makes the system more sensitive	Reduce to improve

*My thanks to Robert Simonis for helping me simplify these rules.

TABLE 11-1   The Six Rules of *Kanban*

include carrying three types of finished goods inventories and each one will be statistically calculated to minimize the total volume yet maintain a high level of customer service (in this case the level of customer service will be 99 percent on-time delivery). Hence, the total number of finished goods *kanban* is the sum of these three stock volumes, divided by the container size.

No. of *Kanban* = (Cycle stock + Safety stock + Buffer stock) / Container size

Kanban Circulation

The kanban system is very flexible, and many types of *kanban* can be used. Likewise, as long as they follow the basic rules of *kanban*, they can be used in a large variety of ways. However, the majority of *kanban* follow a standard pattern. Let's follow a *kanban* as it is circulated (see Chap. 14, App. E, which shows the *kanban* flow on the value stream map for QED Motors). Since lean thinking usually works best if we start at the customer and work backward, let's do just that.

Since Rule 3 says that the product has *kanban* attached, when the customer comes for his pickup, the *kanban* are removed and placed in a *kanban* post. From here, the *kanban* are picked up, normally by a materials handler, and transported to planning, or ideally they go directly to the *heijunka* box in front of the production line. If they go to planning, they generally do little with the *kanban*, but they like to stay in the loop. From planning, the *kanban* are sent to the front of the production line per the information on the *kanban*. The *kanban* are then placed in the *heijunka* box, a load leveling tool. From here, the production workers withdraw the *kanban* from the box in sequential order, and the process then produces the product in the quantity listed on the *kanban*. The *kanban* has just served to be a production work order and is infinitely superior to any MRP/ERP/SCM-type system to trigger production. The worker then attaches the *kanban* to the products made and they are placed in the designated spot, ready for pickup. On his normal circulation, the materials handler picks up the products, with *kanban* attached. The *kanban* tell him exactly where to deliver the products—normally this is the storehouse, which completes the cycle. The *kanban* have moved a distance and have consumed time by:

1. Transportation to, and time in, planning

2. Transportation time to and time spent waiting in the queue, the *heijunka* box

3. Time spent in the production line

4. Time used to deliver the finished goods

The sum of these four times is the replenishment time.

How Do We Achieve Process Improvements in a *Kanban* System?

Process improvement in a *kanban* system is accomplished by the reduction of inventory. The total inventory in a system is the number of *kanban* multiplied by the number of parts per *kanban*. So we can reduce either the number of parts per *kanban* (very useful in a two- or three-bin *kanban* system, for example), or in the number of *kanban*. This can be achieved by:

- Reducing any of the four replenishment times or reducing the pickup volume by the customer, this is usually achieved by increasing the pickup frequency. Reductions in any of these items will reduce cycle stock inventory.

- Reducing the variation in the production rate, which allows safety stock reductions.

- Reducing the variation in the customer demand, which allows buffer stock reductions.

What Does the *Kanban* System Really Do?

Think for a moment about a perfect stockless (almost) manufacturing system. It would have a cell where all the necessary processing steps are connected with zero inventory between stations, one-piece flow, operating with 100 percent availability, and 100 percent yield, and hence the steps would operate in total synchronization. We would simply tell the operators to keep one unit of production in the finished goods inventory and if the customer came and removed a unit, then and only then would we replace it. In this system, with 100 percent on-time delivery, once the customer withdrew an item, it would signal replenishment, and in total synchronization all stations would spring into action and another would be produced, almost instantaneously. The perfect pull production system. Once a customer arrived, product was ready; however, If the customer did not withdraw a product, no production would occur. One hundred percent on-time delivery, with no overproduction, a near perfect lean system. This, of course, would only occur in a perfect—therefore, non-real—system. Unfortunately, we mortals need to deal with the realities of life.

These realities of life include several issues.

First and foremost is the issue of variability. Did we not say it was inevitable? Since perfect synchronization is not possible, 100 percent on-time delivery and zero overproduction, are also not possible. Though these ideals might be ones to shoot for, they are typically impossible and many times impractical. Variations always exist in rate, quality, people, machine, and environments. They are inevitable and omnipresent. All this variation creates, guess what? You got it, inventory. So to compensate for the variation, we need some buffers. This causes our total inventory to rise and Little's Law tells us our lead time will increase, which likely will cause us to hold even more inventory as finished goods.

> "**D**on't be fooled. The system optimum is not necessarily the sum of the local optima."
>
> Unknown

So how do we reduce the inventory—that is, avoid overproduction of both the local (WIP) and finished goods, bringing inventory to its minimum—and still supply the customer with high levels of on-time delivery?

Either task can be done simply, but doing both simultaneously—and well—is the trick of a good business system.

And that trick is *kanban*.

The essence of *kanban* is twofold.

First, it is direct communications to produce material—in other words, to supply the customer. It is the pull signal to produce. Once the product is withdrawn by the customer, at that moment the *kanban* tells us exactly what the customer is using, and hence what the customer will need later. This *kanban* is sent as fast as possible to the production line. In essence, the *kanban* system is doing the "talking" to the production system, telling it to produce because some product has been removed. This system easily bypasses all the accounting and planning systems that tend to not only delay this signal but also add variability along the way. The *kanban* system is dealing real-time with the realities of what is happening on the line. The planning systems deal with what the programmer believed should be happening. I can say with certainty, that when it comes to triggering production, with the minimum lead planning time, no planning system can come close to *kanban*. In this manner, the *kanban* system not only assures supply to the customer, but does so with the minimum planning time.

Second, *kanban* creates an absolute limit on total inventory. Since each *kanban* represents a certain amount of stock, and the number of *kanban* are strictly controlled and limited, this creates an upper limit on the inventory. We will show in Chap. 12 that this inventory limitation is a key factor in making a pull production system function. By utilizing pull production, we minimize overproduction. Furthermore, the continuous improvement aspect of *kanban* works to further reduce this overproduction.

What Is Value Added Time?

Total time—all the time it takes to produce the product, which is made up of:

- **Value-added time**—the time the customer is willing to pay for; when form, fit or function also improve

- **Non-value-added time**—which is waste. It comprises:

  - Pure waste—Activities that can be eliminated or reduced immediately.

  - Necessary waste—Activities that cannot be reduced immediately due to the present work rules or technology.

Finished Goods Inventory Calculations

Cycle Stock Calculation

Cycle stock, you will recall, is the stock that is the volume you need on hand to take care of the normal demand pickups by your customer. Hence, we will calculate the cycle stock volume to be the production rate multiplied by the replenishment time plus some arbitrary safety factor we will call Alpha. A sample calculation of replenishment time is shown in Table 11-2, which quantifies the display shown in Fig. 11-1.

For example, if our typical daily shipment is 1400 units per day. Presume *takt* is 1 minute, so production time is then 23.3 hours. The time the *kanban* cards are in planning is 24 hours, and delivery time (due to the material handler's frequency) is 2 hours. In our typical queue, we have 16 hours of demand in front of this order and we use an Alpha of 0.05. With this, the replenishment time is as shown in Table 11-2.

Planning time	24 h
Queue time	16 h
Production time...1400/60	23.3 h
Delivery time	2 h
Total time	65.3 h

TABLE 11-2   Replenishment Time (RT)

Thus, the cycle stock volume is (65.3 h × 60 units/h)(1 + 0.05) = 4114 units. If, for example, there are 50 units to a box, the cycle stock inventory would be 83 containers, and if *kanban* were used, we would have 83 *kanban*. So we could have as much as 83 *kanban* of finished goods in the cycle stock inventory. This is very unlikely, but if the customer failed to make a pickup for a few days, this would be the maximum volume stored on-site. At that time, since there are only 83 *kanban*, each *kanban* would be attached to a box that is held in cycle stock inventory. Since all 83 *kanban* are attached to boxes, none work their way through the system to the *heijunka* board where they would trigger production—hence, no more product would be made.

Buffer and Safety Stocks Calculations

For both buffer and safety stocks, the same logic and methodology is used. In both cases, you use historical information to calculate the variation. We then determine an acceptable level of on-time delivery, normally 99 percent, and to obtain this, we need a z score of 2.33 Sigma for a one-sided test. Therefore, we need 2.33 Sigma volume of stock to assure 99 percent on-time shipments in the case of buffer stocks. Now the required volume to protect the supply can be determined. The difference between the calculations is that for safety stocks you use the data, which depicts the internal variation, usually the production rate to the storehouse. On the other hand, for buffer stocks you use the external variations that are typically the effects of demand fluctuations, plus delivery variations.

Let's calculate the safety stock for the preceding case:

The production data for a 30-day period are listed in Table 11-3.

Day	Production	Day	Production
1	1460	17	1480
2	1410	18	1350
3	1390	19	1450
4	1300	20	1250
5	1390	21	1370
6	1450	22	1400
7	1400	23	1390
8	1410	24	1480
9	1420	25	1450
10	1460	26	1400
11	1410	27	1350
12	1380	28	1310
13	1370	29	1380
14	1400	30	1510
15	1390	Ave	1400
16	1420	Std Dev	59.0

TABLE 11-3   Production Data

With this information, you can quantify the need for safety stock due to production variations. The standard deviation is 59 units. To cover ourselves for this variation to a 99 percent certainty, we can carry 2.33 Sigma of stock, or 138 units. Practically, we would carry three boxes; two boxes of which would only be 100 units. In the final result, that would mean we might short an order as much as 1 percent of the time. If we examine these data, the variation for this 30-day period is a high of 1510 and a low of 1250. If the pickup is 1400 units and that day we made only 1250, we would need an extra 150, to complete the order, this is exactly what we have, so our system worked nicely to assure on-time delivery.

Buffer stock is calculated the same way, except that we use the external variations. These are normally caused by demand changes by the customer. The demand data for a 30-day period is listed in Table 11-4.

If we wanted to cover the demand variations to 99 percent certainty, we would carry 2.33 Sigma of stock, or about 485 units. That would be ten boxes. In these 30 days, we actually had a high demand of 1800 which then could be covered by the normal production of 1400 plus 400 of the 500 units in buffer stocks. So again inventory management philosophy assured that we could meet on-time delivery for the customer even when they were the source of the variation.

In addition, many people like to mingle these stocks with no segregation of buffer from safety stocks. This is not a good idea for two reasons. Since the reasons for demand variations are independent of supply variations, it is normal to calculate the two inventories separately. Also, since the use of either buffer stock or safety stock inventory triggers immediate corrective action, it is worthwhile to keep them separate since the respective corrective actions typically solve dramatically different types of problems.

Day	Demand	Day	Demand
1	1400	17	1200
2	1400	18	1600
3	1800	19	1400
4	1400	20	1000
5	1500	21	1400
6	1000	22	1400
7	1800	23	1400
8	1500	24	1000
9	1200	25	1600
10	1400	26	1400
11	1600	27	1400
12	1400	28	1000
13	1600	29	1400
14	1400	30	1600
15	1400	Ave	1400
16	1400	Std Dev	208.0

TABLE 11-4   Demand Data

Stock Type	Theoretical Need (At 2.33 Sigma Coverage)	Practical Volume on Hand	# of Boxes; # of *Kanban*
Cycle	4114	4150	83
Safety	138	150	3
Buffer	485	500	10
Total	4737	4800	96

TABLE **11-5**   Finished Goods Inventory

The Total Inventory Situation

The total inventory is summarized in Table 11-5.

Three other aspects of lean inventory management make this a bit complicated.

- First, FIFO inventory management is a lean tool. Stock rotation gets complicated with three types of stock for each part number.

- Second, each time product is withdrawn from the buffer or safety stock, it signifies an unusual event and triggers a formal corrective action.

- Third, it is very common for the *kanban* cards for safety and buffer stocks to be a different color than the cycle stock inventory. It is normal for buffer to be orange and safety stock *kanban* to be yellow. Red *kanban* are sometimes used as a signal of an emergency situation.

So How Good Is the System We Just Calculated?

First, with our inventory management philosophy, we saw that it operated at virtually 100 percent on-time delivery. Now we need a metric to work toward minimizing inventory. Such a metric for inventory can be calculated in many ways, but inventory turns is a common one. For instance, if we have an inventory of 4800 units and make 1440 units per day, we have 3.33 days on hand. If we work 261 days per year (5 days per week plus holidays), we then have 78 annual turns—that's pretty good. But we can still improve it quite a bit.

Inventory Reduction Efforts

The responsibility for inventory reduction efforts normally falls to the planning and the purchasing organizations to execute. For finished goods inventory, planning is normally assigned the task to keep inventory turns up and also to ensure high levels of on-time delivery. In most cases, on-time delivery is the more important metric, and inventory turns of finished goods takes a back seat. Hence, finished goods inventory rise to high levels. Furthermore, since inventory is usually needed to cover variation, for example, variation in production rates, and reduction of that variation is not within the control of the planning department, the response by planning is predictable, automatic, and almost justified. They ignore the causes that are driving inventory up and simply add inventory until they feel comfortable that shipments will not be missed … for whatever reason.

What is missed in the normal plant is the understanding of why the specific levels of inventory are held. In a phrase, there is no management philosophy on inventory management.

With this methodology of calculating cycle, buffer, and safety stocks, not only is the amount of inventory understood but the reasons why these volumes are needed are understood. Furthermore, it is much easier to assign responsibility of inventory reduction to the group that can actually make an impact on that specific inventory creating process.

In this particular case, if we wished to reduce inventories—and we do—it is easy to see that the largest contributor to overall inventory is the cycle stock inventory. Furthermore, the largest contributor to the cycle stock (review the replenishment time calculation) is the planning time. Why does it take 24 hours for the *kanban* to be massaged in planning? Very likely these cards are sitting in an in-box waiting to be processed. In most good systems, planning is completely bypassed and *kanban* go from the storehouse directly to the *heijunka* board. If we could do that here, we could eliminate 24 hours of replenishment time, cycle stock inventory would shrink to 2601 units (that's a reduction of 30 *kanban*) and total stock would now be 3200 units (or 2.22 days) and we would improve from a very good 78 turns to an outstanding 118 turns—all this at no cost and no risk.

Think of all the work that would be needed to reduce ALL the variation associated with the supply and ALL the variation associated with the demand, so we could eliminate ALL the buffer and safety stock—and still you would only reduce 13 *kanban* total. Yet by allowing the *kanban* system to do what it is designed to do, and bypass planning, we eliminated 30 *kanban*. Sometimes our own processes and procedures are the source of huge wastes and need to be addressed (see the discussion on Policy Constraints in Chap. 18).

However, do not lose sight of the goal: the total elimination of waste. Make no mistake about it, all this inventory is waste. It is total non-value-added work. But it is necessary for the time being. We would do away with even this minimum inventory if we could, but it is only the "least-worst" available option—for the time being.

Kanban Calculations

The basic *kanban* calculation for cycle stocks is:

No. of *Kanban* = (Replenishment Time × Production Rate) (1 + Alpha) / (Container Size)

Make-to-Stock versus Make-to-Order Production Systems

Background

Many lean systems use a make-to-stock production system, at all steps in the process. The beauty of a make-to-stock system for finished goods is that it virtually assures 100 percent on-time delivery as long as you have a good inventory management policy.

In a make-to-order system, no finished goods inventory exists at all. The finished goods are made only after the receipt of an order. In a make-to-stock system, by definition there is inventory.

So When Does Make-to-Order Make Sense?

In a lean system with a fairly stable demand, there are times when an entire family of parts is made in one production cell. For example, let's say the family has 30 models, but only 5 models comprise 90 percent of the total production. We might call them A models or runners. The other 25 models we will call strangers. It may then be advantageous to produce the runners on a make-to-stock system and produce these strangers on a make-to-order basis, holding only the cycle stock on hand. In so doing, you forego much of the resources it would take to hold the buffer and safety stock inventory for these 25 strangers. To accommodate the variation, you would then buffer these cycle stocks with a time-buffer strategy such as a plan to work a little overtime when it is needed. This is fairly common, easily calculable and very often a good, Lean, business decision.

The other time make-to-order makes sense is in the job shop—that is, the extremely high-mix production situation, which is usually low volume as well. The key problems with a make-to-order system are that very often you really do not know the demand volume or the due date, until the order is in your hands. Most of these orders are unique, so to be competitive most job shops either have a great deal of invested capital or long lead times—each of which creates a problem of its own.

What Is the End Result of It All?

The typical Lean Manufacturing system is a make-to-stock system—that is, it normally has a finished goods inventory, with a sound policy for the management of the inventory, to assure supply to the customer. It is also a pull production system, that is, production is only triggered by customer consumption, to avoid the waste of overproduction. However, this finished goods inventory, although it is necessary, is still waste and, as such, we wish to eliminate it. As we get better at removing the variation and then reducing the inventory, you can see that the optimum condition would be to have no inventory at all, which would be a make-to-order system. The logical extension of a fully matured make-to-stock system is therefore a make-to-order system. The catch is, to remove all the inventory, we need to remove all the variation, but since variation is "the inevitable differences...," we can not remove it all. Interestingly enough, the perfect system would be a make-to-order system with no inventory and the lead time would have to be zero. Of course, that is impossible, but it is interesting.

Chapter Summary

Becoming Lean is not synonymous with JIT, for JIT is only a part of becoming Lean. To affect inventory reductions, it is important to understand that inventory is created largely due to the variation that exists in the manufacturing system and that this variation, at some level, is inevitable. Consequently, we want to reduce the variation to a minimum, which will then allow us to reduce the inventory without hampering customer service. This inventory reduction lets us not only save money but also allows us to reduce our lead time and hence become more flexible and responsive as a business. Though a number of inventory reduction tools exist, *kanban* is one of the most powerful in the House of Lean and must be applied totally—following all six rules—to be really effective.

CHAPTER 12

Lean Manufacturing Simplified

Aim of this chapter … to explain the philosophy of Lean, which drives the goals and culture; the foundational aspects of quality control on which it is built; and the strategy, tactics, and skills utilized in the quantity control to become Lean. The House of Lean (see Chap. 24) is a descriptive metaphor in graphic format that will assist you in understanding how all these aspects work together to describe the mature Lean Manufacturing system.

The Philosophy and Objectives

At the heart of Lean is its philosophy, which is a long-term philosophy of growth by generating value for the customer, society, and the economy with the objectives of reducing costs, improving delivery times, and improving quality through the total elimination of waste.

> **"A**fter World War II our main concern was how to produce high quality goods...... After 1955, however, the question became how to make the exact quantity needed. **"**
>
> T. Ohno

The Foundation of Quality Control...The Foundational Issues

Strategy

This foundation of high quality has two strategies. First is the training and development of the workforce. Second is the effort to make all processes stable and capable of meeting customer needs. It is a strategy designed to achieve high levels of delivered quality.

> **P**oint of Clarity   The Toyota Production System (TPS) is a quantity control system.

Tactics and Skills for Quality Control

It is not a well-known fact that the primary purpose of all the lean tactics and skills (often called the tools of Lean) is to highlight problems in the manufacturing system. (Technically they may be either problems or symptoms of problems, although we often call them all problems.) Their primary purpose is to cause these problems to become "visible" so countermeasures can be implemented. For example, *poka-yokes* give you immediate feedback on a system creating off-spec product. *Kanban* tell you if inventories are changing. *Heijunka* and production-by-hour boards tell you if you are meeting the plan or not. *Andons* tell you if the equipment is performing properly. OEE metrics tell you if we have a problem with availability, quality, or cycle-time performance. The list goes on but their primary purpose is problem identification and exposure.

People

People and the proper handling of people—including training, career planning, and the commitment to a job—are forever at the heart of the TPS. The culture of Toyota is built on the people, and the company makes few compromises in this area. Some of the basic needs to execute the TPS are covered in the following subsections.

Multiskilled Workers   Multiskilled workers are required to staff the production facilities, for two major reasons. First, to achieve process improvements, it is often necessary to reduce or change the elements of the work. This in turn often requires a redistribution of the work. Second, work cells are often designed so they can be operated by one, two, three, four, or five people, depending on changes in demand. If the workers are not multiskilled, the dynamics of Lean are lost. Multiskilled workers are at the heart of flexibility in Lean Manufacturing. The TWI (Training within Industries Services) methodology of training for highly repetitive tasks is clearly the best proven method.

Problem Solving by All   Problem solving by all has been a hallmark of the TPS since its inception. Workers are expected to solve simple problems, and the TPS incorporates a time trigger regarding the escalation of problems and the involvement of others. Here we have the very revolutionary concept of line shutdowns initiated by the line worker himself. To really give justice to allowing the operator to shut down the production line is a book in itself. But just for fun, let's touch on one of the topics here—that is, how problems are perceived differently within the TPS compared to the typical attitude many companies have toward problems. In a normal Western plant, problems are seen as a nuisance and even a sign of failure of management, engineering, or even the worker himself. Hence, problems become a thing to hide and shrink away from. No one wants to accept the resultant blame handed out, and so many problems go unresolved even though they are obvious to many. This is commonplace, even today, in most facilities where we work. However, within the TPS, problems are viewed as a weakness in the system and an opportunity to improve the system and make it more robust. Guilt and finger-pointing are avoided and problems are addressed and solved.

Now, let's do a little exercise in imagination. First, envision several problems within your organization. Think of a production problem that has persisted for a while, maybe one that people feel a little uncomfortable talking about. No one else is around, so be honest!

Now ask yourself, "What must we do as a company to remove the root cause of this problem?" Do not be surprised if myriad answers come to mind, few of which are really doable.

Next, ask yourself, "What must we change in our company so this problem will never appear again?" and you will get an idea of the deep cultural change needed to alter the attitude toward problems in your company.

Although problem solving is done by all, it is a myth, quite frankly a popular myth, that during, his job, on the line, the worker is engaged in *kaizen* activities or even problem solving of any depth at all. Quite frankly he is fully loaded with his work as he is producing to *takt*. Within Toyota, when they have excess personnel, they will often assign line workers to do *kaizen* activities and problem solving but this is the exception rather than the rule. More often than not, the line worker may highlight a problem, but the team leader, group leader, or some other support person is the mechanism used to solve a problem.

Problem solving is at the very core of a lean transformation; for that reason Chap. 6 is dedicated to that topic.

Understanding of Variation   An understanding of variation is a topic almost skipped in Ohno's book. Yet this topic *is* the topic of problem solving, process improvement, and inventory reduction, to name just a few. So why is it missing form Ohno's writings? Well, after some thought I've concluded that he had both a deep understanding of, and an ability to manage, variation reduction to such a level that it was simply obvious, it was second nature, to him. And Ohno—if he has a weakness—sometimes does not state the obvious. Do not slight this topic. It is at the heart of your company's survival and nearly all the lean tools require an understanding of, and reduction in variation, to work.

Leader Standard Work   The purpose of Leader Standard Work (LSW) is twofold. First, it is a policy deployment technique. Second, it is used for personnel development. It is created for all leaders, from the Corporate CEO down to each team leader. It is a documented format that is constructed with structured work such as routine reports, routine checks, and spot check of metrics along with routine audits and periodic structured contact with each worker. There are also unstructured tasks, including responding to problem solving and personnel matters or an unscheduled customer visit. The portion of structured time may be as much as 60 percent for a team leader, 40 percent for the production manager and yet only 25 percent for the plant manager, and 10 percent for the CEO. All LSW includes a feedback loop with your respective supervisor. This then creates a opportunity for continual review of the work being done and the degree to which it is being done. This is an excellent forum for both job redesign and performance feedback. For this reason, LSW must be dynamic in nature. The danger of LSW is that managers lose sight of its objectives, and it then becomes mismanaged without the dynamics, the feedback, and the proper focus. It then becomes just highly structured micromanagement—and is worse than a total waste.

> **P**oint of Clarity LSW is to improve policy deployment and accelerate individual development.

Stability

OEE   OEE (Overall Equipment Effectiveness) is the primary measure of production effectiveness. It can be used for value stream or individual work station performance evaluation. Good value stream OEE is one of the key precursors to the implementation of Lean and is the product of these three important operational parameters:

1. Equipment availability
2. Quality yield
3. Cycle-time performance

To calculate OEE, you will need five parameters. First is the planned production time for the line. Second is the unplanned line downtime. Third is the line cycle time, or cycle time, of the bottleneck. Fourth is the total production including scrap, and fifth is the total amount of salable product. Let's say we have the following data:

- Planned production time is 20.5 hours. It is 24 hours less 1 hour per shift for lunch and breaks, and less one-half hour for planned preventive maintenance.
- Unscheduled downtime was 1.5 hours.
- Design cycle time is 30 seconds per piece.
- Actual total production was 2020 pieces with 50 rejects, yielding 1970 pieces of salable product. This then allows us to calculate:
  - *Availability* = Total uptime divided by total planned uptime = A = (20.5 – 1.5) /20.5 = 0.927
  - *Quality Yield* = The total of salable production units divided by the total production = Q = 1970/2020 = 0.975
  - *C/T Performance* = Total units produced, good and bad, divided by the volume, which should have been produced during the actual uptime at the design cycle time = P = 2020/[(20.5 – 1.5) × (3600/30)] = 0.886
  - *OEE* = A × Q × P = 0.801

This is a way to express how our production facility is performing, and then prioritize our problems and allocate our resources. In this example, OEE is 80 percent. The losses can be stated as about 2.5 percent due to quality issues, 7.3 percent losses due to availability issues, either materials delivery or machinery downtime, and we are losing 11.4 percent due to the line not performing at the design cycle time. It can be a very good picture of plant performance and will allow management to focus on the appropriate goals for improvement.

OEE like any tool can be misused and abused. First, it has a definite life span in a lean initiative. Once the value stream has a good quality yield, good availability, and cycle losses are low, OEE loses its effectiveness. Its major problem is that it is a lagging measure. In a mature lean facility, you will find that the entire concept is replaced by Rapid Response PDCA (Plan, Do, Check, Act), something many facilities will struggle for some time to attain. For example, in a mature facility, each quality defect will trigger real-time PDCA and be resolved in real time. With this mechanism in place, there is no need for OEE. There are at least two other problems worth mentioning. This is an easy metric to understand so it is not uncommon for the management team to use it as a metric to compare one-value stream with another or even one plant with another. This is a mistake of galactic proportions. All OEE metrics should compare themselves to themselves…and in no other way! The objective is improvement—not some form of disingenuous comparison and competition. The second is to try to come up with an OEE metric for a complex plant. OEE will work for a work station, a value stream, or even a plant if a plant is also a value stream. The problem is that with stations or value streams in series parallel, some managers have tried to convert this to a mathematical model that does not exist in reality and create a plant OEE. I have never seen this work. The math is the math but the logic is unsound. For example, I have 2 feet and each foot has 5 toes. I also know that a foot is 12 in. Combining these two comparisons I now have 1 ft = 5 toes = 12 in. or 5 toes = 12 in. Hence each toe is 2.4 in. The math is perfect! But do you buy into this? If you do, then go ahead and calculate a plant OEE.

MSA   MSA stands for Measurement System Analysis. It is the statistical calculation of the variation in the measurement system and applies to both attribute and variables data. MSA must be done on all measurement systems. The most common use of MSA is as a precursor to doing a capability study on a product characteristic or a process parameter. It is crucial to understand the variation in the measurement system since it detracts from the capability performance of the process. Frequently, process performance can be improved simply by working on the variation in the measurement system.

Cp and Cpk   Cp and Cpk are the industrially accepted measures of process performance. They are both called process capability indices. Several good books describe how to calculate Cp and Cpk, but one major point of understanding must be accepted—specifically, Cp and Cpk have no meaning if the process does not exhibit process stability—that is, process predictability. Process stability is best evaluated using a control chart and is absolutely necessary for lean transformations to be implemented. Nothing is more basic to successful lean implementation than process stability.

Availability   Availability is the concept that the production process shall be capable to produce product, when it is scheduled to do so. High process availability is a necessary characteristic of a process ready to be leaned out. Low process availability is almost always a sign of an unstable process. Usually, low availability is associated with machinery downtime or the inability to deliver quality raw materials to the production line.

Cycle-Time Reductions   Cycle-time reductions are very important to lean implementations. It is best to work hard on cycle-time reductions prior to implementation of a lean transformation. This helps stabilize the process and then the quantity control issues are more easily managed. However, often during a lean implementation, cycle-time reduction opportunities are discovered and they usually translate directly into higher production rates. These cycle-time reductions are truly the "low hanging fruit" of lean implementations. Any time a cycle-time reduction can be achieved, the resultant extra production is the lowest cost product you can make. Basically, you are transforming the cost of raw materials into the value of the finished product.

Standard Work   Standard work, as defined by Ohno, has three elements:

1. The cycle time
2. The work sequence
3. The standard inventory

However, it is a much misunderstood concept. In his book, Ohno says, "…I want to discuss the standard work sheet as a means of visual control, which is how the Toyota production system is managed."

Notice he uses two interesting terms. First, he uses the term "visual control," and second he says it is how the TPS is "managed." He does not say, "this is how the TPS is operated." He is very specific, so do not be confused. This explains why, when you enter a Toyota facility and see the standard work sheet at a work cell, it is not facing the operator. Rather, it is facing the aisle so it is available to the supervisor, the engineer, and the manager. Standard work is not used by the line operator but by the team leader, engineer, or manager so they can audit the work, understand the status of the process, and provide assistance if the process is not performing as designed. The standard work chart is part of the concept of transparency and is there for visual control by the management team. It is a myth that the standard work chart is made for the operator.

Transparency   *Transparency* is the concept that the performance of the process or the entire line is able to be "seen" simply by being on the floor. It is not generally a set of charts that will allow this—to the contrary, it is a set of visual controls such as *andons*, *heijunka* boards, and space markings that make the process performance "transparent." Where transparency is implemented properly, a manager can determine within 1 or 2 minutes if his process is performing as designed—and if the process is deficient, the manager can quickly discern the problem areas. The more lean transformations I am involved in, the more I appreciate the power of transparency. At first, I thought of it mostly as a way to "explain the floor operation quickly" so problem solving and other actions could be performed in real time. This it certainly does. However, over time I have come to appreciate the manifest power of "transparency done well," which goes well beyond floor level process understanding and creates some amazing results. First, it shares the information. I have found it is almost universal that many decisions are not made at the floor level, not because of a skill deficiency, but rather due to a lack of information. Hence when transparency is mature, it is empowering and leads to others, besides the supervisor, being able to make decisions. This, in turn, frees up the supervisor for other tasks. Second, it "opens up" the culture. A key factor in "closed cultures" (see Lean Killer No. 9 in Chap. 2) is secrecy and not sharing information. Third, it feeds the natural motivator of "being in the know." So it is empowering at all levels, improves the culture as well as works to motivate the workforce. That is a "triple crown" of benefits that is hard to beat. Transparency is manifest in our process improvement mantra… .

- Create the standard—make it visual
- Train the standard—make it visual
- Execute the standard—make it visual
- Reflect and Improve

5S   *5S* is a set of techniques, all beginning with the letter "S." They are used to improve workplace practices that facilitate visual control and lean implementation. The 5Ss in Japanese and English are:

1. *Seiri*—Sort
2. *Seiton*—Set to order
3. *Seiso*—Shine
4. *Seiketsu*—Standardize
5. *Shitsuke*—Sustain

TPM   *TPM* [Total Productive (*not preventive*) Maintenance] is a revolutionary approach to the management of machinery. It consists of activities that are designed to prevent breakdowns, minimize equipment adjustments which cause lost production, and make the machinery safer, more easily operated, and run in a cost-effective manner. In most plants, wishing to implement a lean transformation, we find that equipment availability is a large source of the process losses and often, the largest of the three losses in the OEE metric. TPM is therefore a powerful tool to improve overall performance of the plant. It is generally defined as having the following five pillars:

1. *Improvement activities*, designed to reduce the six equipment-related losses of:
   - Breakdown losses
   - Setup and adjustment losses
   - Minor stoppage losses
   - Speed losses
   - Quality defects and rework
   - Startup yield losses

2. *Autonomous maintenance*, which is an effort to have many routine activities performed by the operator rather than the maintenance department.

3. *A planned maintenance system*, which is based on failure history. This is not timed maintenance. Instead, it is based on historical evidence.

4. *Training of operators and maintenance personnel* to improve operations and maintenance skills.

5. *A system for early equipment maintenance* to avoid the loss that occurs upon new equipment startup.

Process Simplification   *Process simplification* is a basic concept but is frequently overlooked by most. It is the idea of eliminating and simplifying steps in the production process. This is one of the most powerful variation reduction techniques you can employ.

Sustaining the Gains   Sustaining gains is the concept that once a process improvement is achieved, the next step is to standardize it. Thus, we want to institutionalize the gains so they will be there forever. We then want to build on this gain. It is curious that almost everyone knows this, but almost no one does it, not even modestly. In my work with over 200 companies, I can't give you one example of any company that does this well. This is such a crucial topic; it warrants its own treatment, Chap. 15.

Quantity Control

Strategy

The quantity control strategy has two "pillars": *jidoka* and just in time (JIT).

Jidoka

Jidoka is a revolutionary 100 percent inspection technique, developed by Toyota. It is done by machines not men, using such techniques as *poka-yoke* (error proofing), which will prevent defects from advancing in the system by isolating bad materials and/or implementing line shutdowns.

It is also a continuous improvement tool because as soon as a defect is found, immediate problem solving is initiated, which is designed to find and remove the root cause of the problem. In the design case, the line does not return to normal operation until it has totally eliminated this defect-causing situation.

This powerful concept has been in place at Toyota since its inception as an automaker. In the Toyoda family, it was first implemented in 1902 when it was applied to looms to trigger automatic shutdowns when a thread snapped. Since then, *jidoka* has been continually evolving to higher levels of sensitivity. It is truly a revolutionary concept. (Read about the impact of *jidoka* in Chap. 21.)

A great deal has been written about *jidoka* in cultural terms, with such topics as the interworking of men and machines, which allow the machines to do the repetitive simple checking and let men do the higher-value work, such as problem solving. Ohno called it "autonomation," and he speaks of it in terms of "respect for humanity." Here is also where the revolutionary concept of "shutting down the line by the operator for production problems" is also manifest.

JIT

Just in time (JIT), on the other hand, is designed to deliver the right quantity to the right place at precisely the right time.

Tactics and Skills for Quantity Control

The *Jidoka* Pillar

Poka-yoke   *Poka-yoke* is a series of techniques, limited only by the worker's, supervisor's, and engineer's imagination. The purpose of *poka-yokes* is to achieve error proofing of a process activity and thereby make the process more robust. *Poka-yokes* are also used in the inspection process to achieve 100 percent inspection. There are two types of inspection *poka-yokes*: those that control—that is, shut down—the process or isolate the product upon finding a defect, and those that warn the operator via an *andon*.

Five Whys   Five Whys is the cornerstone of the TPS problem solving effort. The "5 Whys" technique is simple enough in concept. However, it will not work unless those using this technique have both expertise and experience in the problem area. They must fully understand the cause–effect relationships to utilize this seemingly simple technique. The check on the "5 Whys" is the "Therefore" technique.

Kaizen   *Kaizen* is the concept of improving a process by a series of small continuous steps. Often times these improvements are small and hard to measure; however, the accumulated effect is significant. Over the years, *kaizen* has evolved to mean improvement.

CIP   CIP stands for Continuous Improvement Process or Philosophy. Many can talk about it, but few can show a process flow chart for their CIP, and even fewer can adequately measure it. An example of a continuous improvement flow chart is shown in Fig. 12-1. In addition, the TPS advocates the concept of *yokoten*, which concerns extending the process improvements to other locations, as well as other similar applications, as part of CIP.

The JIT Pillar

Takt   *Takt* is the design process cycle time to match the customer's demand, normalized to your production schedule. It is the key calculation used when we synchronize supply

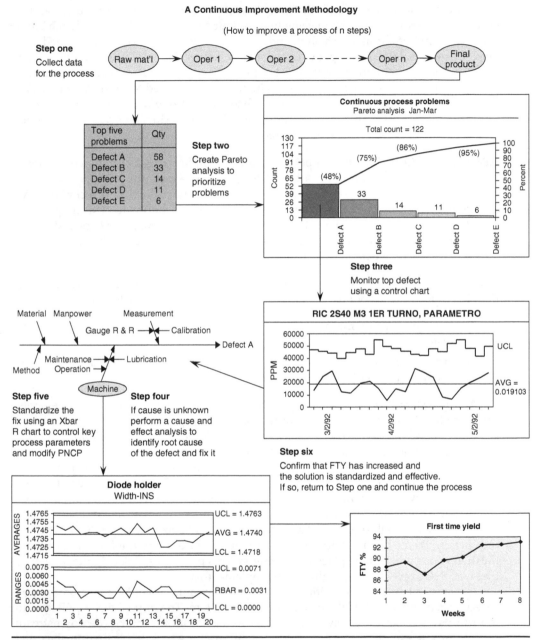

FIGURE 12-1 Continuous improvement flow chart.

to the customer. *Takt* is calculated by dividing the available work time by the product demand. The system is then designed to produce the product at this rate. If we produce at a cycle time higher than *takt* (hence, under-produce), we will not be able to supply the customer demand. If, however, we produce at a cycle time lower than *takt* (overproduce), we will either increase inventory or idle the line to stop the overproduction.

Both of these are wastes—recall that the #1 waste is the waste of overproduction. For example, if we run our operation using two 10-hour shifts with a 30-minute lunch break and two 15-minute breaks each shift, and run 5 days per week, holidays included, and need to produce 500,000 units per year, our *takt* would be: $\{(365 - [52 \times 2]) \times (2 \times [10 - 1]) \times 60)\}/500,000 = 0.56$ minutes or about 34 seconds. That is, to stay in step with the customer's demand, considering our work schedule, we will need to produce one salable unit every 34 seconds. Consequently, since there are losses, our production cycle time will need to be shorter. For example, if OEE was 0.80, a production cycle time of about 27 seconds (0.8×34 seconds) would be required.

Balanced Operations   Balanced operations are a simple industrial engineering technique to have all operation steps—of a cell, for example—operating with the same cycle time. It is the first step in synchronizing the internal production. This technique, not unique to Lean Manufacturing at all, is designed to avoid the waste of waiting. However, this technique places a large emphasis on the ability to standardize operations so we can avoid variation in the process. If any step in a process has high variation, that step will naturally unbalance the entire line or cell.

Pull   Pull systems are production systems that are designed to minimize overproduction, the most grievous of the wastes. Pull systems have two characteristics:

- They have a maximum inventory volume—for example, when using a *kanban* system.
- Production is initiated only by a signal from the customer, and that only occurs when some inventory has been consumed.

A pull system is one in which the customer, the next step in the process, removes some product that then is the signal for the upstream step to produce. For example, for some reason the finished goods inventory of our customer is full and the entire complement of *kanban* cards are attached to the finished goods in the storehouse. Since *kanban* cards are the signal to produce, and they are all attached to the finished goods, production has stopped. Hence, our overproduction is limited to whatever we have in finished goods for the cycle stock, buffer stock, and safety stock. However, when our customer arrives and withdraws product, then the *kanban* cards are removed and circulated back to the production cell, signaling that production is authorized to begin. Once it starts production, the cell will produce only that volume dictated by the *kanban*, and these finished goods will then be placed in inventory. This process of replacing the inventory that was withdrawn is specifically named replenishment.

A manufacturing system with a limit on the maximum inventory, and production based on replenishment, is the essence of a pull system. The opposite of a pull system is a push system. In a push system, there is no maximum inventory, the downstream process produces until it is told not to, usually by the scheduler. It then pushes that product onto the next step whether the next step needs the production or not. Hence, on the production floor, there is no maximum control on the WIP (work in process), so WIP can grow uncontrollably. With this uncontrolled growth of WIP, lead time will grow, with resultant quality, delivery, and cost problems escalating.

Minimum Lot Size—One-Piece Flow   Minimum lot size is a means to reduce lead times. By reducing production lot size and transfer lot sizes, the process proceeds much faster.

Two benefits are achieved. First, we reduce the lead time for the first piece through the process. This benefit is usually felt in quality responsiveness. If the first-piece lead time is reduced, and there is a problem with the product, this information is fed back to the problematic station more quickly. The problem can be resolved more quickly, and if rework is required, fewer items will need to be reworked. (For a dramatic example of this effect, see Excalibur Machine Shop in Chap. 13 and The Bravo Line in Chap. 21.) The second benefit is that the overall product will be completed more quickly, reducing production lead time for the lot. Minimum lot size, with the ultimate being "one-piece flow," is the key to plant flexibility and product supply responsiveness.

Flow   Flow is the concept that parts and subassemblies do not stop except to be processed, and then only for value-added work. It is more of a concept to be attained than a reality. It is the primary tool used to reduce production lead time. The typical technique is to design the process so that as little inventory as possible exists at each work station, and the work stations are synchronized as close as practical. The design ideal is a multi-station cell with no inventory between work stations. The Ideal State we seek is one-piece flow with 100 percent value-added work only. This Ideal State is frequently not possible, at least initially, because there are obstacles to flow. (The Seven Obstacles to Flow are detailed in Chap. 13, with a case study in Chap. 21.) For example, let's review a process running multiple products in which one of the steps is a large machine, say a press, which must undergo changeovers between production runs. To avoid stopping the process during a changeover, an inventory buffer is built up both before and after the machine, so the rest of the production process can continue to run while the press is undergoing the changeover.

All the items in these buffers will arrive "too early" to be just in time. However, considering all the options, creating a buffer is the least-waste-generating choice for the process, so it was selected. This does not create the ideal system but it is the economically practical answer. In every case, if it is not currently possible to eliminate all inventory in a process, then the next best solution is to design a system with the minimum amount of inventory. The amount is calculated, and posted at the work station as a maximum. Whenever the upstream process has produced that maximum volume of inventory, the upstream process must stop production to avoid the waste of overproduction. *Kanban* is just one system that is used to avoid overproduction. Most other systems used to minimize inventory are based on limiting the physical storage space that the parts may occupy. This is simple and creates a very good visual management tool.

Lead-Time Reductions   Lead-time reductions are the essence of waste reduction in Lean. They give the process both the maximum flexibility and maximum responsiveness to changes; especially changes in demand either in quantity or model mix. Read about lead-time reductions in Chap. 13, with a specific case study in Chap. 21, which shows how you can break through the obstacles to flow and significantly reduce lead time.

Leveling   Leveling is spoken of in two terms. First, leveling is the concept to maintain a consistent nonvariant rate of production over time. It is also a waste reduction technique called model-mix leveling, that calls for the simultaneous production of multiple products, or models of a product, from a given production line. To do otherwise is to create a batch in the system. We have already stated that Lean is a batch destruction technique. In a perfect world, we should level production to the individual production unit level. In practice, this often is not practical and sometimes not desirable. Consequently, we will frequently level based on the packaging requirements. That is, if we package 60 units in one carton, we will run 60 of that model and then switch production to another model. For example, if a certain manufacturer produces 50 models of a given product, all in equal volume, and he has the ability to run all 50 models, one piece at a time, it would be easy to implement perfect model-mix leveling. However, let's say he packages 60 units to a box and the cycle time is 30 seconds, so it takes 30 minutes to fill a box. If this operation is run with perfect leveling, then at the packaging station there are 50 boxes being simultaneously filled, and every 25 hours a large batch of finished goods needs to be transported. If, however, the process is leveled so that one box is run at a time—this is called a "pitch"—then there is only one box at a time being filled? Quite frankly, this system of producing one pitch at a time makes the downstream handling more "level" and also makes the *kanban* system much easier to use. Considering the current conditions, leveling to a pitch is normally the optimum for any lean system.

Kanban   *Kanban* is the revolutionary practice of using cards, for example, to smooth flow and create pull in a lean system. It is also a continuous improvement tool. The cards represent and account for all the inventories in the system. By controlling the number of *kanban* cards, we control the inventory. *Kanban* is a technique used to control inventory, minimize overproduction, and facilitate flow. The *kanban* cards are used to trigger replenishment. This will make the system more responsive to customer demand and shorten lead times because the signal comes directly from the customer and triggers replenishment. For a *kanban* system to be effective, all *kanban* rules must be rigorously followed. The Six Rules of *Kanban*, from *Toyota Production System*, *Beyond Large-Scale Production* (Productivity Press, 1988), are as follows:

1. Later process goes to earlier process and picks up the number of items indicated by the *kanban*.
2. Earlier process produces items in a quantity and sequence indicated by the *kanban*.
3. No items are made or transported without a *kanban*.
4. Always attach a *kanban* to the goods.
5. Defective products are not sent on to the subsequent process. The result is 100 percent defect-free goods.
6. Reducing the number of *kanban* increases their sensitivity.

Cells   Cells are work areas that are arranged so the processing steps are immediately adjacent to one another. This lets parts be processed in near-continuous flow either in very small batches or in a one-piece flow. This, in turn, allows minimization of the wastes of transportation and inventory—in this case, WIP. The most common shape is the "Inside U" cell, which minimizes walking distance when standing operators are used. Cells have some natural advantages over the classic assembly line. First, the ability to use people for more than one activity in a cell allows the control of demand variations by staffing differently. For example, if a six-person cell were to cut production by 50 percent, it is commonplace to then staff the cell with only three people and have each person work at two stations. This, of course, requires worker cross-training, but that is a staple of Lean Manufacturing. Second, cells are much more flexible. For example, in place of a 20-person assembly line, if we use four- to five-person cells, we have a much greater model-mix capability without creating large batches and without having large time losses due to changeovers. But the coolest aspect of cells is that, although it is a very well kept secret, cells can be a natural variation reduction device. Cells are a very interesting topic; see Chap. 19, Cellular Manufacturing for more details on cells.

SMED/OTS   SMED/OTS stands for Single Minute Exchange of Dies and One Touch Setups. SMED technology is a science developed by Shigeo Shingo and is designed to reduce changeover times. The problem is simple. Any machine that has long change-over times must have an excess capacity to account for the downtime of the changeover. Furthermore, to supply the rest of the downstream process during the changeover, a large batch must be stored up. Any effort to reduce the changeover times also reduces these two forms of waste: excess capitalization and overproduction. ("Single minute" means a single-digit number of minutes that is less than 10.) In actuality, the objective is to reduce the changeover time as much as possible. In some refined cases, the change-over is handled by having multiple fixtures on the same basic machine, and by simply throwing a switch, the changeover is made. This is called One Touch Setups (OTS), or sometimes One Touch Exchange of Dies (OTED). In his writing, Ohno refers to three basic elements of JIT. They are pull systems, operating at *takt* time with continuous flow. Those may be the big three but JIT is seldom practical without some application of SMED technology. It is a major batch destruction technique. The basic procedure of SMED is simple; it is a three-stage process:

1. Separate internal from external setup
2. Convert internal setup to external setup
3. Streamline all aspects of the setup operation

When an SMED application is first undertaken, we have found the best tool is the simple Gantt chart, showing all the steps in the changeover. Gather the knowledgeable people on the changeover, and then list all the changeover steps. Categorize them as internal setup, external setup, or internal but can be external; also list the conditions to make it external setup.

This is the basic starting point. From here you delete any unnecessary steps and simplify any steps you can. Next, you convert as much internal setup into external setup so it can be done with the machine running. With only internal work left, the technique is generally to create as many parallel paths as possible. At this point, you can get involved with intermediate and holding jigs, automatic adjustments, and a huge volume of imaginative approaches to shorten the changeover time.

SMED and *poka-yokes* are two of the lean techniques that are truly for the imaginative. This combination is a powerful set of tools to use as we reduce lead times and more fully utilize our processing equipment.

Much has been made of making a video of a changeover. I support this and have found it to be useful, but generally it is best to do it after you have applied SMED techniques at least once. The reason is this: When you apply SMED, the entire process will change, so it is not very worthwhile to view the old process. You will get some minor improvement ideas, but the majority of the ideas come from the development of the Gantt chart referred to earlier; for this step I can find no adequate substitute. However, there is one large benefit to be gained from making a video: Watching the old technique is usually humbling if not downright funny, and feeling a little humility as well as a good laugh are both good for the soul. At any rate, doing a video is easier than it used to be, so I do not discourage it completely.

The application of SMED technology is a key batch destruction technique and should not be underestimated in terms of its potential. It is one of the major efforts that must be undertaken if Lean is your objective. For further study, I suggest you go directly to the author of the tactic, Shigeo Shingo. He has written two major books. One is *A Revolution in Manufacturing: The SMED System*, (Productivity Press, 1985), his landmark book on the topic. In his another book, *A Study of the Toyota Production System*, (Productivity Press, 1989), he expanded his coverage on parts of his SMED system. He has refined his three stages into eight techniques. It is good reading for the lean professional.

Cycle, Buffer, and Safety Stocks   Cycle, Buffer, and Safety Stocks is the threefold approach to inventory management used in Lean Manufacturing. Each of the three types of stock is calculated and marked separately. The common way to separate the stocks is to use color-coded *kanban*. For example, white cards are used for cycle, yellow for buffer, and orange for safety stocks. Red is normally reserved for emergency runs. Consequently, when a colored *kanban* shows up at the *heijunka* board, the production people are aware that something is abnormal and they usually have a specific protocol to follow. This description of inventory management is focused on finished goods inventories, but the concepts also apply to WIP inventories. Inventory volumes need to be reviewed periodically to assess possible waste reduction opportunities. Remember, any inventory beyond what is required to protect the supply to the customer, which is the next operation, is unnecessary waste.

Cycle stock is to account for the necessary inventory built up between customer pickups.

Buffer stock, on the other hand, is the inventory kept on hand to cover the variations associated with causes external to the plant, including demand changes and such items as transportation variations.

Safety stock is that inventory kept on hand to cover the variations internal to the plant, including line stoppages, raw material stock outs, and anything else internal that hampers the ability to deliver the customer's demands.

Chapter Summary

The heart of Lean is its philosophy of long-term growth generating value for the customer, society, and the economy with the objectives of reducing costs, improving delivery times, and improving quality, all through the total elimination of waste. The key foundational strategies that support this philosophy are the investment in people and the stability of the processes that then yield a system that will produce a high-quality product. On this foundation of high quality is built the strategy of quantity control. The quantity control strategy is supported by two substrategies: *jidoka* and JIT. All the strategies and substrategies are supported by a broad range of tactics and skills. Together we refer to the strategies, substrategies, tactics, and skills as the Tools of Lean. In Chap. 24, we include a House of Lean that shows how all the tools of Lean work together to execute the objectives, which is simply "better, faster, cheaper," through the total elimination of waste.

CHAPTER **13**

The Significance of Lead Time

Aim of this chapter … to explain the history of lead time as well as the power and meaning of lead time in a business. In addition, we will go through a case study, the story of Excalibur Machine Shop, focusing on lead-time reduction efforts.

Originally, I was not even sure if I should include a special chapter on lead time. Not that lead time isn't important—it's an extremely important lean concept. But lead time is discussed in several chapters, including Chaps. 11, 14, 21, and 22, and although it is an extremely important topic, I didn't want to overdo it.

However, lead time, when it comes to useful information in Lean Manufacturing is like onions in an omelet or cheese on lasagna—you just can't have too much of it. And so we have this chapter.

To cover lead time, we will:

- Look at its history
- Explain the benefits of lead-time reductions
- Review one case study
- Explain the seven techniques used to reduce lead time
- Explain why lead time is the "key" measure of leanness

Some History of Lead Time

Until the JIT (just in time) movement got some traction in the United States in the late '80s, there wasn't much talk about lead time at all. One of the early books on JIT was *Zero Inventories* (APICS, 1983) by Robert Hall, and although he put together a good review of how the Japanese manufacturing industry had changed, he made virtually no mention of lead time. There was, however, an interesting treatment of lead time in *The Goal* (North River Press, 1984) by Goldratt and Fox. In addition, Richard Schonberger, in his book *World Class Manufacturing* (The Free Press, 1986), has some excellent information on lead time and lead-time reductions. These books were from the mid-1980s. But after Ohno and Shingo published their books in the United States, and *The Machine that Changed The World* (Rawson Assoc. 1990) by Womack, Jones, and Roos was published, the topic of lead times became more prevalent among lean professionals. This was about 1990. However, many people did not know, beyond the conceptual level, what lead time really was, and few knew how to calculate it. Lead time made for good talk, but it wasn't really understood.

Then, in 1998, Mike Rother and John Shook published *Learning to See* (The Lean Enterprise Institute, 1998). I consider it a landmark book, not only because of the concepts they explored, but also because they showed the world how to do value stream mapping (VSM). In the process of teaching VSM, they also taught readers how to reduce two lean metrics to numbers. These two metrics are the percent value-added work and lead time.

> **"I** often say that when you can measure what you are speaking about, and express it in numbers, you know something about it; but when you cannot express it in numbers, your knowledge is of a meager and unsatisfactory kind... **"**
> Lord Kelvin

Benefits of Lead-Time Reductions

As a Business Advantage

In his book, *The Toyota Production System, Beyond Large Scale Production* (Productivity Press, 1988), when asked what Toyota was doing, Ohno was quoted as saying "All we are doing is looking at the time line, from the moment the customer gives us an order to the point when we collect the cash … And we are reducing that time line by removing the non-value-added wastes." For Ohno, lead-time reduction is a key method in improving cash flow in the company. It clearly has this business advantage in addition to its advantages in the manufacturing system.

As a Manufacturing Advantage

Two key characteristics that few businesses measure in any form but that all businesses want are flexibility and responsiveness. They go hand-in-hand. With one, you also get the other. I know of no company that measures manufacturing flexibility or responsiveness and posts it with the other business metrics, but it is critically important nonetheless. Ask any planner what they would like to have more of, and second only to accurate forecasts is the ability to quickly change plans and still meet delivery dates. There is a good example of how reducing lead times helped the Bravo Line in Chap. 21. Read that now if you would like. It is made clear how lead time and lead-time improvements were turned into business advantages with huge gains. Two lead times are of critical importance and we will elaborate further on them. They are as follows:

1. First-piece lead time is the time it takes for the first piece to finish and be ready for packaging. The primary benefit of this metric being short is that, typically, the last quality inspection is done just prior to packaging and so this is the response time it takes to confirm that either quality is good, or we need to change the process.

2. Shipment lead time is the time it takes to complete the entire shipment. This, of course, is the key metric used in planning.

Table 13-1 shows the data from the Bravo Line. Take a look at the flexibility and responsiveness factors achieved by shortened lead times.

Time Impacts	First-Piece Lead Time, Cell 1	Shipment Lead Time
Original case	232 min (3.9 h)	149 h (6.2 days)
After lean improvements	6.5 min	28.4 h (1.2 days)

TABLE 13-1   Bravo Line, Lead-Time Improvements

How might the advantage of flexibility work for you in your business? Well, let's say the customer calls up and wants to change the model mix. In the Original Case, we have to tell the customer that we have a batch in the production cell now, it will take about 6.2 days to clear the line, and then right behind that we can run their request which will run another 6.2 days, letting us ship it in 12.4 days. Now, any planner who values his life will add some fat to that because, if you recall, this line did not always run to schedule. So the planner will promise something like 15 days and probably will not sleep well until the shipment leaves.

On the other hand, with the leaned out process, he tells them it will take 2.4 days, and since it runs on schedule more often than not, he not only tells them we will ship in three days, but he confidently delivers that message. But life is not always that kind. Problems can arise in the best of systems. In the short lead-time situation, if the current production is delayed, thus holding up the request from the new customer, it is known in one day, making some countermeasures possible. In the long lead-time case, it may take a week for the problem to surface. This is yet another type of flexibility inherent in a short lead-time production system; the ability to respond to abnormalities more quickly.

Please return to Chap. 10 and the section entitled, What is Lean? Here you will get a good dose of just what we mean when we say it is "emotionally much Leaner." That planner can proceed with confidence and, quite frankly, he will sleep better. Those examples abound in a lean facility.

Responsiveness and flexibility are the life blood of a typical job shop. For them, these advantages can be achieved through the reduction of lead times. We have worked with a number of job shops and taught them the benefits of lead-time reduction by using lean techniques even though the application of lean techniques are just as applicable to the manager who "knows his stuff" (Chap. 2, Lean Killer No. 1).

Excalibur Machine Shop, Lead-Time Reductions

A look at the Excalibur Machine Shop will give us some insight as to the applicability of these principles to a job shop using batch-type operations.

The Background

We were hired to train Excalibur in lean principles. Although they knew little about the TPS (Toyota Production System) or lean principles, they thought it might help them with some of their manufacturing problems. They described their problems as:

- Labor efficiency was only 56 percent compared to their goal of 80 percent minimum. This was a comparison of bid hours for a job compared to actual hours worked.

- They frequently had quality issues. No job went through without rework; most jobs had two or three episodes of rework.
- Even if they started jobs with plenty of time, they always seemed to need overtime to complete jobs. Even after working overtime, 35 percent of the jobs had unscheduled expedited freight charges that cut deeply into profit margins.

We took a plant tour and it was obvious that lean techniques were something they desperately needed. Around their problems we designed a four-day training curriculum in Lean. During the class, we also did a group project. The task was to apply the lean principles to one of their manufacturing applications. In the class they got so excited about what we had accomplished on paper, they wanted to apply it to the floor. They extended our contract and we went to the floor.

The Product

This plant was a metal fabricator. The product they had chosen to lean out was a junction box used in the telecommunication industry. The external dimensions of the box were 36 in. high by 24 in. wide and 8 in. deep. It was made from 12-gauge precoated steel. In addition to the 80 rivets and 40 screws per assembly, there were 32 items on the bill of materials.

The Process

The CNC Punch Press—Turret

The process consisted of using a Computer Numerical Control (CNC) punch press to stamp out a topside assembly consisting of the top and seven other smaller parts. The large sheets were manually loaded onto the punch press; each sheet made five assemblies. While it was cycling, the operator would separate the prior sheet, remove the protective coating, segregate the parts, and load them into containers to be transported to the deburring operation. After 100 assemblies were completed, the operator would transport the production to deburring, change the setup, and produce the bottom-side assembly. The bottom-side assembly consisted of the bottom and four other smaller parts. These parts were handled the same as the topside assembly, with protective coating removal, segregation, and placement into containers for transportation to deburring. After 100 assemblies had been produced, these too were transported to deburring and the operator would make a changeover to his next product. The machine cycle time was 10 minutes, or 2 minutes per assembly, for both the top- and bottom-side assemblies. The changeover time for this product was 36 minutes.

Deburring

The work in process (WIP) from the Turret was then transported to deburring. With automatic deburring machines, the cycle time per assembly was 36 seconds and no changeovers were required. The deburring operator was very lightly loaded. He would deburr a batch of 100 topside assemblies and transport them to the Press Break for bending. He would wait for the bottom-side assembly to arrive and then deburr and transport these to the Press Break.

The Press Break

At the Press Break, the operator would bend the pieces to the assembly and although there were more pieces and more bends to the topside assembly, both the topside batch and the bottom-side batch took 1.2 minutes. Even though there were several change-overs per batch, the changeover time was trivial since the tools were already loaded and the operator only needed to change the program in the computer's database.

The Assembly Cell
The completed WIP was placed in a holding area, awaiting time on the assembly line, which had eight work stations. When all the raw materials were available, the product would be scheduled on the assembly line.

The Analysis

The Lead-Time Chart and Minimizing Lot Sizes
To begin the project, we had completed a lead-time chart in the classroom, as shown in Fig. 13.1.

In this case, a large lot was 100 units and the batch lead time was 1220 minutes, or over 20 hours of production time. This batch of 100 was also about two months demand for this particular cabinet. We wanted to reduce the batch size and, of course, we would like to go to one-piece flow, but one-piece flow with one-piece transfer lots was not possible at this time. Their use of large multipurpose machines made this impractical. But we still wanted to reduce the batch size. We selected a batch size of 20 because there were 20 finished units to a pallet and they would never produce less than a pallet, they said.

This would increase the number of changeovers for the punch press from two to ten for the 100-unit batch. Recall that the CNC punch press was used for cutting out both the topside assembly and the bottom-side assembly units. The CNC punch press was operated by one machinist who handled loading and unloading, as well as did the changeovers single handedly.

Balancing the Assembly Cell
Next, we reviewed the operation of the assembly cell and completed a balancing study. There were eight operators in the cell and although there were only 26 minutes of work per unit, there was a bottleneck at station 4 lasting six minutes. Even though we only had 26 minutes of work per unit, with 8 operators and a bottleneck of 6 minutes, we needed 48 minutes of paid work per unit; a full 80 percent excess labor cost. Nearly all operations were manual, so by rearranging some work we were able to balance the work and reduce the bottleneck to 4.5 minutes. With this new constraint time, we only needed 26/4.5 or 5.8 people. We decided that although it would be possible to use only six operators in the assembly cell, we would start with seven; still a reduction of one person. We modified the work stations and work instructions and were ready in the assembly cell.

Reducing the Changeover Time at the CNC Punch Press
Next, from our prior work in the classroom, we thought there were opportunities in the changeover times for the CNC punch press. If we could not reduce the changeover time, we would add almost 300 minutes to the lead time. Again, in the classroom we did an SMED (Single Minute Exchange of Dies—i.e., quick changeover) analysis. Recall that the changeover time was 36 minutes. In this 36 minutes, we found over 11 minutes of external work; this left 25 minutes of internal work to analyze. It was just 25 minutes of changing tools for the most part, so we acquired two additional operators and put them to work. One operator was the deburring operator who was lightly

loaded, while the other one was the person freed up from the assembly cell. We just divided the 25 minutes equally among the three people and gave them the necessary training to change out the tools on the punch press. This allowed us to perform the tool change-outs in three parallel paths and so we reduced the changeover time to just over 8 minutes, on paper. However, in our planning we used 10 minutes as the time for a changeover. Actually, when we did the first changeovers, they took about 10 minutes. With three people all gathered around this punch press, access was an issue. Although we were confident we could make further reductions, we made no more changes at this time. Notes were taken on possible time reduction opportunities and saved for future use.

Making the First Run

Starting Production

We were now ready and started the production. It went well and we were able to produce small lots (per Fig. 13-2, notice the change in scale from Fig. 13-1).

Quality problems surfaced and we could "see" the type of quality problems they were experiencing. During this initial run, we also learned why these quality problems had such a large effect on their labor efficiency.

Encountering Quality Problems

When we started up the assembly cell, which was about 2.5 hours into the run, we found a problem on the topside assembly. One small hole had been made, on the CNC punch press, with the wrong recess dimensions. Luckily, we found this after only 40 units had been produced. We were able to rework the 40 units offline without disrupting the assembly cell. On the next changeover of the CNC punch press, we installed the correct tool. In addition, we found that one bend had been missed on the bottom-side assembly at the Press Break. These were returned to the Press Break and finished. The overall loss of time at the assembly line was about 1 hour.

The Process Smoothes Out

Other than that, the process proceeded smoothly. About three weeks later, they needed to make another run of 100, using the same small lot production philosophy, and it went flawlessly. In both cases, they were extremely pleased.

The Results

Labor Efficiency Skyrockets

In the first run, using smaller batches, their labor efficiency was calculated to be 118 percent, and in the subsequent run three weeks later, labor efficiency rose to 146 percent. Go figure out. (I did a paper *kaizen* study and found that with no capital, this labor efficiency metric could be improved another 40 percent to over 180 percent.) As so often is the case, we redesign the system, improve it to create a new present case condition, and then again find further opportunities to improve. The cycle seems endless—and it is!!

Changeovers Are Increased, with No Out-of-Pocket Costs

It is interesting to note that there was some resistance to increasing the number of set-ups on the CNC punch press since they stated that this tied up labor and did not

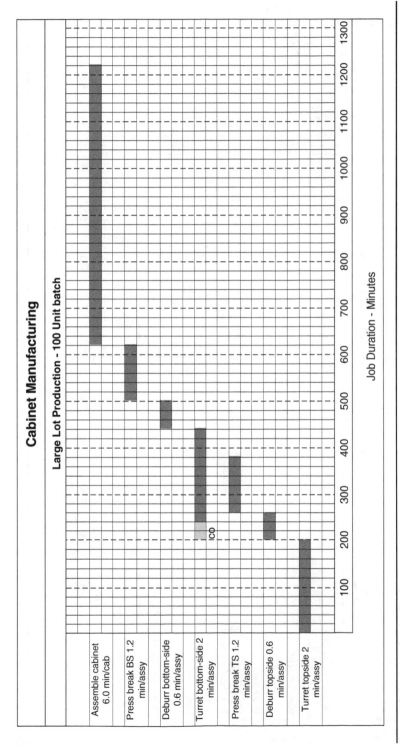

FIGURE 13-1 Large lot production.

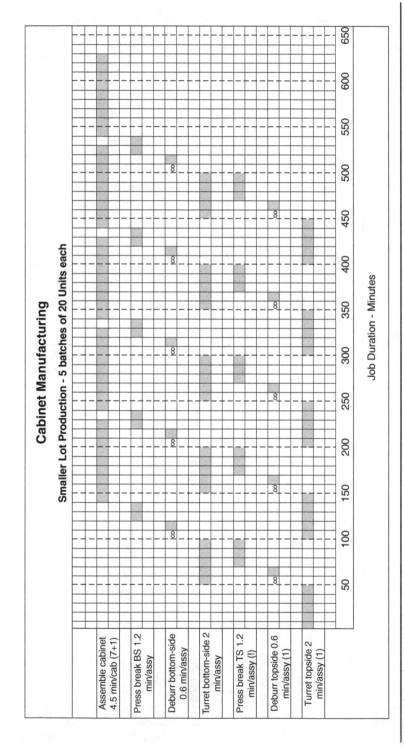

FIGURE 13-2 Small lot production.

increase production. They considered it waste. However, a quick review of some related operating practices showed the following.

First, in an attempt to find quality problems early, they had instituted a practice of doing 100 percent inspection of the first production piece from the punch press. This was done using an automatic optical inspection tool. In addition, it was their policy to stop producing after the first sheet until the quality tech gave them first-piece approval. This inspection process step was to take 12 minutes. However, the inspection was often a holdup and took far longer than 12 minutes. I reviewed some records and found that over the last two months, the inspection delay averaged 46 minutes (above the 12-minute inspection time allocation). Everyone was aware of this, and no one liked it, but the bottom line was that there was no action or even a proposal to reduce it. They simply accepted this time delay but would baulk at the time delay caused by another setup.

Second and even more pointedly, the CNC punch press was only scheduled to run 73 percent of the time, maintenance included, and it was staffed 100 percent of the time. So they were paying the same for it whether it was running or not. In this case, at this time, changeovers were free on this machine, yet they baulked at doing them.

So it is easy to "see" that they have some serious wastes in the system that are larger than the cost of a changeover, yet these go "unseen." This is not unusual and we find this in many businesses. Often, it is due to the paradigms these businesses live with, and these paradigms typically go unchallenged. The paradigm is not the problem—we will always have paradigms. The problem is the unwillingness to challenge the status quo. Herein lies the advantage of having a consultant (or *sensei*) to assist you. He will "see" these opportunities much quicker than the people within the business will see them. Consequently, your consultant will bring forth opportunities that you might be blind to. Your consultant will help you in your "Learning to See." This is invaluable help and the following paragraphs will describe how that advantage was turned into solving business problems and, in a phrase, making some money.

Lead Times Are Dramatically Reduced

Table 13-2 shows the key lean metrics of first-piece lead time and batch lead time.

First-piece lead time, a key measure of responsiveness, is cut by 75 percent, and total time to deliver the completed lot is cut in half. All of this was achieved by simply:

- Reducing the batch size
- Applying SMED
- Balancing the assembly cell
- Implementing a type of *jidoka*

We were able to do it faster, better, and cheaper. And the most encouraging lesson from this story is, we are just getting started!!

	First-Piece Lead Time	Total Lead Time
Large lot, 100 units	620 min	1220 min
Smaller lot, 20 units	140 min	630 min

TABLE **13-2**   Lead Time Changes, Bravo Line

How Did the Shortened Lead Times Positively Affect Those Concerns Stated Earlier?

First, for labor efficiency, they had rather liberal estimates for the needed work; and then both planned and performed poorly compared to the estimates. Labor efficiency on this lot was 118 percent compared to the 56 percent they had been achieving. Labor efficiency more than doubled—and at no cost! For them, that meant they could do the work with less than half the manpower. I would call that a huge gain!

Regarding their quality problems, it is clear they had huge problems with rework. The shorter lead-time model allowed them to correct the problems. For example, with the quality problem on the CNC punch press, we were able to catch the problem after the production of only 40 units; 150 minutes into the run. We were able to promptly feed-back the quality problem and fix the problem with the changeover procedure and this also led to reworking a much smaller volume. In the large lot production situation, a review of the schedule shows that the CNC punch press would finish the entire lot after 460 minutes and probably be on another job when the problem was found in assembly after 620 minutes. In this case, there is no possibility to correct the problem and our only choice is to manually rework all 100 units. Whatever the quality problem, whether it is scrap or rework, the long lead-time model only exacerbates the magnitude of the problems. Again, the shortened lead time allowed us to correct the problems because they could be found in a timely fashion and reduced the labor to rework the problems. That was a huge gain for this company since you will recall that it is common to perform rework two or three times, on each job!

As for the scheduling concerns, this by now should be obvious. **It is very difficult to plan your way out of problems that you have managed your way into.** A review of past production for this model showed that in no case were they able to start and complete this product in the same calendar week! Even with their large lot approach, this job should be done in less than three shifts. It is easy to see that with the large lot production model, quality problems, and the delays associated with correcting them were bound to be large and very punishing in terms of lead time. Yet in the reduced lot size we were able to produce the lot in less than 11 hours. It was done the same day! We reduced the theoretical time to produce from 22 to 11 hours, and we reduced the actual time to produce from over seven days to less than one day. I would also call that a huge gain!

How Well Did We Implement the Lean Concepts in This Batch Operation?

Let me give you a word of both caution and encouragement. We were not able to get the essence of "Ohno's" lean system incorporated into this production process since we could not really set up pull systems, operate at *takt*, and flow one piece at a time—and we may never be able to, considering their business.

> **P**oint of Clarity Do not let the things you cannot do prevent you from doing the things you can!

However, in the end we were leaner than when we started because we were able to produce the same product using less labor, less space, and do so with a much shorter lead time.

Regarding the Applicability of Lean at Excalibur

So although this is not a textbook example of the application of the principles of the TPS, Excalibur Manufacturing clearly was able to reduce the waste in their process and produce a superior product in a more flexible and more responsive manufacturing system. They were able to apply the lean principles and produce their products with less waste. I call that a huge success! What do you think?

How Lead Time Works for a Job Shop

There is one last point to make about lead time in a job shop environment, and it is absolutely critical to understand. Unlike the typical automobile supplier, job shops often do not have a promise of three years worth of work on the horizon; usually they live from job to job. Typically, their jobs are competitively bid, and very often quoted delivery time is a crucial decision-making criteria, second only to cost in the final bid analysis. If you are not able to deliver on time, even with the low bid, you will lose the job. So lead time is crucial to these entrepreneurs. Although short lead times will not guarantee success, long lead times will almost certainly guarantee failure. Short lead time is equal to future business to these dynamic businesses.

Lead Time as a Basic Tool in Variation Reduction

Reducing the lead time is not only a result of reducing variation, it is also a variation reduction technique of extreme power. Reducing the lead time directly allows inventory reductions in the process. This reduces exposure to environmental factors and possible damage and deterioration.

Also, there is the variation introduced to the process as a direct result of the planning process. I find the planning process to be a huge source of variation in its best form. Anyone who has tried to manage a production floor using MRPII (Manufacturing Resource Planning Two) or any of the other planning models will attest to this. It is not practical. The shop floor is operating at a speed that the typical planning processes are not able to achieve—even under the very best of circumstances. Usually, the planning cycle has a weekly update, but the floor is changing hourly or faster. It makes no difference if the planning process is updated daily; it still isn't fast enough to manage the production floor.

In addition, the farther out you need to project your plan—that is, the greater the planned lead time is, the more uncertain the plan is. This uncertainty is simply variation by another name. To make sure all contingencies are covered, planners pad their estimates and add just a little, "just to make sure" or "just in case" (JIC). Now we have a JIC planning process trying to guide a JIT production process. When the "padded" plan is given to the tier 1 suppliers to cover their contingencies, they "further pad" their estimates, creating another layer of JIC. It proceeds this way down the entire planning chain. This adding over time is amplified, and the further out the schedule goes, the more the amplification. This happens along the supply chain and is a major source of variation. Or simply put, the shorter the lead time, the more accurate is the forecast.

Planning programs such as MRPII are still needed to do long-term planning and raw materials handling. They also function well as an interface to accounting, for example. But make no mistake about it, to those who do not understand a JIT system, MRPII makes sense. To those who understand JIT, MRPII is a large waste-generating tool—when it is used to trigger production on the floor.

> **P**oint of Clarity   Lead time is the basic measure of being Lean!

Techniques to Reduce Lead Times

Seven basic techniques can be employed to reduce lead time and improve flow.

Reducing Production Time

Reducing production time is a combination of:

- Eliminating unnecessary processing steps
- Reducing production defects
- Changing the current conditions so those processing steps which are currently necessary, but not valued added, can be eliminated

Reducing Piece Wait Time

Piece wait time is reduced by balancing, so the flow is synchronized.

Reducing Lot Wait Time

Lot wait time is the time that a piece, within a lot, is waiting to be processed. To reduce lot wait times, shrink lot sizes and level the model mix. The goal of minimum lot sizes is one-piece. When lot wait time is reduced, first-piece wait time is also reduced. This time is often overlooked but is incredibly important. There will always be quality and production issues, and these issues must be uncovered quickly so they can be solved. First-piece lead time is often the key to being responsive to quality problems.

Reducing Process Delays

Process delay is the time an entire lot is waiting to be processed. Often it is called queue time. To eliminate this, we must level production quantities and processing capacity and synchronize the flow in the entire plant. The most common causes of these delays are mismatched capacities and batch production. This can also be caused by lack of synchronization and by transportation delays.

Managing the Process to Absorb Deviations and Solve Problems

Many sources of deviation increase production lead times, such as machinery breakdowns and stoppages for quality problems, to name just a few. All these deviations cause inventories to rise, and inventories are the nemesis in Lean Manufacturing—we want to reach zero inventory if we can. Where we have variation in the system, do not add inventory. Instead, attack the variation. One of the key tools to manage the process is the concept of transparency. If the condition of the process is transparent, then Rapid Response PDCA (Plan, Do, Check, Act) can be performed.

Reducing Transportation Delays

One-piece flow, synchronization, and product leveling all place emphasis on transportation, which (if you recall) is a waste. To reduce this waste, several strategies can be employed. *Kanban* is the first thing most people think of, but *kanban* has inventory and creates a second delay, the delay of information transfer, so it is a double waste in itself. Thus, try to avoid *kanban* systems by instead using close-coupled operations, such as those used in a cell.

Reducing Changeover Times

Whenever a machine has multiple uses, we must changeover between production runs. To maintain production before and after the machine, we install inventory buffers that, while they allow continuous production, slow down the overall flow. Consequently, if we reduce the time for the changeover, we can reduce the inventory both before and after the machine. This lead-time reduction and productivity technique, known as SMED, will often have dramatic effects on inventory reduction and consequently improve lead time and flow significantly.

Why Lead Time Is the Basic Measure of Being Lean?

Short lead times and lead-time reduction is such a basic tool in Lean that you will find it to be a strong measure of leanness. In addition, if a company has short lead times, several other inferences can be drawn about that company. Almost without exception, you will find that they have:

- Good inventory management
- Good quality
- Good delivery performance
- Good machine availability
- Good problem solving
- Low levels of variation
- Stable processes

Chapter Summary

Lead time as a metric was not discussed much at all until the book *Learning to See* (Lean Enterprise Institute, 1998) heightened people's awareness of this powerful concept. The business benefits of improved responsiveness and flexibility due to reduced lead times are now well understood. The story of the Excalibur Machine Shop cited here showed in detail how reducing lead time enhanced their production as well as improved their ability to find and solve problems. In addition, this is a good example of how lean tools can be applied in the job shop environment. The seven techniques to reduce lead time were explained, highlighting not only that lead time is the key metric to evaluating whether a facility is Lean, but that those firms with short lead times have other natural competitive benefits as well.

CHAPTER 14

The Path to Lean—The Five Strategies to Becoming Lean

Aim of this chapter ... to cover in a simple format, the five strategies to make a value stream Lean. In support of the strategies, we will detail the diagnostic and analytical tools used to reduce the seven wastes. We will cover the *takt* calculation, the time study, the balancing analysis, the spaghetti diagram, and value stream mapping. I have attached five appendices to this chapter, one to cover each technique.

Overview of the Path to Lean, the Lean Transformation Strategies

The Overall Lean Approach

The overall approach to make a value stream Lean consists of five key strategies. They:

1. Synchronize supply to the customer, externally.
2. Synchronize production, internally.
3. Create flow.
4. Establish pull-demand systems.
5. Standardize and sustain.

Of the five strategies, all but standardization will be detailed in this chapter. To standardize and sustain is an extremely large topic and will be covered in Chap. 15 to follow.

The Diagnostic and Analytical Tools

To apply the five strategies, there are five basic diagnostic tools you will need to use in evaluating a value stream. They are:

1. The *takt* calculation (see Appendix A)
2. The basic time study (see Appendix B)

3. Balancing analysis (see Appendix C)

4. A spaghetti diagram (see Appendix D)

5. The present state value stream map and the future state value stream map (see Appendix E)

Eliminating the Wastes

The goal is to apply the five strategies using the five diagnostic tools to eliminate the seven wastes, which are (remember TWO DIME):

- Transportation
- Waiting
- Overproduction
- Defects
- Inventory
- Movement
- Excess processing

Implementing Lean Strategies on the Production Line

Strategy 1, Synchronize Supply to Customer, Externally

Conceptual Discussion

To synchronize externally is to supply the product to our customer at their needed demand rate, normalized to our production schedule. We want to supply all the customer needs but we do not want to overproduce and create excess inventory. These tools allow this balance to be achieved.

In order to properly synchronize to the customer, we need to meet the contractual volume demand and, in addition, we will need to handle the normal variations in both supply and demand. In a mature make-to-stock production system, with good raw materials supply, reliable production equipment, stable cycle times, and high-quality yields, our supply variation should be low. However, we will still have supply variations; therefore, we will need a safety stock inventory to compensate for these variations. In addition, there will be demand variations to contend with if we wish to be synchronized to the customer. This variation will require buffer stock inventory.

Tools Used

- **The *Takt* Calculation** will allow us to understand at what rate the customer will normally wish to have product supplied. This is the basic starting point for all production rate calculations. This is often referred to as rate leveling or heijunka. We want to avoid the ups and downs of normal production and rather stabilize the rate.

- **Cycle, Buffer, and Safety Stocks** are inventories, but they are the definition of necessary inventories. Cycle stock is necessary to assure normal pickup deliveries are in place, buffer stocks will handle the demand variations, and safety stocks will take care of internal supply variations. In this way, we will ensure we meet demand, but have the minimum inventory on hand. These inventories are designed to handle normal variations in both supply and demand and therefore allow the production process to stay at *takt* and remain as stable as possible. (Chapter 11 has an example set of inventory calculations.)

- **Leveling of Model Mixes or Products** is used when more than one product is made on a given production line. The goal of leveling is to avoid making a batch of model A and then a batch of model B, and instead make both products, simultaneously, one at a time at the demand rate of the customer. We are trying to synchronize externally to the demand rate of the customer. To level production in amount and by model mix, we will frequently use a *heijunka* box. If leveling is not achieved, cycle, safety, and buffer stocks must be much larger and, of course, we want our inventory to be a minimum.

Wastes Reduced

Overproduction   Overproduction is the waste targeted here. However, when overproduction is reduced, all other wastes are reduced as well, especially the waste of inventory. In addition, this strategy to synchronize externally is the key to on-time delivery. It will allow smooth production line operations so the line can produce at a constant rate, using the safety and buffer stocks to take up the supply and demand variations. In conjunction, it will allow the supply to be both more flexible and more responsive.

Summary of Synchronize Externally

To establish the production rate at *takt* is absolutely crucial. This, coupled with the establishment of lean inventories will allow you to maintain supply to your customer and run the process at a level and stable rate. This is always the first, and the most important, step. It is often difficult to redesign the work stations for leveling of the model mix since it often requires changeovers, and so on. If that is the case for you, then do the model-mix leveling later, but make sure you are producing at *takt* and also have inventories set up to protect your supply to the customer. A word of caution is in order. If you found that your production rate was unstable while doing the systemwide evaluations, carry extra inventory. Seek advice from your *sensei* as to how much. This inventory then becomes the protection that will allow you to both assure supply to the customer as well as reduce the variations on your line from your planning systems.

Frequently, you will want to install a *heijunka* board and a production *kanban* system right away. You may have *kanban* training scheduled for later. However, implement *kanban* here anyway, if at all possible. Have your *sensei* make the calculations and do a short training for just those involved in this system. This type of JIT training will often be necessary and it is not unusual to improvise from time to time like this.

Strategy 2: Synchronize Production, Internally

Conceptual Discussion

To synchronize production internally is to divide the necessary work in processing steps such that each processing step takes the same time. The ideal is that all processing steps perform at a cycle time equal to *takt*. The following lean tools are used:

Tools Used

- **Balancing** is done by completing the basic time study and then designing the work at each work station to be the same. Normally, some accommodation is made for OEE (Overall Equipment Effectiveness) to account for production losses caused by availability issues, quality dropout, and cycle-time losses. The end result of balancing should be work stations that are synchronized.

- **Standard work** is the technique used to review the performance, including the cycle time of a production process, production cell, or a production work station. It is a management key tool in evaluating and assisting the production process to achieve synchronized production.

Wastes Reduced

Waiting is the key waste removed, and while inventory is often reduced, the goal is one-piece flow.

Summary of Synchronize Internally

The key tool used to synchronize internally is the basic time study coupled with the balancing study and chart. The balancing chart will show at a glance three major aspects of the process.

1. The relative cycle times of each step: the balance
2. The waste of waiting due to the imbalance
3. The process bottleneck

From this balancing chart, the work begins and comprises two major steps. First, the production cycle time must be calculated. It is normally the *takt* time multiplied by the actual OEE, if that is known. So if *takt* is seven minutes and OEE is 0.80 or 80 percent, the cycle time of the process will need to be 5.6 minutes. With this known, it is now time for the second task, which is to design all work stations to have a 5.6-minute cycle time. With these two steps we have balanced the process steps (synchronized internally) to a production cycle time that will achieve *takt* (synchronized externally).

In the application of this strategy, it is best to go to the end of the value stream and work backward, as much as practical. For example, if there are three work cells in series that comprise the value stream, work on the cell closest to the storehouse first. This order, of starting at the customer and working backward in the flow, should also be employed with the following two strategies of creating flow and installing pull-demand systems.

Strategy 3: Create Flow

Conceptual Discussion

The concept of flow is such that we do not want the production units to stop, except for value-added work. The flow concept has both overall measures and local measures. The local measure would be cycle time. That is the increment of time between consecutive production units. If work is done, one piece at a time, it is also the processing time at the work station. The overall measure of flow is production lead time. It is the overall time it takes for a unit to complete the entire production process. In every case, if we can reduce cycle time or if we can reduce lead time, we will make process improvements. The eight obstacles to flow include:

1. Inventory
2. Batches and batch processes
3. Distance
4. Any defect-creating process
5. Variation
6. Process steps with mismatched cycle times
7. Changeovers
8. Non-valued-added work steps

Creating flow is "the basic condition" and this strategy has some strategies of its own, including:

- Rate balancing of all steps in the value stream from the customer all the way through the raw materials supply
- Removal of inventory
- Reduction of distances between stations
- Elimination of defects, which we call *jidoka*
- Elimination of non-value-added work

> **P**oint of Clarity   These eight obstacles to flow will be the focus of the majority of your lean improvement activities from this day forward.

Tools Used

- **Minimum lot sizes** with the ideal being one-piece flow.
- **Cells** and other techniques to close-couple process to achieve short transportation distances and one-piece flow.
- **SMED** (Single Minute Exchange of Dies, quick changeovers) to reduce changeover times and the needed inventory to sustain production.
- *Jidoka* (see the section at the end of this list).
- **Problem solving by all,** for the elimination of defects and to achieve process improvements. The goal should be Rapid Response PDCA (Plan, Do, Check, Act).
- **CIP (Continuous Improvement Philosophy) and** *Kaizen* to organize the problem-solving activities.
- **Five Whys** is the key problem-solving tool used.
- **Reduction of variation** is a key tool used in inventory reduction.
- **OEE** is a key metric to use in prioritizing if **quality yield, availability**, or **cycle-time performances** must be addressed to achieve and increase flow rates.
- **Availability** improvements through the use of **TPM** (Total Productive Maintenance).

Jidoka   *Jidoka* is not only the most important strategy to implement, it is also one of the most difficult. However, it is the one I find most often slighted. The danger is that early in the implementation, other effects will be much more pronounced in terms of achieving goals. There is always significantly more quantity control reduction to be made in reducing the wastes of inventory, batches, and transportation than the waste caused by defects. Hence, these aspects get more attention, at the expense of the *jidoka* concept. In addition, most companies, although not very effectively, have been working on defect reduction for some time. However, I have yet to see even one company that had a *jidoka* concept in place—at this point in the implementation—that even remotely resembled the TPS model. However, later, this weak *jidoka* system will undermine literally all other activities. I cannot stress enough that *jidoka* must be given top priority, even though the current gains may not seem to warrant it. Return to Chap. 4 and the section titled, "The Implementation of *Jidoka* Is *Always* a Fundamental Weakness" for further information.

Wastes Reduced

- **Transportation** is reduced by the reduction of distances traveled by using cells, for example.
- **Waiting** is reduced because the production lines, while flowing, have little downtime.
- **Overproduction** is dramatically reduced on a local basis since local inventories and batches do not need premature replenishment.
- **Defect** reduction is the objective of *jidoka*.
- **Inventory** reduction is achieved by several of the techniques, including problem solving, SMED, and minimum lot sizes, to name just a few.
- **Movement** is reduced when transportation is reduced and distances are reduced. In addition, since availability is up, there are fewer instances of reassignment and wholesale personnel movement.
- **Excess processing** is reduced as all non-value-added activities are reduced.

Summary of Creating Flow

This strategy is simply removing all inventory possible, moving process steps as close together as possible, and eliminating non-value-added work—plus, the most important aspect of implementing *jidoka*.

Strategy 4: Establish Pull-Demand Systems

Conceptual Discussion

Pull systems have two characteristics. First, they have a fixed inventory, so the cycle stock, plus the buffer and safety stocks need to be determined. Second, they are activated when product is removed and this signals the upstream process to produce—no signal, no production. All *kanban* systems provide this function. However, for some simple systems such as pull systems within a close-coupled cell, for example, the most effective pull signal often is the "*kanban* space."

With a *kanban* space, when the customer removes the upstream production, the customer has "opened" the *kanban* space—this is the pull signal. Afterward, the upstream process produces more product, but not before. It is the perfect "take one make one" system. Operationally in pure pull systems, it means you do not send anything anywhere. If it leaves, someone came to pick it up. However, it is not possible to have a pure pull system in all cases. Wherever there is inventory, this inventory will delay the pull signal from the customer. This is the basis used in *kanban* design.

In *kanban*, the *kanban* card, for example, is removed from the product as product is consumed. The card is then placed in a *kanban* post and the *kanban* card is then transported back to the *heijunka* board to signal replenishment. Second, since we cannot always use pure pull signals, the pull signals need to be time responsive to the needs of the customer. The time it takes—from the receipt of the signal by the customer (product is removed and the *kanban* card is placed in *kanban* post) until the replacement product arrives at the storehouse—is called the replenishment time. This is effectively "pull signal delay." We strive to minimize pull signal delays. Simply stated, we work to minimize replenishment time.

The perfect pull system is "take one, make one." For more information on replenishment time, refer to Chap. 11, the section "Finished Goods Inventory Calculations" and Fig. 11-1. The opposite of a pull system is a push system. In a push system, product is made at an upstream station and then "pushed" to the downstream station independent of the need of the downstream station. Push system allows local machine optimization at the expense of overall system optimization. They overproduce and create not only excess finished goods inventory but excess WIP also.

Tools Used

- *Kanban* is the second most important tool used in creating a pull system. The most important tool is training. It is critical that all employees understand the concept of pull production. For example, in a close-coupled manufacturing cell, there are no *kanban* cards, but pull is practiced fully.
- JIT (Just In Time), support of all types; including not only JIT pick ups and deliveries; but also JIT maintenance, JIT problem solving, JIT decision making, and especially JIT staff planning support.

Wastes Reduced

Overproduction and inventory (in the form of WIP) is reduced.

Summary of Pull-Demand Systems

In most pull-demand systems, we will establish a signal to produce and then work to reduce the time in the replenishment cycle.

Almost surely, if you have central planning or even a local MRP program designed to do the scheduling, this will need to be altered. This will be a huge cultural change. To bypass this planning step is normally not very hard technically, but it must be done carefully since you will uncover all kinds of non-lean activities going on. You will find that the planning program does not work as designed and requires a great deal of human interaction to make it work. This human interaction is often just variation by another name.

The first thought most engineers have is to implement *kanban;* however, their first thought should be training. It will be a huge cultural change to get people out of the practice of delivering when in reality they are just pushing materials to the next cell. It may be some time before you can set up good delivery loops with materials handlers to take care of WIP. Consequently, often in the early stages of implementation, operators are still moving goods from one station to the next. This needs to be changed from a mentality of delivering and pushing to a mentality of picking up and pulling.

For raw materials delivery, and finished goods pickup, many facilities already have a skeleton *kanban* system in place. If that is the case, teach these people the proper use of *kanban* and implement *kanban* properly—employing all six rules.

Early on, get the planning people involved so they can begin the integration of your planning system with the new lean tools. They have much work to do.

Chapter Summary

The process used to make a value stream Lean consists of implementing five strategies, which are:

1. Synchronize supply to the customer, externally
2. Synchronize production, internally
3. Create flow
4. Establish pull-demand systems
5. Standardize and sustain (we will discuss in Chap. 15)

To apply the first four strategies, the basic diagnostic tools include:

- The *takt* calculation
- The basic time study and balancing analysis
- A spaghetti diagram
- Present state value stream map
- Future state value stream map

The goal is to apply the strategy, using the diagnostic tools to eliminate the seven wastes, which are:

1. Transportation
2. Waiting
3. Overproduction
4. Defects
5. Inventory
6. Movement
7. Excess processing

Appendix A—The *Takt* Calculation

Background

Takt is the rate at which a customer would pick up a product if he picked product up uniformly during the day, while you produced it uniformly throughout the day. It is the true one-piece flow, pull-demand concept. The key waste that the *takt* concept strives to avoid is the waste of overproduction, the greatest of all wastes.

The equation for *takt*, or *takt* time, is the *available work time* divided by the *customer demand* for that work time interval.

An Example Calculation

For example, let's say we produce two 10-hour shifts and each shift includes a 30-minute lunch and two 10-minute breaks. So the available work time is 20 hours – (2 × 50 minutes) = 18.33 hours. Our normal work schedule is 5 days per week and we have nine holidays, so our work year is 365 – (2 × 52) – 9 = 252 days per year. The customer has a contractual agreement to purchase 500,000 units per year.

To calculate *takt*, let's use the work week, since that is a common planning interval (If you use the day or the month, the answer will be the same—try it).

- Available time is 18.33 h/day × 5 d/wk = 91.67 hours/wk, which equals 330,000 seconds/wk
- Customer demand is 500,000 units/yr ÷ 52 wks/yr = 9615 units per week
- *Takt* = 330,000 seconds/wk ÷ 9615 units/wk = 34.3 second per unit

In simple terms, we need to produce one good unit every 34.3 seconds to stay in step with our customer's demand. This is the synchronization time, to synchronize supply externally. If it is met, the first strategy has been executed successfully.

How to Handle Model-Mix Leveling

It is common on many cells to produce several models of the same basic production unit. These models, taken as a whole, are often referred to as a family of products because they use many of the same parts and many of the same processing steps. In that case, the *takt* equation remains unchanged. It is still available work time divided by customer demand but must be calculated for each model. The complication is not in the *takt* equation; rather, it is setting up the cell so the units can be produced simultaneously.

Frequently, the concept of model-mix leveling is avoided. The typical logic used to avoid doing the leveling goes like this, "Since my customer comes for his pickup on Friday, and as long as I have the entire shipment made by then, it makes no difference whether I make the models at a uniform rate during the week or in a batch, just as long as I have all the models completed by pickup day."

However, let's look at a specific example. For example, you produce five models, A,B,C,D, and E and it takes exactly one day to produce the contractual volume of each model. Your work schedule is five days per week and his pickup is first thing Monday morning. So you make A on Monday, B on Tuesday, and so on, finishing with E on Friday, and all five models are ready for pick up the following Monday. This causes spike demands of raw materials as the various models—that is, batches go through the system, but under normal circumstances it does not sound too compelling to force model-mix leveling. But what if something abnormal happens? For example, you get a call on Wednesday and your customer says, "By the way, we want to change our pickup for next Monday. We will not need Model A, B, or C but still need the normal weekly volume but with a mix of 50 percent D and 50 percent E. Well, you now have produced only A, B, and C and the customer only wants D and E. In this case, if you work Saturday and Sunday you still can't meet their demand changes. However, had the model mix been leveled, you could just switch to D and E, and since you already have some on hand, you could work on the weekend and have everything the customer wanted for pickup on Monday.

The fact is, if model-mix leveling is properly employed, the business is just more flexible and more responsive. These are two huge business advantages that are hard to come by and serve you well in the competitive battle to survive and prosper—and leveling augments both.

How Does Cycle Time Relate to *Takt* Time?

Cycle time has many meanings, but generally people mean one of two things, one relating to the product, one to the process. Production cycle time is the time interval between two consecutive production units at the end of the production process. Process cycle time is the amount of time the unit is being worked on at any given production step. If the process cycle time in each processing step is the same, we say the process is balanced; it is synchronized internally. However, this cycle time must not only be synchronized, it must be synchronized to *takt* to stay in compliance with strategy number one: synchronize externally.

This has practical limitations since sometimes the line is not available to produce because of machine failures, stock outs, cycle-time problems, or defective parts. If the production process would be designed to operate at *takt*, then each problem mentioned earlier would result in a customer supply shortage and necessitate overtime or some other countermeasure. Since, it is practically impossible to avoid all these problems, we normally calculate the desired cycle time to be:

Cycle time = *takt* time × OEE or in this case if OEE = 0.88 (OEE is Overall Equipment Effectiveness, defined in Chap. 12), Cycle time = 34.3 seconds × 0.88 = 30.18 seconds

Hence, we would design the production system to be synchronized to 30 seconds cycle time. We would design all stations to perform to 30 seconds, or stated another way: For a one-person work station, we would balance the work so each work station has 30 seconds of work. Now when we have the normal problems of production, manifest by our OEE = 0.88, we will still end up producing the equivalent of one good unit each 34.3 seconds. When we calculate the cycle time thusly, we now are compliant to both strategy number one and two.

OEE is a manifestation of the reality that all problems can't be fixed right now, but we still have to supply the customer. A measure of a lean system is the difference between *takt* time and cycle time. Among other things, this extra time is the waste of manpower that we must pay for because our system has losses. To improve systems, a first pass is always focused on OEE and the three losses of quality losses, cycle-time losses, and availability losses. When we are able to reduce the losses of OEE, we are now able to produce more using less:

- Less space
- Less manpower
- Less capital
- Less raw materials

OEE is a very descriptive and powerful metric.

How Do We Handle Variations in Supply and Demand?

Having this cycle time to *takt* time relationship will take care of our production problems in the long run, but what about the short-term problems we might have? Those short-term problems are usually of two types: underproduction and short-term increases in demand. Say, for example, that the customer calls on Monday and informs you that for their Wednesday pickup they need twice their normal volume. Normally, there is no way to handle this with cycle-time changes, so we hold buffer inventory for exactly this situation: demand variation. Take a second example in which your customer picks up daily and you have a problem with one of your parts and can't make production. How do you make the shipment? We will hold a volume of safety stock to cover this variation. Hopefully these do not happen often, but in both cases the answer is that we protect the supply to the customer with finished goods inventory.

Appendix B—The Basic Time Study

Background

The most fundamental tool to synchronize flow and analyze the work is the basic time study. The Zeta Cell Time Study, shown in Fig. 14-1, is included for your reference. The study can be done in many ways, but several cycles should be observed, and then a rational decision made on the times to use in the study. Notice that, although the average value is employed, several times where the average was overstated due to problems found during the study, a lesser number was used. In addition, those problems encountered were taken care of with *kaizen* activities.

The Time Study

The form used has a place for the:

- Process step number
- Flow chart identification
- The work element
- Eight time studies
- Summary statistics
- Final time selected

It is important that the times measured are for the work only and do not include the wait times. If you wish to catalogue wait time, simply modify the form to have a work column and a wait column for each cycle.

It is important to evaluate several cycles. In this case, we did eight. Ten is very common, and five is too few.

The Time Study Analysis

In this case, we put the data into a tool I call a Time Stack of Work Elements, see Fig. 14-2, which shows the cumulative cell time on the left, with the work steps defined and notation of which operator performs the work. Here, there was one operator per work station.

These data are then used in the balancing study. See Appendix C.

Process to Monitor						Rayco 43-27			Date	3/9/2005, 2 shift				
Station:		Zeta Cell		Done by:				J. O. Bengineer						

Step No	FC Id.	Work Element	Cycle 1	Cycle 2	Cycle 3	Cycle 5	Cycle 6	Cycle 7	Cycle 8	High	Low	Range	Average	Final
1	10	Cut bracket	3	4	3	2	5	11	3	11	2	9	4.4	3
2	20	Assy bushing (3)	11	10	13	12	13	19	12	19	10	9	12.9	12
3	30	Install o-ring and clip	9	6	6	8	7	8	7	9	6	3	7.3	7
4	40	Place in jig, glue	7	8	9	11	10	10	9	11	7	4	9.1	9
5	50	Press in magnets (2)	4	5	6	5	4	7	17	17	4	13	6.9	6
6	60	Insert o-rings, cap, grease	14	12	12	13	19	13	14	19	12	7	13.9	13
7	70	Install support	7	8	7	8	8	9	7	9	7	2	7.7	8
8	80	Install o-ring and clip (2)	6	7	8	9	23	7	8	23	6	17	9.7	8
9	90	Apply epoxy, 3 locations	12	13	15	14	14	14	13	15	12	3	13.6	14
10	100	Install control capacitor	7	8	9	9	8	8	7	9	7	2	8.0	8
11	110	Apply epoxy, topside	7	6	5	9	6	5	5	9	5	4	6.1	6
12	120	Install retainer ring	9	8	9	8	9	7	8	9	7	2	8.3	8
13	130	Install covercap	6	7	7	8	6	7	7	8	6	2	6.9	7
14	140	Unload/load machine (2)	2	3	3	2	12	3	3	12	2	10	4.0	3
15	150	Apply final sealant (1)	22	14	15	28	14	15	16	28	14	14	17.7	15
16	160	Final test, wrap leads	16	19	17	18	22	17	18	22	16	6	18.1	18
17	200	Package	12	10	28	12	13	11	12	28	10	18	14.0	12
18														
19														
20		Total	154	148	172	176	193	171	166	193	148	45	168.6	157

Notes	1	Gun required unplugging hence long times, place on PM program
	2	Long cycle time was due to dropped parts, attention to details
	3	Long cycle times were due to dropped parts, operator needs surgical gloves
	4	Hard to do study, so much inventory and lots of movement plus lots of wait times
	5	Numerous units dropped on the floor
	6	Transportation times not taken
	7	
	8	

FIGURE 14-1 Zeta cell time study.

Time Stack of Work Elements, Zeta Cell					
Time in seconds	Activity	Original Plan	Time in seconds	Activity	Original Plan
80					
79					
78					
77	Apply epoxy 14 secs	Operator 5	157	Package 12 secs	Operator 10
76			156		
75			155		
74			154		
73			153		
72			152		
71			151		
70			150		
69			149		
68			148		
67			147		
66	Install O-ring and clip 8 secs	Operator 4	146	Final test, wrap leads 18 secs	Operator 9
65			145		
64			144		
63			143		
62			142		
61			141		
60			140		
59			139		
58	Install support 8 secs		138		
57			137		
56			136		
55			135		
54			134		
53			133		
52			132		
51			131		
50	Insert O-ring, grease insert, place in cap. 13 secs	Operator 3	130		
49			129		
48			128		
47			127		
46			126	Apply final sealant 15 secs	Operator 8
45			125		
44			124		
43			123		
42			122		
41			121		
40			120		
39			119		
38			118		
37			117		
36	Press in magnets 6 secs	Operator 2	116		
35			115		
34			114		
33			113		
32			112	Unload-load machine 3 secs	Operator 7
31			111		
30	Place in jig, glue 6 places 9 secs		110		
29			109		
28			108		
27			107	Install cover cap 7 secs	
26			106		
25			105		
24			104		
23			103		
22	Install O-ring 7 secs	Operator 1	102		
21			101		
20			100	Install capacitor retainer 8 secs	Operator 6
19			99		
18			98		
17			97		
16			96		
15			95		
14	Assembly bushing to bracket, install cap. 12 secs		94		
13			93	Apply epoxy 6 secs	
12			92		
11			91		
10			90		
9			89		
8			88		
7			87	Install control capacitor 8 secs	
6			86		
5			85		
4			84		
3	Cut Bracket 3 secs		83		
2			82		
1			81		

FIGURE 14-2 Time stack of work elements.

Appendix C—The Balancing Study

Background

A balancing study is completed to see how well the actual work elements will fit into the desired cycle time. It is easy enough to calculate the desired cycle time, but often the work elements do not allow a perfect distribution of work. In addition, at the Zeta cell the design was an outside U cell, with stationary operators so the operators could not be moved as a technique to balance the work. Instead, the work needed to be moved to the operators. Nonetheless, we came up with a nice balance for a first pass (see Figs. 14-3 and 14-5).

The Present Case Balancing Graph

On the balancing chart, you can quickly see three things: how much time is wasted, the degree of balancing achieved, and the bottleneck.

- First, the vertical distance from the *takt* line to the station cycle time represents the waiting time, which is time wasted for that work station.
- Second, by comparing the heights of the bars, you can see at a glance if the process is unbalanced and what results rebalancing will yield.
- Third, the highest bar is the bottleneck, which is now obvious.

The balancing chart for the original ten-operator layout looked like that shown in Fig. 14-3.

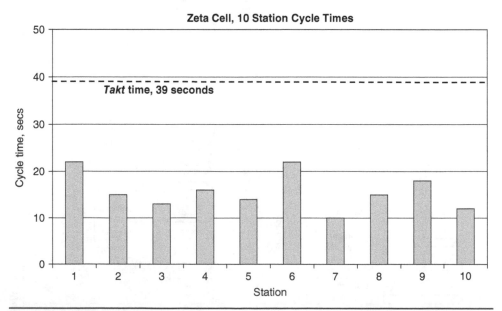

FIGURE 14-3   Zeta cell balancing graph.

			Zeta Cell, Original Design			New Balanced Proposal		
			10 Operators			5 Operators		
Step no	FC Id.	Operator No.	Work element	Final time	Time per operator	New design, operator No.	Time per operator	
1	10	1	Cut bracket	3		1		
2	20	1	Assy bushing (3)	12	22	1	31	
3	30	1	Install o-ring and clip	7		1		
4	40	2	Place in jig, glue	9	15	1		
5	50	2	Press in magnets (2)	6		2		
6	60	3	Insert o-rings, cap, grease	13	13	2	35	
7	70	4	Install support	8	16	2		
8	80	4	Install o-ring and clip (2)	8		2		
9	90	5	Apply epoxy, 3 locations	14	14	3		
10	100	6	Install control capacitor	8		3	28	
11	110	6	Apply epoxy, topside	6	22	3		
12	120	6	Install retainer ring	8		4		
13	130	7	Install cover cap	7	10	4	33	
14	140	7	Unload/load machine (2)	3		4		
15	150	8	Apply final sealant (1)	15	15	4		
16	160	9	Final test, wrap leads	18	18	5	30	
17	200	10	Package	12	12	5		
			Total	157	157		157	

FIGURE 14-4 Zeta cell rebalancing calculations.

The Present Case Analysis

Let's analyze the chart for the three simple reviews available from this type of a graph, specifically.

This line balance chart shows graphically the time wasted, which is the distance from the top of the bar graph for any station to the *takt* line—that is, if the production could be made at *takt* rate. You will find in Chap. 22 that this cell is performing so poorly it takes over two shifts, rather than the design of one shift, to produce the volume. So even though this graph shows clearly there is a huge time waste, it also grossly understates the waste. The actual wasted work time is much worse than this chart shows, and this chart is very bad! (The actual paid time was over 78 seconds per unit × 10 operators or 780 seconds, yielding 623 seconds per unit of waste; over 80 percent wasted labor time.)

The balance is poor since the tallest bar is 22 seconds and the shortest is 10. This is not good balance; the cycle times should be very similar. The bottleneck, the longest cycle time, occurs twice. Both stations 1 and 6 are 22 seconds.

The Redesign: Synchronize to *Takt*

Next, we rebalanced to *takt* by redistributing the work at the work stations. Recall that there were 157 seconds of work and we need a *takt* of 39 seconds, which yields: 157 seconds of work ÷ 39 seconds/station = 4.02 theoretical stations. With one person per station, we can't have 0.02 persons, so we will check out five stations, which gives us 157 ÷ 5 = 31 seconds per station. And if OEE is 90 percent, we could design for a cycle time of $39 \times 0.90 = 35$ seconds. This will work as a starting point. So our balance will be based on five operators and a design cycle time of 35 seconds.

With that in mind, we redistribute the work as shown in Fig. 14-4.

Redesigned Case, Balancing Graph

Figure 14-5 shows the Zeta cell balancing graph, redesign.

The Redesigned Case: Analysis

To analyze this line balance chart, we find:

- The waste is much smaller using 5 people with balanced operations than in the case using ten persons. Useful work is 157 seconds per unit, paid work is 39×5, or 190 seconds per unit, so waste is 33 seconds per unit.

- The balance is reasonable. Actually, it's quite good considering the restraints we had to deal with. The longest cycle time is 35 seconds, right at max design, and the shortest is 28 seconds. Clearly an improvement in balance, and quite frankly pretty good.

- The process bottleneck is now station 2 at 35 seconds.

The Results

The entire story of the Zeta Cell is told in Chap. 22. Read it there and see how employing the Four Strategies created the process gains shown in Table 14-1.

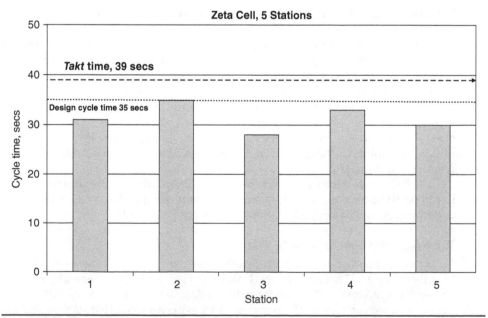

Figure 14-5   Zeta cell balancing graph, redesign.

Metric	Original Case	Leaned Process	% Improvement
First-piece lead time	4.5 h	9 min	97% reduction
Batch lead time	20 h	8.5 h	58% reduction
Space utilization	425 sq. ft.	160 sq. ft.	62% reduction
Operators per cell	10	5	50% reduction
Labor costs/unit	15 min/unit	3.19 min/unit	79% reduction

TABLE 14-1   Process Gains Summary

Appendix D—The Spaghetti Diagram

Background

The spaghetti diagram, see Figs. 14-6 and 14-7, is a simple yet powerful tool to visualize movement and transportation. When the transportation paths are seen, it is often easy to spot opportunities to reduce these wastes. A spaghetti diagram is normally hand-drawn on a simple floor layout. The example here is of the movement of a motor as it progresses through the processing steps.

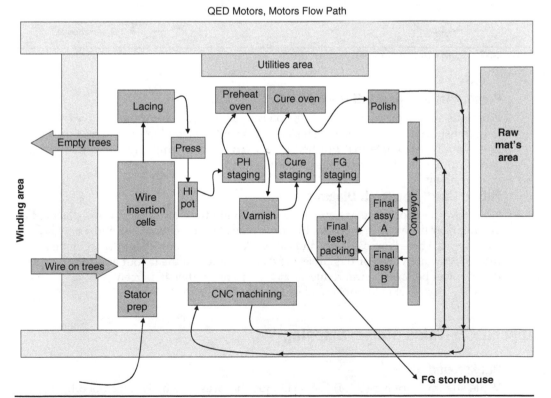

FIGURE 14-6   Spaghetti diagram, QED Motors, before.

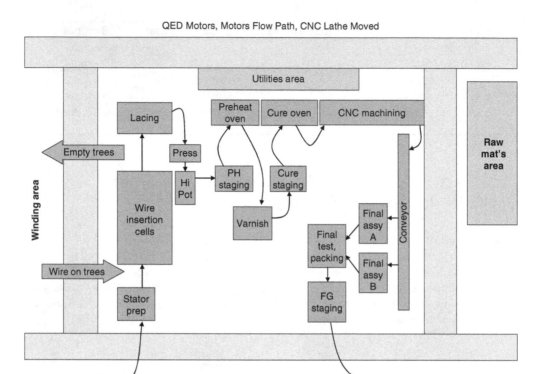

QED Motors, Motors Flow Path, CNC Lathe Moved

FIGURE 14-7   Spaghetti diagram, QED Motors, after.

Present Case: Spaghetti Diagram

From this simple diagram (Fig. 14-6), you can see that the motor makes an excessively long transportation from the Polish step to the CNC Machining process and then back to Assembly. Also note that finished goods staging is a long ways from the finished goods storehouse. See the changes shown in Fig. 14-7.

Future Case: Spaghetti Diagram

Note how the floor space has opened up by moving the CNC Machine and placing FG Staging near the aisle. In this case, this floor plan should be reviewed and a new layout considered. There is excess space and the work areas are not well laid out. On the list of future *kaizen* activities for this project is an item to revise the layout and for no capital the engineer believes he can reduce the floor space another 46 percent.

The story of QED Motors is covered in detail in Chap. 22.

Appendix E—Value Stream Mapping

Background

Value stream mapping (VSM) is a technique that was originally developed by Toyota and then popularized by the book, *Learning to See (The Lean Enterprise Institute, 1998)*, by Rother and Shook. VSM is used to find waste in the value stream of a product. Once identified, you can work to eliminate it. The purpose of VSM is process improvement at the system level.

Value stream maps show the process in a normal flow format. However, in addition to the information normally found on a process flow diagram, value stream maps show the information flow necessary to plan and meet the customer's normal demands. Other process information includes cycle times, inventories held, changeover times, staffing and modes of transportation, to name just a few. The typical VSM is called a "stock to dock" or "door to door" value stream map since it normally covers the information and process flow for the value stream at your facility. VSMs can be broader and cover any part or the entire value stream. However, functional responsibility often precludes the ability to take actions on these larger value stream maps. If actions are not the result of the value stream map, then the map has lost most of its effectiveness.

The key benefit to VSM is that it focuses on the entire value stream to find system wastes and tries to avoid the pitfall of optimizing some local situation at the expense of the overall optimization of the entire value stream. The strength of VSM may also be its weakness. It is not uncommon to find large wastes in cells, for example, which are not detailed on VSMs. If this is the case, large wastes can go unnoticed. This is a problem to those who only use VSMs in their battle to find and eliminate waste. VSM is only one tool in the battle for waste reduction, and to truly attack wastes, many tools are required.

VSMs detail information and the two key metrics highlighted are value-added work and production lead time.

Value Stream Mapping (VSM) Common Sense

First, VSM is a tool to assist you in the battle to reduce waste. Don't get too hung up on the rules. The general ideas will guide you nicely, but if there are some relevant data you think would be helpful but you haven't seen it on some primer on VSM, don't hesitate to use it. The objective is waste reduction—don't lose sight of that. Make the VSM do what you want it to do. See Figs. 14-8 to 14-10.

Second, we often deal with three types of value stream maps, Present State value stream maps (PSVSM), Future State value stream maps (FSVSM), and Ideal State value stream maps (ISVSM). Each has its own merits and problems.

The PSVSM is just what it says. However, a word of caution: finish them. Very often I see early efforts of PSVSM get derailed by starting to make changes, even before the mapping is done.

Next are the ISVSM. Quite frankly they are the best and worst of VSM. They are the best because they ask you to think about the ultimate in waste removal, which will achieve the best percentage of value-added work and the absolute shortest lead times. They are the worst because they are the least defined—and if you are not careful, a waste of time. For example, what technology do you use for the Ideal State? Do you have unlimited capital? Can you move the plant? These are serious boundary questions that must be answered. Well, my answer to that is to simply state the assumptions on the ISVSM, and go from there. I have seen a number of seemingly impossible ideas reach fruition because someone spent some time making an ISVSM. The trick is to stretch the boundaries of your paradigms, but still have some possibility of achieving, at least in part, the Ideal State.

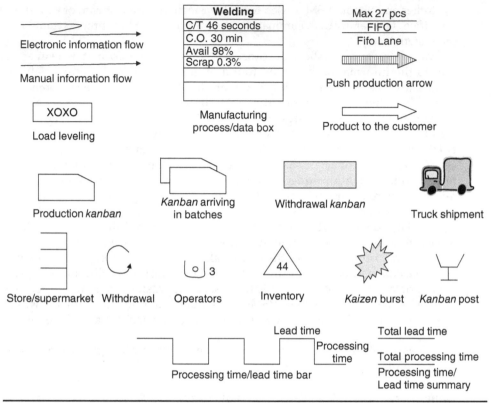

FIGURE 14-8   Typical value stream mapping icons.

Finally, the real driver to most improvement activities is the FSVSM. They should not be "ideal." Rather, they should be "achievable," and generally in a reasonably short, say 3- to 6-month time frame. My experience is that I have yet to see a FSVSM completed as it was drawn. Halfway through the *kaizen* activities, new ideas emerge, a new PSVSM is made—along with a new FSVSM—and the cycle starts all over again.

Third, the VSM is designed to be a tool to highlight activities. In leanspeak, we call them *kaizen* activities, for waste reduction. Once highlighted, the purpose of a VSM is to communicate the opportunities so they may be prioritized and acted upon. Hence, the prioritization and action must follow the VSM. Otherwise, it is waste just like any other waste.

Fourth, who should make values stream maps? Well, since the purpose is action, those involved in the action decisions need to make the maps—or at a minimum be on the team that makes the map. There is no rationale for having a VSM specialist make all the maps—that is, counterproductive. The benefits of VSM do not come solely from the creation of a map. Instead, they come from the interaction of the people making the maps, with the process and making the observations on the floor, which are necessary to gather the information for the value stream maps. In short, there needs to be a management presence in the value stream map construction process.

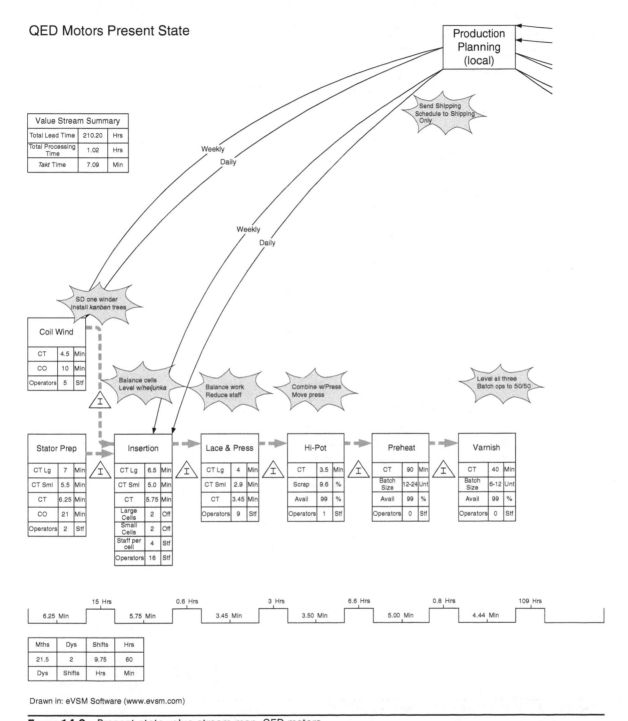

FIGURE 14-9 Present state value stream map, QED motors.

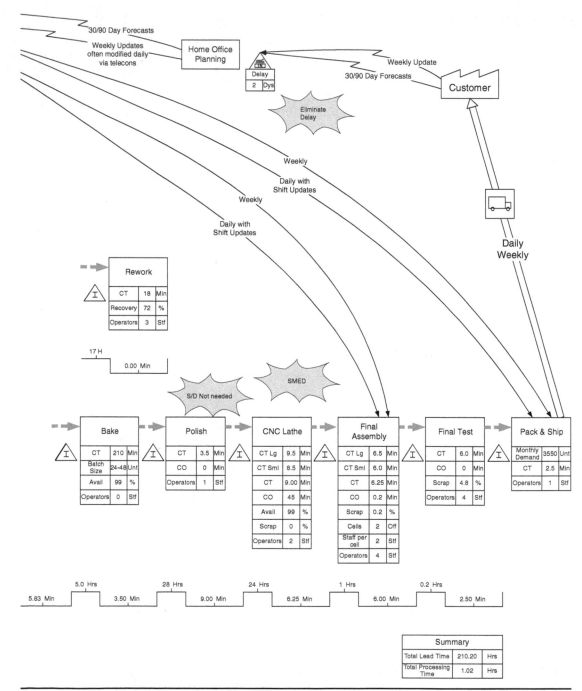

FIGURE 14-9 Present state value stream map, QED motors. (*Continued*)

QED Motors Future State

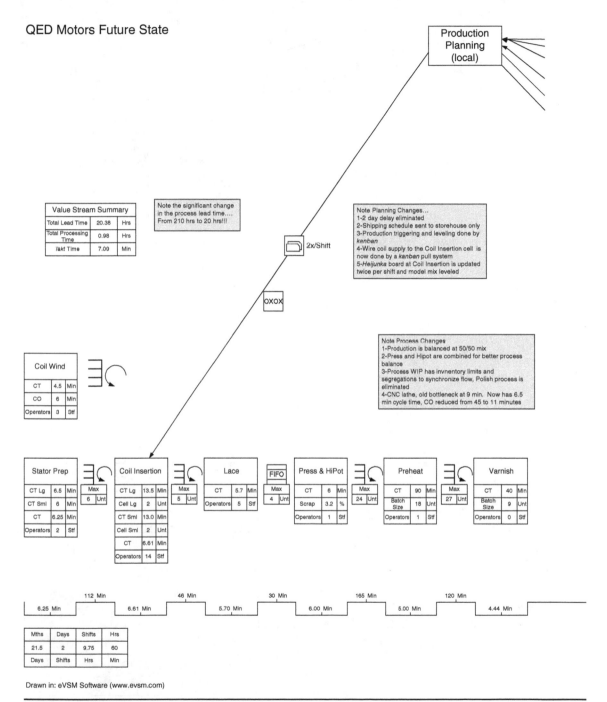

FIGURE 14-10   Future state value stream map, QED motors.

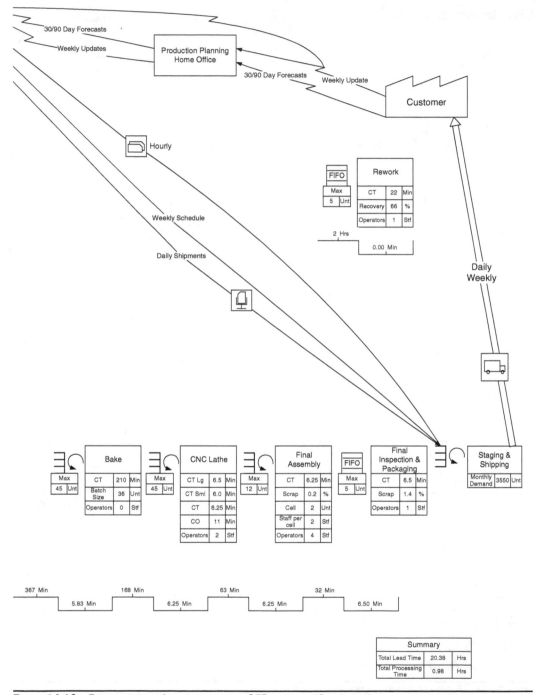

Figure 14-10 Future state value stream map, QED motors. (*Continued*)

Fifth, how do I start to make my VSM? Well, walk the value stream, starting closest to the customer and working backward up the value stream.

Sixth, should these VSMs be handmade or computer generated? I make mine by hand. Most serious practitioners make them by hand. There are a few advantages of having them on software, such as transportability, quick update on calculations, and easier use in presentations. However, what I have seen is that the people then tend to get stuck in the office, which totally undermines the benefit of the VSM. On balance, I believe handmade—in pencil—is a clear choice.

Two value stream maps are included. The PSVSM and FSVSM from QED Motors. Read the story of QED Motors in Chap. 22 for all the details and see if you can make the map from the process information … everything is there.

CHAPTER 15

Sustaining the Gains

Aim of this chapter... to explain the importance of sustaining the gains and its nature as a key foundational topic. We will discuss its importance and its application, especially the more powerful techniques of product and process simplification. Unfortunately, for most of us applying lean tools today, the product is designed and the process in place, so we have to work with what we have. Seldom is it practical to change the product, so we are left with improving the process. Because of this, we will address how we can sustain gains in these circumstances.

Why Is It so Important?

It seems almost intuitive that to sustain gains is a key issue in any business. Why make the process improvements only to lose them over time? If the gains are sustained, over time the net effect is much larger, of course. The two techniques used to sustain gains include maintenance and standardization of improvements. Maintenance is the ability to restore equipment to the original condition so the status quo can be restored. Standardization, on the other hand, is the ability to get all people, machines, and methods to continue to do what has once been shown to be effective, be it the status quo or a process improvement. Hence, to sustain the gains from our process improvements, we need to at least standardize the gains achieved. (See Fig. 15-1.)

So why doesn't everyone do it? Sustaining the gains is something we consider to be of the greatest importance. This is what separates the companies that are prospering from all the others. Every company is encountering and solving problems—there is nothing particularly unique about that. Some do it better than others, some are more efficient, and some are more effective than others. However, how many of these companies are spending time solving problems that had been solved before? This cultural characteristic of spending lots of time on fixing problems, only to have them reappear and be resolved, is incredibly common. It takes great discipline to check, double-check, then review and finally audit to make sure the last problem was solved—and solved for good. It is simply a lot sexier to move on to the next problem and solve that one. This pattern—the pattern of fixing problems without institutionalizing them—is the pattern of most cultures; it is the pattern of a culture of "fire fighting." It cannot be the pattern of a lean culture.

> **"T**he only effective approach to quality is to make it part of your culture, that is our aim.**"**
> **The motto of Quality Consultants**

"Sustaining the gains," is an area in which our company, Quality Consultants, specializes. If the changes are not institutionalized, there will be natural degradation.

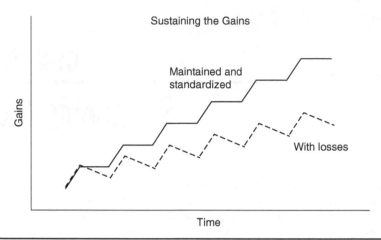

Figure 15-1   Sustaining the gains.

Remember that the entropy of the world is increasing; all systems need maintenance; nothing is forever. This can be said in many ways, but they all mean the same thing: If progress is your desire, then you need to fix it only once and fix it correctly.

One of the real enemies of sustaining the gains is the rapidly changing world we live in, as outlined in the following list.

- First, to survive, we must have a culture of continuous improvement, which means we need to change internally.

- Second, if our own internal efforts of improving processes and reducing costs with those resultant changes are not enough, customers will in addition forever ramp up the required standards of their suppliers.

- Last but not least is the issue of employee turnover—the most difficult type of change of all.

When all three change factors—internal changes, external changes, and employee changes—act on a business, I cannot give a single example of even one firm that is excellent at sustaining the changes needed to prosper.

The first two enemies to sustaining the gains are not avoidable, so as a business we must manage the personnel in such a way as to minimize turnover. The issue of employee continuity is crucial to being able to sustain the gains. I know of several firms that are adequate at sustaining the gains, and they have one major item in common: a high level of employee continuity. In terms of numbers, they have developed a culture that holds on to the top management for an average of 16 years, supervisors for an average of 12 years, and hands-on workers for an average of 7 years.

Now look at the example of the House of Lean (Chap. 24), and you will find that "Sustaining the Gains" is a foundational issue. In fact, it is more than a foundational issue; it is *the* foundation of all foundational issues. It almost seems infantile to point out that we must continue to do the good things that allowed us to progress; however, it must not only be pointed out, there must be an entire system of activities in place to ensure that we continue to do those things that allowed us to progress. I often find it almost insulting to discuss this with top managers, but it is the most basic of problems. I can't give you one example of any company that does this well. I can give you some examples of several who are adequate, but not one that does it really well.

How do I measure that? It is simple: "Do the problems, once solved, stay solved?" Actually, this is a two-part issue. Can they solve problems? This is a major weakness with most companies (as was discussed in Chap. 7). So, problem solving is a precursor to sustaining the gains. Among those companies that have good problem-solving skills at all levels, they must now institutionalize those solutions. Embedding these changes as part of the normal fabric of doing business is not easily accomplished. Most managers recognize that sustaining the gains must be done, but put forth only a minimal effort, mostly because they do not make it a priority to ensure that the changes are institutionalized. Instead, they move on to the next issue.

I said that most managers do not make it a priority to sustain the gains. In my practice, I have found many who will agree that institutionalizing the gains is important but a scarce few who can put it together in a simple flow chart that shows the actions they must take to sustain these gains. Most managers will readily agree that sustaining the gains is a process, but not many can make a process map that describes it.

Most readily agree that it is important, and just as rapidly they skip right over it. Yet others have an idea of what to do, but are not willing to put forth the effort—which is frequently substantial. Once again, the reluctance to sustain the gains was predicted by those men of yore when they said:

> "Men stumble over the truth from time to time, but most pick themselves up and hurry off as if nothing happened."
>
> Sir Winston Churchill

> "Opportunity is missed by most people because it is dressed in overalls and looks like work."
>
> Thomas Edison

We are not a voice in the wilderness decrying the need to standardize and institutionalize improvements. This concept of sustaining the gains is deeply embedded in Lean and also in Deming's philosophy. Read Deming's landmark book, *Out of the Crisis* (MIT CAES, 1982), and spend a few minutes on his 14 Obligations of Management and his 7 Deadly Diseases. Deming is clear when he states the following three points:

- We must sustain the gains.
- We cannot make progress without sustaining the gains.
- Ultimately, we cannot survive without sustaining the gains.

How Do We Know There Is a Loss?

Transparency

In regards to transparency, refer to Chap. 24 of this book, which shows that transparency is one of the foundational issues. It is the concept that lets you "see," in real time, right now, what is happening in your process, letting you make a determination as to whether anything has changed or needs attention. *Andons*, 5S, tool shadow boards, and *heijunka* boards are all examples of transparency. For example, by just glancing at a *heijunka* board, we can tell if production is ahead, behind, or on schedule. We can make that evaluation right now in real time. There is no need to check the information center. It is not necessary to check with the storehouse or some computer database; the *heijunka* board will allow us to "see" it directly, at this moment. Today, it is more common to hear the phrases visual management or visual system used in place of transparency. However, I prefer transparency because it better captures the concept, as you will "see" in the example later.

Transparency Misunderstood

A great deal of the original concept of transparency has been lost. Most of what is commonly displayed as transparency is really visual management, and it is very long on the "visual" and all too often short on "management." Unfortunately, most of what is posted is not consistent with the original concept of being able to "see" what is happening in the process. A common example of "visual management" is display boards that show the status of such things as monthly production volumes, the status of problem solving, and preventive maintenance, for example. Frequently, this information is kept in one location, known as an information center. The information center often has process information such as standard work combination tables, standard work charts, and other engineering and training intensive information that generally I find are not valuable on the floor, but make for nice wallpaper at the information center. All of this information is needed, at some time, at some place, and for some reason. What should be kept at the line is what will be *used* at the line. Keep that information there; the rest should be kept where it is needed.

The Ultimate Purpose of Transparency

> **"N**ot everything that is faced can be changed, but nothing can be changed until it is faced. **"**
>
> James Baldwin

Transparency is a very powerful concept when used properly. First, it is a huge communication tool. Having real-time information in a visual format on the floor lets everyone know how they are doing. It allows the floor operator to "check" in real time how he is doing. This in turn creates a sense of accomplishment on his part. Keep in mind this is one of the five intrinsic motivators. Second, transparency is a huge benefit in delegation of authority. As I mentioned earlier, very often the reason that decisions are not made at the floor level is not because the people do not have the talent to make the decision; rather they do not have the information. Transparency is one large tool that can be used to overcome this roadblock and allow greater delegation. Third, transparency is a very simple way to clarify goals and objectives. It is a powerful combination of traits to have trained floor personnel armed with clarity of purpose making your products. Fourth, transparency facilitates the execution of Leader Standard Work (LSW). Clear visual displays make it easy to "check" on the status of the process whether it is done by the team leader, the group leader, the production manager, or the plant manager...all will gain from the transparency. However, the ultimate purpose of transparency, starting with situation assessment is Rapid Response PDCA. Recall that PDCA stands for "Plan-Do-Check-Act." This is the iterative improvement cycle that is inherent within the *kaizen* improvement process. As part of PDCA, first it is necessary to determine something has changed, such as a production rate flagging. Next, it is necessary to:

- Plan a corrective action or countermeasure (plan).
- Implement it (do).
- Confirm whether it is successful or not (check).
- Determine if additional actions or thoughtful inaction is appropriate (act).

The cycle then starts all over again; hence, it is an iterative process. Thus, it is critically necessary to have some information to be able to:

- Discern something has changed.
- Confirm that the countermeasure was or was not successful.

Furthermore, to perform PDCA in a "Rapid Response" fashion, or in leanspeak, to do that in a JIT (just in time) fashion, requires JIT information. This availability of JIT information is what transparency is all about. In a transparent system, these data are available JIT and JIT problem solving can then ensue.

An Example of the Implementation of Transparency

For example, in a high-speed electrical manufacturing plant with multiple lines operating at 6.65-second cycle times, it was hard to tell, except after several hours of operation, if the lines were producing to the plan. Even once we got a grasp on the production, if there *were* problems, we didn't have any idea where the problems might be, so JIT problem solving was something that wasn't possible.

In addition, the technicians, line leader, and area supervisors were always busy. They were always correcting something; however, the stories of what was done and why they were done were filled with generalities and terms like "It was not operating quite right."

These actions and attitudes are clear signs of a number of things, and in this case let me point out three of them:

- Fortunately, the workers, as well as the management, were both motivated and trying very hard to be engaged. They were trying to do the right things, and they were doing work that resembled their job description.
- Unfortunately, they were doing many wrong things.
- Yet more unfortunately, even when they did the right things, there was inadequate feedback of information to confirm that something positive had been done.

At this time, the relevant information to assess the production rate, which was readily available on the floor, included *andon*s; rejected product segregated into collection bins; and production boards that covered a full day's operation.

The *andon*s had no recording, only warning features. Hence, they were not useful for problem solving unless we were right there when they were activated or cleared. The scrap data was sorted and recorded hourly, but actual scrap was very low and not a factor in low production.

The production-by-hour (PBH) boards had hourly and cumulative production goals with actual production numbers entered at the top of each hour. These goals were calculated based on the hourly goal of 600 units per hour, taking into account lunch and rest breaks. There were 21.5 available hours per day, so the daily production goal was 12,900 units. This hourly goal was met nearly 50 percent of the time, but the daily goal was met less than 3 percent of the time. A review of the previous month's data showed production of 9330 units per day, or 434 units per hour: a full 27 percent below goal.

As part of their management system, there was a daily production meeting on the floor, run by the line supervisor. It lasted about 15 minutes and had a good agenda, but when the production shortage was discussed, which was a topic nearly every day, the answers were—well, amazing is the best word I could use. First, they were very general, and in almost every case where a specific problem was discussed, it was decreed to be solved. Unfortunately, these same problems would reappear later, and it did not seem odd to anyone that, although these problems had already been solved, they still reappeared. Quite frankly, during this part of the meeting everyone was on "autopilot." It was apparent they had accepted these "amazing" explanations for so long that they sort of believed them themselves. Yet day after day the production remained below goal, and they were forced into working 7 days per week to meet a plan for 5 days. In this case we could conclude that the production data were inadequate—there was no "transparency" of the production data. We could neither understand if we were on plan, nor could we solve problems when we *did* understand them.

After a meeting with the top management, we decided that taking on the low production was our number one goal. It was the key reason this product was highly unprofitable. The first thing we did was try to break down the problem into solvable pieces. We asked three questions. Is the production shortage due to:

1. Quality losses?
2. Availability losses?
3. Cycle-time losses?

We had the quality information, and at this step of the process, quality losses were insignificant. Our segregation bins gave us all the real-time information we needed. We could "see" that quality losses were not the answer to our low-production problem. Or in leanspeak, our transparency regarding quality yield was adequate for this issue.

Now we had to determine if availability was an issue. A quick check showed that material stock outs were virtually nonexistent but the technicians were working on the machines seemingly all the time. Significant downtime occurred, but we neither knew how much of it there was nor did we know what was causing the downtime. We had serious concerns here but had no information at all about the availability losses. Our transparency in this instance was not only inadequate, it was nonexistent, but we did have the *andon*s.

Next, we looked into cycle time. Other than a few time studies done by the engineers, no data on cycle time was available. The measured cycle time was advertised to be 6.0 seconds. If the process performed at this cycle time, production should be 600 units per hour—the hourly goal referred to earlier. No measurement of cycle time—of any kind—was done on the floor. The transparency about cycle time was similar to availability information. It was inadequate and practically nonexistent.

With this review, we decided to implement an OEE (Overall Equipment Effectiveness) program. (See the description of OEE in Chap. 12.) OEE information would allow us to begin the understanding of production losses and segregate the information into quality losses, availability losses, and cycle-time losses. Forms were made, training was done on gathering and entering data, and the information was set up to manage these data with feedback after each shift.

We soon found out that the losses were about 20 percent on cycle time and almost 10 percent of availability losses. Although OEE is always lagging information, it was valuable information and told us we needed to work on both cycle time and availability. However, when it came to understanding and improving the process in real time, we were no closer than a shift away from good information, so we needed to improve that.

Nonetheless, we set up some goals, created an improvement plant, and the first thing we attacked was cycle time. We did a controlled study and found the cycle time was really 6.65 seconds, which surprised everyone. It meant the goal of 600 units per hour was not even attainable. The 6.65-second bottleneck was a manual operation in the welding process. This welding machine was operated by a robot that had a microprocessor with a small display screen. Using some great imagination and innovation, the engineering supervisor found a way to program the microprocessor and display the cycle time for the manual operation. The operator now had real-time information regarding cycle time. Immediately, the cycle time began to drop and stabilize, and likewise production increased. The drop in cycle time was amazing. In less than two weeks, it had improved to 5.5 seconds. We implemented several *kaizen* activities to improve the work station even further. One of these activities was to program the microprocessor to display the average cycle time. It told what the average was, how many units were produced with that average, and when the averaging had started. We retrofitted the station with a reset button and—*voilà!*—we had an excellent piece of transparency. Ultimately, after a few *kaizen* events, we were able to achieve cycle times of 4.6 seconds, consistently. Furthermore, now the operator, supervisor, and anyone interested could look at the display and see how the process was performing. Literally, we could find the instantaneous production rate. This was an imaginative solution that added significantly to the transparency of the system. And although the OEE was a lagging indicator, we now had a "leading indicator" (or almost so) of total production rate.

However, remember that availability losses were about 10 percent. The OEE information was valuable in helping us understand the losses, but not very helpful in solving many of the problems due to the lagging nature of the information. Nonetheless, we were able to reduce some losses. Check sheet information logs on white boards were installed on the *andon*s to gather information at the time the *andon* was activated. This helped distill the information further and made it available for future use. We now had a recording function for the *andon*s. However, the review of the OEE information showed that 90 percent of the availability losses were due to machine adjustments. A quick review showed that most adjustments could not be explained and were just apparently "tinkering" by well-meaning technicians. We did some training on common versus special cause variation, clarified some work procedures, and these availability losses immediately dropped like a rock—as did the workload of the technicians. They were not comfortable with this at all, however, since they were highly motivated, very energetic and weren't used to having any spare time. Getting them to "do nothing" was a major cultural issue.

It might be asked how we ever met the hourly goal of 600 units when the process cycle time would not support it and our average production was only 433 units per hour? Well, the answer is both simple and revealing.

The simple part is this. The way production was measured was to count the production of trays that each held 120 units and multiply by the number of trays. No partial trays were counted. So, production was in multiples of 120. A tray of 120 units was the transfer batch. Also, since the line leaders knew that the managers wanted the production goal to be met, it was better to meet the goal some of the time than none of the time. Consequently, at the top of the hour when production was counted, if a tray was nearly filled, the line leader might wait to enter the data. Of course, the next hour the production was even shorter than it would have been if the data had been reported accurately. This 120-unit transfer batch created some problems beyond accounting for production, so we later cut the transfer batch size to 48 units. This not only helped the accounting but reduced the processing lead time, as you might expect.

So how is this so revealing? It was a symptom of the entire facility. Everyone wanted to do well and be seen doing well. As I said earlier, they were a motivated group. And this motivation served us well as we implemented other process changes and improvements.

We could go on, but let's briefly review what was just discussed.

- First, our production system improved. The production rate increased by 42 percent with virtually no invested capital.

- Second, once the improvements had been made, we could now "see" the system status, and if there were problems, we could rapidly find and implement the needed countermeasures. We now had the information to do Rapid Response PDCA. Our system transparency had improved immensely, allowing our problem solving to improve immensely, too.

Transparency Executed ... Following Is a Mantra We Teach on the Floor:

- Create the standard—make it visual

- Train the standard—make it visual

- Execute the standard—make it visual

- Reflect and improve

See how this plays out in "A Story with in the Story."

A Story within the Story

Within this story of production improvements through improved transparency are numerous stories, one of which we will elaborate on. It is the story of engaged workers and a dedicated, motivated management team that was hampered by a certain type of blindness. A type of blindness we all share—to some degree.

In this case, this product was not profitable at all. After 18 months of production, the company was losing 30 percent on this product. The primary problem was the insufficient production rates. Under design conditions, production needed to produce 330,000 units per month to meet the business plan. The management was absolutely frustrated because, as they said, "We have done everything we can to improve the production." They then listed the things they had done to get to the 6.65-second cycle time. The list was long and impressive. It included a great deal of training and efforts to motivate the workers, as well as a long list of changes made to the equipment.

Although they were both sincere and passionate about the statement: "We have done everything we can to improve the production"—they were very wrong. They *had not* done everything they were able to do, as you will "see" in this story.

Specifically, when we measured and posted the cycle time, it dropped from 6.65 seconds to 5.5 seconds, with no other process improvements. Two conclusions are inescapable.

- Once the workers "knew" what management wanted and once the management allowed them to "see" what was really happening, the rate improved by more than 18 percent. In rough terms, that meant we could now produce the same volume in one day per week less. Our fixed costs had now just dropped 18 percent per unit. Isn't that incredible? Consequently, we could conclude, with certainty, that the workers were both capable and motivated to improve the production.

- The second conclusion is a bit more uncomfortable to those in management. That is, the management had not "done all they could do," actually they had done all they knew how to do, but still, they were the obstacle that prevented the improvements.

Let me refresh your memory of some information contained in other chapters. A review and deeper understanding of these comments may assist some managers in their efforts to remove roadblocks that are hindering the ability of others to perform.

Now let me put these two quotes into context with this story.

First, the problem of low production could only be measured in a meaningful way, after a full day of production—a full 24 hours late. Then, they would discern, as was normally the case, that the production rate was low. But how did they numerically describe the problem of low production. They could not tell if the losses were from quality problems, materials delivery problems, machine availability issues, or poor cycle-time performance. So their knowledge of "low production" was, as Lord Kelvin would say, "... of a meager and unsatisfactory kind." Their knowledge *was* meager, to say the least, and it was unsatisfactory

> **"T**ell me how a man is measured and I will tell you how he will behave.**"**
> **Unknown (from Chap. 17)**

> **"I** often say that when you can measure what you are speaking about, and express it in numbers, you know something about it; but when you cannot express it in numbers, your knowledge is of a meager and unsatisfactory kind ...**"**
> **Lord Kelvin (from Chap. 13)**

to solve these problems, for certain. The same could be said about their "knowledge" of cycle times.

Second, they had failed to explain to the operators what the cycle-time goals really were. They had told them to improve production—that is, "work better"; and improve the cycle time—that is, "work harder." They had done this and expected the operators and the process to improve, without giving them any meaningful way to determine if they had done either. They had no meaningful way to measure the performance of the system and the performance of the individual.

Behind each of these requests—"work better" and "work harder"—was some management belief. These beliefs, respectively, were:

- Since we have low production, the workers must not be performing well enough, so we need them to "work better."

- Since cycle time is clearly too slow, operators obviously must apply themselves better, thus they need to "work harder."

Both of these beliefs were wrong, as has been shown. The two exhortations of management were heard, but the operators were virtually powerless to make the corrections they so vigorously sought. What was needed was not more effort by the worker but rather more effort by management. What management needed was a *cold, hard, dispassionate, honest, introspective* review of the system, and especially their role in the system. To their credit, they did this. They did it well and they did it rapidly. Once this was done, the beast of Rapid Response PDCA was unleashed and progress followed immediately. Furthermore, the management learned, at this time, and only at this time, that they had not "done all they could do to improve the production."

This was the case here, and nearly 90 percent of the time, this is the case for everyone.

In a nutshell, there was no lacking on the part of the rank and file workers. As we so often find, the system was deficient. Recall that Deming says that 94 percent of the time (or more) the system is the problem, not the people. Once again we found that the system that management had created was deficient. Hence, to improve this system, the leadership and involvement of management was needed. But, let me not be too critical of this particular management since, quite frankly, I found them to be open, honest, very hard working and sincere in their efforts to improve not only the plant, but all who worked there. They worked long hours and applied themselves fully. There was no lack of trying on their part.

> **"A**dopt the new philosophy. We are in a new economic age, created by Japan. Western management must *awaken to the challenge*, must learn their responsibilities, and take on *Leadership for Change.***"**
>
> (Point 2)
> W. Edwards Deming

Their weakness was inadequate awareness.

We are all blind to certain things, at certain times and in certain places. Often it takes some outside influence to help us "see" those things we are not aware of; those things that are our blind spots. Once this management group was able to "see," we were then able to make great progress.

Deming spoke of this in his writings, so let me quote him here using his 2nd and 12th Obligations of management, both taken from *The Deming Route to Productivity and Quality* by William Scherkenbach (CEE Press Books, 1988).

In this case, as is always the case, leadership was needed to improve this situation. Again, I cannot say it better than Deming, so I wrap up this section with some words of his (Point 10) also from *The Deming Route to Productivity and Quality.*

This story-within-a-story shows the manifest power of transparency. Once the workers could "see" what was wanted through clearer goals, and once they could "see" what they were doing through better metrics, they simply performed better and the system improved. The workers were already motivated; they simply needed the proper tools and then they could make the system perform at a higher level.

This is their story.

A story that is repeated way too often.

At some level, this is your story as well. Don't forget it.

> **"R**emove the barriers that rob the hourly worker of this right to pride of workmanship. The responsibility of management must be changed... .**"**
>
> (Point 12)
>
> **"I**nstitute leadership. The aim of leadership should be to help people, machines and gadgets to do a better job. Supervision of management is in needs of overhaul, as well as supervision of production workers.**"**
>
> (Point 10)
> W. Edwards Deming

Back to Transparency

Just how did the greater level of transparency work for the future activities? First, now the hourly production was more accurate and we could determine if we were performing as planned. At the top of the hour, the supervisor would enter the production rate, which was now within 48 units of being exact; or accurate to within 4 minutes of the schedule. Second, if we were off schedule, we could look at:

- The rejected product segregation bins to see if we had a quality problem
- Look at the *Andon* log to see if downtime had been a problem
- Look at the cycle-time information to see if the process was performing to cycle time

Generally, once we had this information, it was easy to focus our attention on the specific problem machine or issue, for Rapid Response PDCA.

We were not done improving the transparency of this system, but we had been able to improve significantly. Earlier, we were trying to solve problems 24 hours after they happened and having very low success rates. Now we could find out nearly all the data we needed to solve most problems in a 10-minute window. We had made significant progress, and it was largely due to the transparency we had built into the system.

Transparency and Imagination

Transparency is one of those concepts that is truly ripe for development as part of Lean Manufacturing—we are only limited by our imagination. With transparency, along with the SMED (Single Minute Exchange of Dies; quick changeovers) and *poka-yoke*

> **"I**f you see in any given situation only what everybody else can see, you can be said to be so much a representative of your culture that you are a victim of it."
>
> S. I. Hayakawa

technologies, there simply is no end to what we can develop to fully exploit these tactics. Just keep two concepts in mind.

- We want "transparency" to be able to distinguish and tell us that something has changed.
- We want "transparency" to show us the necessary information, immediately at hand, in order to implement Rapid Response PDCA.

Transparency is that way of showing the information such that these two aspects of process diagnosis and process improvements can be accomplished.

What Is Process Gain?

Most process gains are achieved by either reducing the variation in the product, the process, or both. This variation reduction will then reduce waste and produce gains that are manifest as higher yields, shorter cycle times, or greater uptime, to name a few of the typical manufacturing plant gains.

So, to achieve the gains, we need to reduce the variation, and there is a specific approach that can be made to reduce variation. We can work with:

- The product
- The raw materials
- The process equipment
- *Poka-yokes*
- Process procedures

Simplify the Product

The greatest leverage in reducing variation is to simplify the product. In one case I can recall, we were working on improving yield on an electronic control unit (ECU) that had 13 functions. This ECU had over 300 components populating an 8- × 8-in printed circuit board (PCB). The next generation of ECU, although it had 42 functions, required only 46 components and the PCB was 4 × 5 in.

Technology had allowed this design simplification and, of course, the processing equipment was both dramatically reduced and simplified as well. For example, in the solder application process, done with a screen printer, the PCBs were initially printed in a pair—that is, two PCBs per panel were printed. After the design simplification, six PCBs per panel were produced. Before the redesign, there were five placement machines in series. After the redesign, the new version required only three placement machines.

This is just part of the impact, but it is easy to see that more units were produced in shorter times using less machinery with less investment. In addition, space requirements were reduced as was future maintenance. Also, as expected, when the new process started up, initial process yields were greatly improved over the earlier design. This type of product simplification is the most powerful, but in the typical manufacturing environment, unfortunately, it frequently is not possible. More often than not, you have to deal with the product you now have.

Reduce Raw Materials Variation

Often the incoming raw materials have significant variation and this is a problem of varying degree to everyone. However, many manufacturing firms work with the suppliers at the process level to improve their processes. They use tools such as 8Ds, Supplier Certification Programs, and some of the most forward-looking producers have good Supplier Support organizations. Supplier development was a key tool used by Ohno to improve his manufacturing system, but he really did not implement this until nearly 20 years after he began his quest for quantity control. Very likely, this will not be a large issue in your facility for a few years. When it is a priority, luckily, there are a series of good books on supply chain management, so we will not go into detail here on this topic.

Simplify the Process

The second most powerful technique is to simplify the process. Most often, the effort is focused on reducing human interaction. Frequently, this means you will use robotics or other types of automation. Often, and it is especially true for a tier-1 automobile supplier, this type of simplification is not practical. First, there is the cost issue of the capital improvements, and beyond that suppliers are required to jump through a whole series of hoops that tend to discourage these types of improvements. These hoops, known as process validation, which are mandated by the typical automobile customer, are so costly and time consuming that even if the change is warranted, it is often not done due to the rigors involved. In addition, these changes also open the door for the topic of price reductions. All in all, process simplifications of any magnitude, such as converting manual operations to robotics, are not done very often.

Poka-Yokes

Poka-yokes (see Chap. 12) are powerful tools and should be fully exploited before other procedural changes are implemented. I find it curious that *poka-yoke* technology is not used more often. It does not help that I find very little in-house training being done with *poka-yoke* technology, although there are several good references in the literature. Other than a little periodic maintenance required of most *poka-yokes*, I see very little justification for their underutilization. The beauty of *poka-yokes* is in their simplicity and effectiveness. We can give numerous examples of *poka-yokes* that saved $10,000 per dollar invested. That kind of investment is hard to beat.

Standardize the Process Procedures

Unfortunately, the most common type of improvement comes about by improving process procedures. What this generally entails is an attempt to reduce variation through the use of better work descriptions. But in the end we still rely on the human element to perform well. I used the word "unfortunately" in the first sentence for two reasons.

- First, the human element is often the greatest source of variation and we are simply trying to do it better, not really differently.
- Second, it is unfortunate because, often times, even with the most imaginative engineers and managers, that is the best we can do. In fact, this is the most typical approach taken in the manufacturing world to reduce variation.

Short-term improvements are often realized but the larger problem is *"how to sustain these gains?"* That is the subject matter of the rest of this chapter, so we can realistically sustain the most common of our improvement techniques—improving the processes done by humans.

The Prescription

A Five Step Prescription exists on how to sustain the gains. It is not amazing, but once implemented, it is always effective. This prescription includes:

1. Good work procedures
2. Sound training in the work procedures
3. Simple visual management of the process
4. Hourly and daily process checks by leadership and management, LSW
5. Routine audits by management

Now let's look at the prescription, one element at a time.

Good Work Procedures and Standards, Reflective of the Facility Goals

First, we must have good work procedures, including good job breakdown sheets (JBS) to describe the sequence and details of the work needed to be performed along with checkoff lists, startup procedures, maintenance procedures, and standard work to name a few. In short, all the procedures are necessary to run the business and do it efficiently and effectively. There is a lot of good literature on writing good job instructions so I will not belabor this point, except for two points of major concern. The work instructions and standards:

- Must be written in behavioral terms
- Must be auditable

People Trained to the Standards, Including Being Reality Tested on Performance to the Standard

Next, we need to train the employees on these standards and procedures. Again, there is a lot of good material in the literature; however, the very best I have ever used are the materials developed as part of the Training within Industries (TWI) Services program in 1944. The materials were developed to train workers during World War II. Quite frankly, even though the materials are over 70 years old, they are still the best format for training in repetitive jobs I have ever seen. You can get the original TWI documents in pdf format from several sources using just an Internet search. You can also find them in Word format, likewise on the Internet. There are a number of excellent TWI trainers available to guide you through the Train-the-Trainer program material and are available to assist you in deploying this training in your organization. Toyota was trained using these methods when they were revitalizing Japan following WWII and even today the training model used at Toyota follows the TWI model almost verbatim. Whatever training model you use, the training, once completed, must be evaluated using a practical test done at manufacturing conditions, like in the TWI format. Let us say, for example, we have trained operators to visually inspect parts for attribute characteristics, prior to final packaging. These operators need to be tested—in this case, with an attribute Gauge

R&R study—and it must be done at line conditions, at *takt* time. Written tests and video training are okay, but in the final analysis the purpose of this training is behavioral modification, and that modification must be tested; *there is no substitute for this.*

Simple Visual Checks to See If the Standards Are Met...Transparency

At the jobsite, we now need simple visual checks to see if the objectives are being met. This is a key element of transparency. If the operator is to meet a specific cycle time, how can he tell? Is there a visual readout that shows his performance? If not, how does he know? If routine maintenance needs to be done, can that be audited by the manager walking by? In this case, the transparency allows the operator to discern if his process is performing properly and respond accordingly if it is or is not. The visuals in this case, used in this way, are a short-term "check" on the process result and consequently a "check" on the hypothesis; that is, "if we do these specific things, we will expect this specific result."

Leader Standard Work Done by Leadership and Management, Done Weekly, Daily, or Even Hourly

In my experience, most companies do the first three steps with some degree of efficiency, and here in steps 4 and 5 is where the system to sustain the gains often breaks down. A key element of *Hoshin–Kanri* (H-K) planning is daily, weekly, and monthly management review. More and more, I find it less common for even midlevel managers to go to the floor daily. These trips to the floor are absolutely necessary if continuity is the concern. The countermeasures to this problem are the implementation of both H-K planning and LSW. One purpose of LSW is policy deployment and here the transparency will be used as a process "check" to see if the cell, line, or value stream in performing on a longer-term basis. The operator is checking each cycle, each pitch, or each hour.

The leader/manager is checking each hour, each shift, each day, or each week to see if the process is performing as designed. He is also making a "check" on the process rate but on a longer interval. In so doing, he is making a longer-term "check" on the hypothesis; that is, "if we do these specific things, we will expect this specific result," in the longer term.

> **P**oint of Clarity   Managers do not get what they EXPECT, they get what they are willing to INSPECT.

Managers must be aware of actual operations and operating conditions; they must not only be lean competent they must be "lean aware." Also being on the floor keeps them in touch with the people. It helps the manager evaluate not only what is happening on the floor but helps him evaluate the information he gets from others, which is generally the bulk of his information. And finally it helps the manager learn about the people, about the process, and about the product. It is a critical part of Deming's 14 Obligations of Management. Toyota thought it was so important they gave it a name and made it a basic principle in The "Toyota Way, 2001," (The Toyota Motor Corporation, April 2001) their guiding document to all employees. It is, "We practice Genchi Genbutsu...go to the source to find the facts, to make correct decisions, build consensus and achieve goals at our best speed."

Sometimes the manager may have a specific agenda, sometimes he may just want to observe to see what happens—but there is no substitute for his or her presence.

The most common management practice is to show up on the floor only when there is a problem. Consequently, when the workers see the manager, they know that something is amiss. There is nothing like this to create an atmosphere of concern which then evolves into a culture of fear and secrecy. The manager must be on the floor because it is part of his normal job—not only to investigate when things go wrong, but also to investigate when work is going well. He can then tell that the methods to standardize are working.

In addition, I have found the very best at this go to the floor for two other reasons. Really good managers have an ability to go to the floor and just listen to the rank and file workers and be able to learn about the process—directly from the horse's mouth. It is the "Ask a question and then just listen" skill that only a few managers possess. Another technique that pays high dividends is to have on your calendar the birth dates, company service dates, and wedding anniversaries of all your employees. These provide reminders to go visit the workers on a more personal basis. Then the manager cannot only get to know these people better, he can stop by and discuss their first day at work, for example. This type of touch needs to be sincere, and if it is not your style, don't do it, you will only come across as phony and, hence, gain nothing. However, if it is your style, it pays huge dividends. In the future, these people will be far more willing to "tell it like it is" rather than the "what I think you want to hear" mode so common in most plants.

Routine Management Audits...to Teach the Managers, to Check the System

Again, the H-K planning model includes the concept of management audits. The general paradigm of audits is to check the system to see if it is working, and most auditors are not really happy unless they can find a few things done wrong. It seems to quench their thirst and convince them that they have done a good job. In my experience, nearly all auditors have that desire: to catch people doing things wrong. Those audits are not really helpful, much like the standard itself. On the other hand, the objectives of these Routine Management Audits are much different and are twofold:

- To teach management
- To check the ability of the system to meet the policy

> **P**oint of Clarity   Seldom is the person working in the system, the real problem...nearly always the real problem is the system which the person is working in.

It is my experience that the major opportunity for improvement in manufacturing lies not in improving the people; rather, it lies in improving the systems. Let me say that again, it is my experience that the major opportunity for improvement in manufacturing lies not in improving the people; rather, it lies in improving the systems. In fact, in controlled studies we have done, we routinely find that 85 to 95 percent of all variation is system created. More simply said, most of the variation is because people are:

- Using the raw materials they are supplied
- Running the machines they are supplied
- Following the instructions they are supplied
- Working in the environment they are given

Consequently, most of the time, when we wish to make progress and when the analysis of the variation is complete, it is the raw materials, machine-operating conditions, work instructions, and work environment that must change, not the people. Keep in mind that the selection of raw materials and machines, the writing of work instructions, and the creation of the work environment are all done by management. These four things largely define "the system." (Recall the definition of variation in Chap. 11: *the inevitable differences in the individual outputs of a system*.) Unfortunately, and all too often when problems appear, the managers do not have the necessary understanding of variation to respond properly, and all too often they inappropriately focus the attention on the workforce when it is these systems which they have created that must change.

The managers, particularly the middle managers, have a lot of emotional investment in these four aspects of the system and do not really want to change them—after all, they created this system. To change this system is then an admission that they had created a defective system. The truth is, no system is free from deficiencies—all can be improved.

All kinds of cultural forces protect the status quo, yet the status quo must change, and frequently the middle managers do not see or do not want to see the changes necessary. For this reason and others, I have found that there is no substitute for top management presence on the production floor. These audits provide just such an opportunity and provide the teaching of the managers and the middle managers as well.

There is yet another benefit that is achieved when the managers perform these audits. Here I do not mean, "make sure these audits get done," I mean to do them. I mean:

- To review the standard
- To compare the actions to the standard
- To draw conclusions and develop corrective actions
- To follow up that the work was completed
- To document the audit

These audits are, by definition, non-delegable. The benefits are tremendous here, and they all have to do with improving communications. Let me state just a couple benefits, but the list goes on.

- It provides a real-life connection of the manager to the floor and the workers know the manager is committed, not just involved. The managers will be seen as real workers, not just shiny pants looking at spreadsheets in their office.

- It provides an understanding, at the worker level, that their problems are understood by top management, and that they, the workers, are not totally insulated by middle management. The workers can easily see that they have a conduit of communication to the top managers.

- By getting involved and changing or accepting the systems they are auditing, they are also communicating and confirming their standards.

Through these audits, system deficiencies can be found and corrected and standards can be confirmed. They complete the PDCA cycle for the system. Few companies do management audits, and even fewer have an annual audit policy. This lack of an auditing system—and the lack of management audits in particular—hinders system improvements and is a major impediment to sustaining the gains.

Chapter Summary

One of the key foundational issues is the ability to sustain the process gains, once they are achieved. This is a key issue to maximize process gains over time, but strangely enough few companies do this well. The first problem to sustaining the gains is to make sure there is, in place, a system to ensure that losses become obvious, when they occur. The technique used to make the losses obvious is called transparency. The two strongest techniques to sustain the gains, once they are found, are through product and process simplification. Although these are strong techniques, for the typical lean initiative, they are not practical. The next best method is to standardize through the use of *poka-yokes*. This is also a very strong technique, but this is also often underutilized. Finally, the most common but least robust technique to sustain the gains is to standardize the process procedure. A Five Step Prescription exists on how to standardize the process. You will not only find this prescription amazing, but once implemented, you will find it to be effective as well.

CHAPTER 16

A Lean Transformation

Aim of this chapter... to first explain the incorrect mindset that guides most implementations to either partial or total failure. Then we will discuss the steps to take at the corporate level to organize for a lean transformation across the entire corporation. You will see the emphasis on the need for management education and learning-by-doing as a means to create lean competent managers and leaders and an Alpha Site to develop and prove the techniques needed. We will then describe the prescription for the individual lean value stream transformation.

> **"F**ailing to plan... is just planning to fail."**
>
> Unknown

You and Your Team Must Have the Correct Mental Model of Lean

Pay attention... read very carefully and thoroughly...focus on the next few paragraphs. If you heed these next few paragraphs, your lean transformation will have a decent opportunity to succeed; ignore the advice herein and you will doom your effort to failure.

Failure to understand this concept has caused innumerable failures in lean implementations. The typical approach as outlined in Chap. 3: with an emotional, hurried decision to undertake a lean transformation; led by a mid-level manager with little lean experience; driven by a staff organization working for the mid-level manager... is the formula for failure. Although this approach sounds very logical, there is a very specific reason it will fail. It is built on the foundation of a very flawed mental model.

> **P**oint of Clarity Lean is not just a new "thing"... it is a "completely new way" to do the things we must do anyway.

Lean Is Not a New "Thing"

Supporting this belief that this system of deployment will work is the underlying mental model that Lean is a new technical "thing" like a new version of Enterprise Resource Planning (ERP), where the stated approach actually may work very well. However, simply stated, for a lean transformation, this is a bad mental model. In fact it is a lethal mental model. If you accept this mental model, it will lead you... logically and disastrously so... to design a lean implementation that is destined to fail.

It is a bad mental model for at least three compelling reasons. First, the ERP conversion metaphor only deeply affects a few functional groups; among them are planning, IT, and to a lesser extent production. But more importantly, it really does not deeply affect management as a group. Implement the new ERP model and managers still manage much like they did, before the conversion. A lean transformation, on the other hand, deeply affects everyone and it affects management the most of all. It makes their job bigger, better, and more influential. At the end of the day, the lean transformation changes everything everyone does in some way or another. Second, the ERP is basically a technical solution and although some job descriptions and training requirements are affected, it really only affects some staff personnel in terms of behavioral modification. The line organization is really not affected very much. They get the production schedule and execute it, and it does not really matter too much if it is from an MRPII, ERP, SAP, SCM, XL spreadsheet, or whatever alphabet soup acronym was used to create it; the production schedule is the production schedule. It basically says who is going to do what next. On the other hand, the lean transformation is not only a list of new and modified technical "things"; in the fullness of time, it will change the way everyone executes their jobs. Top to bottom in the organization, from the CEO to the floor operator significant changes will be made. Also across the entire organization chart from the line management to all the staff groups, all these roles and responsibilities will also evolve and change. And as a direct result of these changes, you will have transformed the entire organization and now you will have a facility that is a better money-making machine, a more secure workplace for all, and the supplier of choice to your customers. Third, as a result, while the ERP "thing" will cause some minor changes, a lean transformation is a total cultural transformation in that not only the behaviors will change but so also will the underlying beliefs, values, language, and artifacts change.

Think of some of the "things" your facility will need to do. These are the "what needs to be done." With or without a lean transformation, for example, you will need to plan, organize, schedule, train, and problem-solve, to name a just short a list of items. Just to survive in the world of manufacturing, you must do these things.

Lean Is More About "How" You Will Do These "Things"

Although much of "what" you do will not change, in the manner of "how" you do each of those things—after a lean transformation—there will be many changes. In many cases you will do them in a "completely new way." For example,

- Top-down autocratic, my-way-or-the-highway planning methods that not only force but also facilitate and accept "numbers gaming" … change to H-K planning, using real data, with extensive use of catchball and job-appropriate follow-up.

- Organizations which comprise functional silos bent on suboptimization using silo-specific goals…change to being organized by values streams working to supply value to the customer and supporting the value-creating workers. Organizations that are "staff driven" and have self-serving staff goals that supersede facility goals…change to being "staff supported" with Lean, focused staff groups being good suppliers and supporting the value-creating line organization.

- Scheduling that used to be this monolithic monthly monster that does not represent reality … changes to real-time scheduling on the floor using such tools as *kanban* and the three-part inventory system.

- Training using the ineffective buddy and word-of-mouth training systems with no decent documentation … change to using the TWI methods with real-time job breakdown sheets which are largely prepared, tested, and improved upon by the workers themselves.

- Problem solving which is more often some mode of firefighting, done only by supervisors and engineers always seeking short-term solutions because the processes are so unstable and chaotic...changes to a system of real-time PDCA which is adept at finding and removing root causes, and it is done by everyone in an environment of predictability.

So, you see, Lean is not just a new "thing";...it is a "completely new way" to do the things we must do anyway. It is this correct mental model that also understands that the "completely new way we do things around here"...is just the operational definition of installing a completely new culture (see Chap. 4).

> **P**oint of Clarity...a culture is "just how we do things around here."

The Lean Transformation...at the Corporate Level

Leadership Growth and Learning by Doing

At the enterprise level the lean transformation consists of two major paths operating in parallel. First is the creation of a critical mass of lean competent leaders at the executive level, starting with the CEO. Education of this group is paramount as they absolutely must guide, model, and support the behaviors they wish to see in everyone else. In addition, teaching by this group will largely be the method to deploy the lean transformation throughout the Corporation. Second, and in parallel with this, we will create an Alpha Site value stream where we will begin a full-blown lean transformation, led by the plant manager and guided closely by the sensei.

The Alpha Site Concept

This will be the place the Lean Leadership Team (LLT) in conjunction with the plant change agent and your *sensei* can use to develop and test the plan and the model you wish to use to transform your entire organization.

Selecting one value stream as an Alpha Site is a countermeasure to "Roll Out Design Error No. 5—Plan the implementation across multiple value streams: or worse yet, across multiple plants," as discussed in Chap. 3. Using an Alpha Site has several benefits. First, a lean transformation is a huge change. Since many of the responses to our changes are unknown and yet others are unpredictable, we wish to undertake the first transformation under a controlled environment where we can observe it carefully and respond quickly if need be. Second, it will be the test ground to train our first group of managers so we can better understand what future training other managers and leaders will need. Third, it is also the site where the LLT can go to observe, learn, and practice the lean skills they so often need. Fourth, by transforming your Alpha Site and allowing your lean transformation model to unfold, this provides a time window of opportunity for the LLT to further develop their lean competencies so they will be better prepared when the Corporation roll out occurs. At that time they will be under much broader pressures to exhibit the Six Skills of Lean Leadership. Fifth, as you roll out your lean transformation to the rest of the value streams, this Alpha Site can be the place you test new concepts.

How to Get Ready, Get Organized, and Get Going at the Corporate Level?

Once it has been decided that you wish to undertake a lean transformation. The first phase is to get organized and get going at the corporate level. This will consist of six basic steps:

Step 1: The CEO should hire an outside resource, a *sensei*, to assist in guiding the transformation (Plan).

Step 2: Select the key top level managers and create a steering committee. This will be the Lean Leadership Team (LLT) (Plan).

Step 3: Grow the LLT (Do).

Step 4: Create an Alpha Site to learn by doing (Do).

Step 5: Evaluate the status of the management lean skill set and status of the Alpha Site (Check).

Step 6: Continue the effort at the Alpha Site and expand to the next value stream when ready (Act).

CEO to Hire an Outside Resource (Plan)

The selection of an outside resource is a very important step. Pick someone who has prior experience in a lean transformation. The advisory function, the *sensei*, cannot be done by a good manager who understands lean techniques. This must be done by a seasoned veteran in lean applications. If you, for some odd reason, have decided not to employ a *sensei* and are doing this with a seasoned manager with solid engineering skills, but who is not seasoned in Lean, go hire someone with experience to assist you. I have never seen a good plan come out of anything but a person seasoned in Lean. In the absence of a *sensei* or a consultant, *quit this effort and move onto something else.* I cannot state this more clearly. The likelihood of partial failure is 100 percent, and the likelihood or total failure is significant. The bottom line is that management is not committed (they will fail questions 2, 4, and 5 in the commitment test).

Select the Key Top-Level Managers and Create an LLT (Plan)

The leader of the LLT is the CEO; the *sensei* is the advisor to the LLT. The CEO is also the leader of the transformation effort (see Chap. 3, Roll Out Design Error No. 2). The LLT will then have five "first tasks":

1. Define, document, and publish their specific task, often this is a Vision and a Mission Statement (Plan).

2. Create, document, and publish a Corporate True North (Plan).

3. Complete a thorough evaluation to determine the viability and need of a lean initiative. This is a countermeasure to "Roll Out Design Error No. 1—Proceed without a factual review of the situation" in Chap. 3 (Plan/Do).

4. The LLT should develop a routine meeting schedule (Plan/Do).

5. The LLT should create a written plan with actions items, responsibilities, and due dates. (Plan/Do)

   a. At all times the LLT should have a plan. Initially it will have three objectives:
      i.   First is management growth and becoming lean leaders.
      ii. Second is Alpha Site development.
      iii. Third is to develop a timing and methodology to roll out the lean initiative to the rest of the corporation.

The purpose of the LLT is to provide the support, guidance, direction, and proper modeling, so the transformation can proceed first at the Alpha Site and then to the rest of the corporation. Since Lean is a process of continuous improvement, it is likely the LLT will become a permanent management group.

Beware.......

At this point, almost always the LLT wants to create some kind of a staff group to support the lean transformation; they might call it the lean support staff, or the lean office. Whatever it is; it is a really—really bad idea. Don't do it. Except for a few technical circumstances, staff support is not necessary and certainly it is not necessary at this juncture in the transformation. Worse than being superfluous, it is destructive in at least three ways. First, as soon as this staff group is formed, the management sees them as the source of teaching and technical expertise for the transformation and they back off and let others do what they themselves need to do. Both of these items, teaching and expertise supplied by the staff group, are a little like using narcotics. They make you feel good for a little while, but will surely destroy you in the long run. Second, for a lean transformation to be sustained, the vast majority of the teaching and direction must come from the lean leaders to their direct reports and from nowhere else. Staff groups can cause great confusion regarding work priorities. Almost always they cause the line organization to lose clarity of purpose. Third, the lean expertise that is truly learned, ultimately comes from doing; we "learn by doing." So if the staff group is the one who is making the changes, they are the ones learning and growing rather than the line production personnel. At maturity, the expertise must reside in the employees at the facility and especially among the management and preferably with the line personnel.

Now, I know this is a "weird" thought and certainly countercultural to a lot of current activities and advice you may get. My advice was acquired from 45 years of industrial experience where I have seen the benefits of good staff support; and likewise I have seen the deleterious effects of staff support that is poorly managed. Lest you think I am "too weird," let me give you some supporting documentation from those wiser and certainly better published than I.

"The unchecked growth and excessive power of service staffs is considered by practically all our foreign critics to to be a serious weakness of U.S. industry, and a major cause of its poor performance," Peter Drucker.

"One reason why Japan has been able to invade U.S. markets successfully in one industry after another, is that in the U.S. staff positions have far outstripped line jobs in number and in influence," Arch Patton.

Schonberger, in his seminal book states: "The wastefulness of staff growth and overspecialization is not to be tolerated any more than inventory buildup and the waste of producing defective products"...he goes on to summarize..."Industry does not need a lot of improvement programs coordinated or run by specialists; production managers and workers can do it themselves" Richard Schonberger from "Japanese Management Techniques-Nine Lessons in Simplicity," The Free Press, 1982).

Grow the LLT (Do)

This will be the first challenge for the management team, and it must start immediately. Rarely, at the initial phases, is there a high level of lean experience at the top management level. This must be corrected as we must create a critical mass of lean leadership. This only comes about through individual growth, and this growth occurs through two mechanisms: awareness and practice. The awareness can come from either study or observation, but the practice is just that, practice.

They Must Become Lean Competent…They Must Become Good Role Models   The practice of lean techniques by the top management is of paramount importance for two reasons. First, these managers will be the teachers to their direct reports. Second, if the top management is not "walking the talk," the transformation will falter. (see Lean Killer No. 1)

> To explain the manifest power of modeling behavior by top management, I wish to share with you a quote from a friend of mine…..."If you want people to do it right, you have to show them how to do it right *every time*. If you want people to do it wrong, you only have to show them once." …Robert Simonis

They Must Focus on Learning by Doing   To learn Lean, you must practice Lean. To practice Lean, you must know the basics of Lean. For those at the executive level who are just starting with Lean, this presents a major challenge. However, each of the executives must practice lean techniques at every opportunity; and the opportunities will present themselves. Some techniques that every executive can utilize and properly model almost at day one include the use of "catchball" in goal attainment, using the Five Questions of Continuous Improvement, practicing *genchi genbutsu*, properly performing management audits, more use of questioning, and less use of directing. Along with learning and practicing the Six Skills of Lean Leadership, there is a litany of lean activities that the LLT can exhibit.

In the very beginning to gain a better intellectual understanding and awareness, a good place to start is to have topical reading available for the LLT along with a completion schedule. Book reviews are another good idea. Also to broaden your awareness, your *sensei* can prepare some general trainings and visits to existing lean facilities. Lean competence at the top levels of management is a significant issue (see Lean Killer No. 1). Additional learning can come about by routinely visiting the Alpha Site as well as attending seminars and workshops. Spare no effort on this topic. These top managers are almost surely very busy but they have chosen to make their organization Lean so they *must* become lean competent. This cannot be compromised. If you are the CEO and sense the management team lacks the commitment to become lean competent, flush this out immediately. They must be given an ultimatum; there is no room for half-hearted commitment. A CEO friend of mine once said it very succinctly when speaking about management engagement, he said, "we are not looking for volunteers."

How to Accelerate the "Learn by Doing"?   The problem that most managers encounter is that to become lean competent, you must "learn by doing." And most managers are not in a position to learn about *kanban* calculations, detailed balancing calculations, or to practice TWI-type training. This stymies them. It need not stop them. For example, they may not be able to design a *kanban* system, but through several methods they can study the benefits and become versed in how *kanban*, for example, can help the business system.

Furthermore, there is still a lot they can do to both "learn by doing" and modeling the correct behavior. As they learn from their readings and visits and they learn from the Alpha Site, each manager will have numerous opportunities to become more lean competent. (A number of suggestions are available in Chap. 2.) The critical factors will be whether they will capture those activities to learn, become more lean competent, and model the correct behavior…or let them slide by because they claim to be too busy. If they are not becoming more lean competent by the moment, then that should be a very special topic for the next meeting of the LLT.

Create an Alpha Site or Value Stream to "Learn by Doing" (Do)

Early on, the LLT should pick one value stream to use as an Alpha Site. The Alpha Site must be carefully selected. First, and most important—by far the most important— factor to consider is that you will need to have, on-site, a plant manager who is committed to making it work. He not only needs to be committed, he literally needs to be ecstatic to have his plant chosen. His energy, drive, effort, and desire will make or break the effort. He will be the value stream change agent and his energy, or lack thereof, will be catching. You cannot accept that someone else will be the change agent. The production manager or quality managers do not have the necessary position power to make the quick decisions which are necessary in the dynamic nature of the transformation which is needed. Second, but clearly second is the issue of transportability. Given that you might have multiple adequate change agents, you may wish to pick a value stream that has some commonality with other value streams. Work done at the Alpha Site is thus more transportable. Third, and a distant third, is the accessibility to the LLT. Quite frankly success is the most important factor and asking the top management to travel a little while they learn is a small cost when compared to failure. At the Alpha Site a full transformation will be designed and executed so you can test all your ideas and uncover the obstacles to transforming your production facilities. The obstacles to the transformation will be a long list of many items. There will be technical issues; there will be personnel issues and often a whole litany of cultural issues. The list will be long but basically it is a huge compilation of the answers to question three of the Five Questions of Continuous Improvement…that is, "What is preventing us from reaching the desired condition?".

Evaluate the Status of the LLT Skill Set and Status of the Alpha Site (Check)

The first item at each LLT meeting should be a reflection on what has happened since the last LLT meeting. Second is an update on the increased lean competence of the LLT and the growing lean competence of management in general. Third should be a report out on all planned activities, especially those activities at the Alpha Site. There should be a discussion of; what worked, what did not work, and what we need to improve upon? The report out should end with a discussion of next months planned activities. The final item should be a reflection on the meeting itself: what worked, what did not work, and what do we need to do better?

Expand the Effort at the Alpha Site and Expand to the Next Value Stream When Ready (Act)

As the Alpha Site work progresses, the effort will be expanded. The next steps will always be decided upon by the change agent with the advice of the *sensei*. And ultimately, when it is clear what needs to be done for a lean transformation and the corporation is both staffed for and emotionally prepared to take on a second value stream, then and only then should that decision be made.

The Lean Transformation and Corporate Goals

At the corporate level there are really no lean goals yet this strikes most managers as odd. More often than not, this is due directly because they do not understand that "Lean is not a new thing, it is a completely new way to do the things we need to do anyway."

I often get asked, "so how do we know if our lean system is working?". To which I respond, "how do you know if your existing system is working?". Generally some thought on that question alone will allow most managers to clarify in their mind that "Lean is not a new thing ... but it is a completely new way to do the things we need to do anyway," so when transformed, your organization will be able to produce, "better, faster, and cheaper." Hence the way you tell if your lean system is working is to measure what your production system needs to do to be competitive, to survive, and prosper. It really makes no difference if you have the old classic mass production facility or the latest and greatest lean facility; it still must meet the same performance metrics to survive and prosper. Generally this can be displayed on a balanced scoreboard of metrics such as safety; production; delivery performance; internal quality; external quality; and productivity coupled with an honest, dispassionate introspective review of how well you are accomplishing your mission and how well you are approaching your vision.

So How Do I, as a Manager, Use This Information?

Well, first do not create some funny goals like "let's do 25 VSMs this year and set an objective to make $100,000 of bottom line improvements per VSM." As logical as this sounds, it is just another version of the "tools approach" described in Chap. 3. It may give you some short-term benefits, but it is a long-term losing strategy. Second, create and manage goals, the True North for each value stream that are first created via *Hoshin–Kanri* methods and second stretch the imagination and creativity of the workforce.

An Overview on How to Implement Lean at a Value Stream

Much like the transformation at the corporate level, implementing at the value stream consists of two major paths operating in parallel. First is the creation of a critical mass of lean competent leaders at the value stream level, starting with the plant manager. Education of this group is paramount as they absolutely must guide, model, and support the behaviors they wish to see in everyone else. In addition, teaching by this group will largely be the method to deploy the lean transformation throughout the value stream. Second, and in parallel with this, we will begin the changes in the value stream led by the plant manager and guided closely by the *sensei*. This will be where the first transformation occurs and is where the managers will learn and practice the lean skills they so often need. These two factors, management lean maturity and knowledge gained from the initial efforts, will catalyze the changes so we can problem-solve our way to the Ideal State.

Our transformation of a value stream will be outlined here for ease of understanding. It will be managed and executed like any large project. The action items for this project will be compiled from two sources. The first set of data is gathered from the "systemwide evaluations," while the second portion will be from the "specific value stream evaluation." Both sets of evaluations will be compiled, prioritized, and lead to action items: *kaizen* activities. Together, in an eight-step process, all the evaluations and action items will be included in our plan and documented on some type of project documentation, whatever you are most comfortable with. We normally use a standard Gantt chart.

This is THE PRESCRIPTION on how to Transform the Lean Value Stream.

(1) Step 1, select and prepare the value stream change agent.

(2) Step 2, assess the status of the Five Cultural Change Leading Indicators.

(3) Step 3, Create True North metrics and a balanced scoreboard.

(4) Step 4, perform the systemwide evaluations.

(5) Step 5, document the current value stream condition.

(6) Step 6, redesign to reduce waste.

(7) Step 7, document and implement the *kaizen* activities.

(8) Step 8, do it all over again.

The Transformation Plan Will Be Documented on a Gantt Chart

We have spent little time in this book on project management. Since this book is written to those in manufacturing, like engineers, performing projects is something we assume you are already skilled in. If you are not, a number of excellent references are available. I recommend *Project Management DeMYSTIFIED* (McGraw-Hill, 2004) by Sid Kemp. At a minimum, your Gantt chart—or project management tool, whatever it is—should have certain data. For example:

- Show all *kaizen* activities.
- Show critical milestones.
- Show completion dates and responsibilities.

Step 1: Select and Prepare the Value Stream Change Agent

There must be a person in the plant who is both committed to making the transformation and someone with enough authority to make it happen. Normally the plant manager is the slam dunk choice for the change agent. He must be prepared by the LLT. The LLT needs to make it clear what objectives they have for this Alpha Site. In *Hoshin–Kanri* planning this is the "what we want accomplished" step. The change agent must be made aware of what resources are available and must be given time to digest and think about his task. Almost without exception he will need both a great deal of advice and guidance from the *sensei*, before he can report back on the "how" he feels he can accomplish this task, if he thinks he can. Once this "what-how" catchball process has been completed, and it likely will take more than one iteration, then the change agent, with the help of his *sensei*, must put together a plan to begin the process of transformation of this value stream.

Step 2: Assess the Five Leading Indicators to Cultural Change

Do We Have the Leadership to Make This a Success?

The change agent, with help from the *sensei*, will make an internal assessment of the strength of the leadership skills on the current staff. This leadership assessment almost always identifies some general training that is required. Also, frequently training on how to create, execute, and follow up on a plan is also required; in other words, project management skills.

Do We Have the Motivation to Be Successful?

Motivation is a complex cultural issue as you read about in Chap. 6. Most managers can determine, at least at the gut level, if the organization is properly motivated for this transformation.

Two very important things can be done, at a minimum. First, one of the great motivators is "to be in the know"—that is why rumor mills are so popular. So, to "get them in the know," the lean effort will need to be publicized and all that Lean is should be open and discussed. That puts a stress on a lot of cultures, which are very closed. Characteristics of closed cultures include lots of secrecy, even about the most mundane of topics, and managers who do not appear to be forthright. For example, in these closed cultures, many questions get avoided and there is an unstated but understood air of secrecy that forbids some questions being asked. Those elements will destroy a lean transformation, so they must be corrected. So tell them what is going to happen, when it is going to happen, and what you expect the results to be. Be sure to get them "in the know," and above all else be open and honest.

Second, for the group to be motivated, the leaders must be motivated. This motivation must be obvious by their actions. If the leaders are not motivated—not just motivated, but literally exuberant about the future of this lean transformation—either you have picked the wrong people or they simply do not know the power of a lean transformation. Quite frankly, in the world of manufacturing, there is no cultural change transformation that can create so much hope as the proper implementation of Lean!

Finally, supervisory training in motivation is a must. The topic is very well outlined in Chap. 6, which can be used as a guide.

Do We Have the Necessary Problem Solvers in Place to Be Successful?

Do we have enough and do we have the needed level of training? Make sure you have earmarked the problem solvers and they know their roles. It is common to have a small cadre dedicated to problem solving. Also, generally assigned to each value stream are what most organizations refer to as process engineers. Train and use these two groups fully. If they are not fully trained, have a plan to get there. It is very common to have a lot of the training done by the *sensei* as you progress. I call this Just in Time (JIT) training. Some offsite training is also common. Whatever the current status, this is a critical aspect because the transformation will only proceed as fast as you can solve the problems—and the problems will come rapidly as soon as you start making the big changes when the five strategies to become Lean are installed on the first value stream.

And above all else, teach everyone the Five Questions of Continuous Improvement and be guided by the information in Chap. 7 as you make this assessment.

Do We Have the Whole Facility Engagement Necessary to Be Successful?

Although most managers equate engagement with "want to" on the part of the employee, they are wrong. Almost always the deficiencies found in engagement are because the management team has not properly made sure the employee knows what to do, how to do it, and has the resources to do it. Typically these aspects of job definition and supplying resources are what limit plant engagement. Be guided by the materials in Chap. 8 as you make this evaluation.

Do We Have the Learning, Teaching, Experimenting Environment to Be Successful?

Almost surely you will find gaping deficiencies in this leading indicator. It will be an area-requiring immediate attention. I have yet to have a client that made this a priority initially, yet following the transformation, this topic is always a focus of management attention.

Step 3: Create True North Metrics and a Balanced Scoreboard

  a. Determine critical process indicators.

    i. Usually safety, quality, productivity, delivery, and cost…in a balanced scoreboard.

    ii. Determine how to measure and display each one.

    iii. Create levels of attainment, "these should be" for each goal with a time frame for achievement.

  b. Deploy these critical process indicators via catchball.

  c. Make them "visual."

  d. Implement Leader Standard Work.

Step 4: Complete a Systemwide Evaluation of the Present State

  a. The Five Tests of Management Commitment to Lean Manufacturing (see Chap. 24)

  b. The Five Precursors to Implementing a Lean Transformation (see Chap. 24)

  c. Perform an Educational Evaluation of the Workforce.

The Role of the *Sensei*

Now it is time for the leader and the *sensei* to make an evaluation of the present state of the entire manufacturing system. They will make three evaluations of the manufacturing system as outlined above.

In this step, the *sensei* is absolutely critical. This step cannot be done by a good manager who understands lean techniques. This must be done by a seasoned veteran in lean applications. If you, for some odd reason, have decided not to employ a *sensei* and are doing this with a seasoned manager with solid engineering skills, but who is not seasoned in Lean, go hire someone with experience to assist you. I have never seen a good plan come out of anything but a person seasoned in Lean.

The First Management Commitment Test

This management commitment test is the first and most important of the commitment evaluations. The most important questions are, "Do we have the level of commitment necessary to make this a success?", and "If we do or do not, what are we going to do about that?".

Before we discuss the evaluation, let's explore exactly what commitment is. (In Chap. 20, we have the story of the Alpha Line, which is a good example of management commitment.) A lot has been written about commitment and many confuse it with involvement. Those walking around, making casual comments, and even helping out are involved. Commitment goes much deeper. The best description I have ever heard uses the metaphor of eating ham and eggs for breakfast. In this case, we can say the chicken was *involved*; the pig, however, was *committed*.

Now, back to the evaluation. This can be done in two ways: quick and dirty or thoroughly. I suggest you do both, for different reasons and at different times. First, you and your *sensei* should do a quick-and-dirty evaluation:

- List the key* (see sidebar in Chap. 5, Leadership on key players) players who are necessary to make this transformation work.
- Make the best evaluation you can, using The Five Tests of Management Commitment to Lean Manufacturing.

Make sure you do your evaluations *based on the behavior you see* from the key players. Don't just go by what they say. At this point, this evaluation may be very rough but it will point out any obvious issues and will get you thinking about this issue. At some point, or at several points, individual or group commitments, or a lack thereof, will be significant issues, and it is best to flush them out early. So an early awareness of problems is a good first step.

From this evaluation, develop potential action items. At this point, this is just an opinion, so treat it very carefully. As you proceed in the early steps of the transformation, take careful note of any commitment issues. Be complete and document the who, what, when, where, why, how often, and how much, for each example. However, unless there are obvious and very serious problems, take no action on this list of items at this time. Save it for later.

The Second Management Commitment Test and Group Dynamics

Next, do a thorough evaluation of the management commitment—this cannot be done initially. For this second evaluation to have significant meaning, there must have been some context developed so the key players can more fully understand what types and levels of commitment the lean transformation will require. It is best to do this evaluation after the key players have been trained in your lean transformation. Also, there must have been several major changes made in the basic operation so the key players can understand, firsthand and by experience, the type of changes required. Generally, this can be done as early as 3 months after kickoff, or as late as 18 months afterward.

This is a crucial time in the program. For the first few weeks of program transformation, particularly if the transformation was sold well by the lean leadership, there will be a lot of positive energy and the majority of those involved will be very enthused about the transformation and its future prospects. This is how nearly all groups form and develop—lots of positive energy and seeming agreement early on. Life seems pretty great during this phase, which we will call the "cocktail party phase."

Unfortunately, sometime after starting the program, some tensions develop due to the necessary and resultant changes. This tension is unavoidable. What happens is that people begin to jockey for position, reassess their positions, and question things like, "How can I do this and still meet my goals?". In short, individual goals conflict with group goals and differences surface. Even if they do not, the worry exists that they might. This creates some conflict that was not present on day one. What started as a conflict-free group with seemingly common goals and a large amount of agreement is now breaking down.

It always happens. It is not only inevitable, it is necessary. This is the second phase of group development: let's call it chaos (a term coined by M. Scott Peck). This chaos has developed because the realities of life have been imposed on the group, and differences that they initially did not envision are popping up all over the place. These differences are now creating problems. Furthermore, having just left the cocktail party, so to speak, they don't really know how to handle the problems.

This is the earliest time we would want to do the second evaluation. Without this chaos and some understanding of the chaos, there is no real context to have a deep and meaningful evaluation. That chaos is often felt as early as 3 months after kickoff, but more typically in the 6- to 18-month period afterward. It is at this time that it is appropriate to do the second evaluation of management commitment. This will be a very sensitive evaluation, and it will require good planning by both the lean leader and the *sensei*. All too often, because it is sensitive and brings up problems, most groups work hard to avoid; it is frequently avoided. Doing this evaluation will be a test of the courage, character, and resolve of the lean leadership and the facility's management.

Just so you are aware, two more phases of group development exist. These are called "emptying" and "community." In "emptying," the group must confront the differences, and then by using honest open communication they must "empty themselves of this issues" and using good problem solving, they can resolve the differences. This is easier said than done, however. After accomplishing this stage, the group can then move into "community." In this phase, the group will be able to effectively and efficiently execute its mission as it performs as a community driven to common goals and driven by common values. There is some very good information available on group dynamics. I recommend *The Different* Drum (Simon and Schuster, 1987) by M. Scott Peck and *The Team Handbook* by Peter R. Scholtes (Joiner Associates, 1988). Both are excellent. (In a lot of literature, Scholtes' for example, these four phases are also referred to as "forming, storming, norming, and performing.")

Since the second commitment evaluation may be more than a year off, I have attached it as Appendix A at the end of this chapter.

The Five Precursors to Lean

Background of the Five Precursors The Five Precursors to Implementing a Lean Transformation have an interesting history. In the late '80s and early '90s, when the first U.S. firms were implementing JIT systems, a large number were encountering problems. Many firms were able to reduce inventory volumes significantly but often other problems developed. Frequently, the production rate would drop—this was the worst and also the most common of the problems. Other less serious issues cropped up as well. Once these unexpected problems surfaced, we would then be asked to assist these firms as they tried to work out of the JIT mess they had so carefully managed themselves into. After a few experiences, we found that the reasons these groups had failed could be classified into a few categories. In addition, about this time Ohno's book and several others became available, which more fully explained the TPS (Toyota Production System) and the JIT portion that so many firms were trying to copy.

Upon reading Ohno's book, and with further study of the TPS, it became obvious that the TPS was vastly superior to the production system in most North American firms, even if these North American firms had already implemented the JIT system.

We quickly realized that the majority of the problems were related to this fact—that is, the TPS is a superior manufacturing system, with or without JIT. We also concluded that when Ohno began his quantity (note: i.e., QUANTITY) control and seriously undertook his JIT effort (in about 1961), his production system was superior, at that time, to the manufacturing systems we are working with now, nearly 30 years later. This was not only sobering to most, it was also depressing.

We understood the problem and began preaching that the failing of a JIT effort was not due to the JIT effort itself, but the fact that the facility did not have the needed foundational elements in place to start a JIT effort. These needed foundational elements we named "The Five Precursors to Implementing a Lean Transformation."

Today we find, without exception that all companies must work on one or more, or often all five, of these precursors to have a fighting chance of implementing quantity control measures. However, these issues need not be totally corrected before quantity control can be initiated—often they can be done simultaneously. Let's say, for example, we have a process that has very high variation in hourly production. Although the stability needs to be addressed and it needs to be one of the first things done, we have seen instances where the solution to the rate variability problem was solved by a *kanban* system. So voilà! We get better stability and quantity control at the same time. This is something that is frequently not seen by the novice, but your *sensei* can give guidance as to:

- Which precursors need to be addressed
- What order these precursors must be addressed
- What amount must be addressed

Each of the Five Precursors should be evaluated and the results of this evaluation will often become a significant part of the transformation plan. A description of the Five Precursors follows. Refer to the Matrix in Chap. 24 for clarifications. (Several references will be made to Level 2 and Level 3, for example; these are defined in the Matrix.)

Stability and Quality   High levels of stability and quality in both the product and the processes are the most basic of standards. Ohno says that flow is the basic condition. It is the foundation of the Toyota Production System (TPS). He says this only because he takes it for granted that process stability is a given. (Return to Chap. 10, and reread the section "The TPS Is Not a Complete Manufacturing System.") Absolutely nothing is more basic to quantity control than process stability. So it is necessary to review all aspects of the product and process stability and make a list of items to include in the goals of your lean transformation. First, evaluate and make sure the quality and the production rate of your product and processes are statistically stable. This can easily be done on a simple control chart—most often an Xbar-R or an XmR chart. Check the stability of the production rate: day by day, shift by shift, and hour by hour. If there are instabilities, put them on the list of items to address. For both the product and the process, check each quality characteristic for both stability and levels; evaluate both. Each can be checked using control charts. List all product and process quality characteristics that are not statistically stable. If they are variables data, list all those that have Cpk below your threshold value, usually most organizations start with a minimum of 1.33. If they are attribute data, work to improve the process to eliminate the need to do the evaluation. If this is not practical in the short term, work to correlate the attribute characteristic to variables data, and then strive to reach the threshold Cpk for this as well. If there are processes that do not meet Level 2 criteria, this constitutes a crisis and should be addressed immediately. The minimum goal for 9 months is to have all product and process quality levels to Level 3 on the matrix; Level 4 can be a goal for 18 months.

Machine and Line Availability   Excellent machine and line availability is frequently a very large problem that has gone unattended for years. For those companies without a formal TPM (Total Productive Maintenance) program, it is common for this factor to reduce OEE by 25percent. In practical terms, this means we must run the line 25 percent more than cycle time would predict, with all the attendant costs of running the line. This problem alone will frequently make a product unprofitable. Usually low levels of line availability are due to two major factors: materials issues (usually stock outs or late deliveries) and machine downtime. Frequently, material issues can be alleviated by a lean transformation and the quantity control aspects of the TPS. However, machine downtime is different and usually must be addressed by a concerted TPM effort. If a TPM initiative is not already in place, almost all firms need to develop a new database to keep track of machine uptime and train the personnel in the use and manipulation of this database. There is some commercially available software for TPM and machinery uptime, although I find most can develop a good Excel spreadsheet and make it quite serviceable. Again, if you find in the evaluation that Level 2 criteria cannot be met, this is a crisis requiring immediate action. Regardless, there is usually a large list of materials and machinery issues that are created by this evaluation. The goal to reach Level 3 should be set for 6 months, with Level 4 scheduled to be attained in 1 year.

Problem-Solving Talent   In several sections we have discussed problem solving and the need for problem solvers. The need for these skills cannot be underestimated; it literally is the vitality of the transformation. Furthermore, to complete this five-part evaluation, the skills of MSA (Measurement System Analysis) and SPC (Statistical Process Control) are required. Hence, anything less than Level 2 is an absolute crisis and must be corrected. If these skills are not available on-site, hire trainers right now. A 6-month goal is to reach Level 3, while Level 4 should be met by 18 months.

Continuous Improvement Philosophy   Mature continuous improvement philosophy is something that is often talked about but infrequently reduced to a process that can be taught to and understood by all. A copy of one we created is shown in Fig. 12.1. If the evaluation is less than Level 2, we again have a crisis and it needs to be taken care of immediately. Reaching Level 3 is a reasonable 6-month goal.

Standardizing   Strong proven techniques to standardize is the foundational of all foundational issues. This skill is so important it received in its own chapter in this book: Chap. 15, "Sustaining the Gains." Anything less than Level 2 is an immediate crisis and must be addressed. Reaching Level 3 is a reasonable goal for 18 months.

As part of this evaluation of the Five Precursors to a Lean Transformation, a list of issues must be compiled. This wish list will usually make up the majority of the effort for the first pass of the lean transformation. After the first 6 months, more and more quantity-control efforts can be implemented, but in our experience most firms have a number of foundational issues, which must first be addressed.

Perform an Educational Evaluation

As you compile the action items that need to be accomplished, with help of your *sensei*, start to compile a list of needed skills; it often is large and the list is dynamic. Some will be general and may lend themselves to group training; however, most of the needed skills will need to be administered JIT to the students.

Specific Skills Training   The systemwide evaluations almost always create a very large list of needed training to teach the strategies, tactics, and lean skills. The composite list will largely follow directly from the Five Precursors to a Lean Transformation. In addition, as you do the assessment of the value stream, training topics will almost surely be found. Once combined, these trainings will almost always include problem-solving training, training on statistical tools, and facilitation training for all in leadership positions. In addition, lean-specific trainings in skills such as line balancing, SMED methodology, *takt*, and *kanban* calculations, to name just a few may also be beneficial. It will be necessary to inventory the needed skills and teach them as they are needed to those using the tools.

Just another word on education and training: It should be focused and JIT. For example, during the transformation, train just those people involved. Often, it is not that simple and some people may need to be trained prior to the transformation of their product; it is never perfect. The point here is to avoid the global mass training of individuals that makes good use of the training resources, yet provides the training either too early or too late. Efficiency of the training organization is not of paramount importance when compared to training effectiveness. If there is long time between the training and the use of that training, a large fraction of the learned material will be forgotten. Consequently, it will not be effective and it is a waste, the very item we are trying to eliminate, not create.

Step 5: Document the Current Condition

Preparing a Present State Value Stream Map (PSVSM)

This document will be used to gather current information of the present state conditions for the entire value stream. This will be a door-to-door PSVSM—that is, we will start at the shipping dock and document the value stream up to the raw materials supply.

Step 6: Redesign to Reduce Wastes

Prepare a Future State Value Stream Map (FSVSM) That Will:

- Synchronize supply to customer, externally
- Synchronize production, internally
- Create flow (including the *jidoka* concept)
- Establish pull-demand systems
- Standardize and sustain

This will analyze current conditions and redesign the process flow to eliminate waste. You will recognize these five steps as The Path to Lean and they are detailed in Chaps. 14 and 15.

Creating a Spaghetti Diagram

This diagram will show the movement of the assembly as it is constructed, and show the movement of both the people and the product. Work to reduce the movement and transportation wastes and free up floor space. Do this on a plot plan, made to scale.

Step 7: Implement the *Kaizen* Activities

- Document all the *kaizen* activities you found in steps 2, 3, 4, 5, and 6 on a Gantt chart.
- Implement finished goods inventory controls to protect the supply to the customer.
- Implement your *jidoka* concept.
- Prioritize and implement all other *kaizen* activities on the Gantt chart.

Step 8: Do It All Over Again

a. Evaluate the newly formed present state.

b. Stress the system.

c. Then Return to Step 1.

Some Clarification on Step 8

As part of the project prescription, you will evaluate the new present state, stress the system, and return to step 1 (or maybe 2).

   To make a system Lean is a never-ending process. Each change brings about a new present state that then gets evaluated for improvement activities, which creates more changes and the cycle starts all over again. On many occasions the system will stress itself through the unexpected appearance of quality or availability problems, for example. Sometimes demand changes will put a stress on the system. All of these are opportunities to improve the robustness of the system. Although it sounds a bit crazy at first, it is wise to stress the system yourself to see what other process opportunities may be present. A typical "stressor" for the system would be to remove a few *kanban* cards and see what the system response will be. The primary tool you will use to protect yourself from system failure will be system transparency. Remember when we said we need to create a culture that embraces change. This may be the clearest manifestation that we have changed the culture, when we start stressing the system to make it better. Recall the metaphor of the athlete in Chap. 6. How did he get better? Isn't this the same concept?

What to Do with the Plan?

Review and Endorsement by the LLT

The plan needs management review, discussion, and acceptance. This should be done in a formal meeting with the LLT. This formal review is done for four reasons:

1. It will show, in one document, what is going to happen and when.

2. It will give top management, the movers and shakers, an opportunity to see the entire effort. They can see and comment on those things in their areas of

responsibility and also those changes outside their areas, but these changes still might affect them. In short, they will have an opportunity to bring up questions.

3. Any plan includes the topics of objectives, timing, and resources. This meeting will allow a check on not only those three topics, but their interrelationships as well.

4. How they respond to the plan will be a reality check on the commitment of the top management. This is most important.

Publish and Follow up the Formal Transformation Plan

Immediately following the meeting, publish the plan and put it into action.
Let the Fun Begin!

Formally Introducing the Lean Transformation

A formal introduction is very helpful to getting started well. Remember, the second requisite skill of leadership—the ability to articulate the plan so all can understand it. It is worthwhile to tell the entire facility, "We are going to make a change and that change is to implement the concept of Lean Manufacturing."

Many facilities make this a monster effort, with special invitations, a formalized meeting attended by all the top management, coupled with meals and motivational speakers galore. I find this degree of effort is not needed. In the end, the most important aspect of selling the issue of changing to Lean is dependent upon the continued actions of the lean leaders and top management. If they talk Lean and do not walk the talk, no amount of up-front selling will work. On the other hand, if the lean leaders and top management do really walk the talk, then no large selling effort is required. Either way, a mega-effort at selling is generally a waste, hence I do not recommend it.

> Demings Point No. 10, Eliminate the use of slogans, posters, and exhortations … for the workforce, demanding zero defects and new levels of productivity, without providing methods. Such exhortations only create adversarial relationships; the bulk of the causes of low quality and low productivity belong to the system and thus lie beyond the power of the workforce. (Sherkenbach, "The Deming Route to Productivity and Quality", CEEP Press, 1986)

Rather I recommend a short to-the-point informational training given to all employees on all shifts. This is typically a PowerPoint presentation describing, why we need to make this transformation, how it will affect you, a discussion of the House of Lean and pertinent details about the transformation schedule. Including ample time for a Q&A session, this usually takes about 90 minutes, and groups can be as large as 40 and still be effective.

Chapter Summary

The book *How to Implement Lean Manufacturing* is summarized in this chapter. First, we need to make sure we have a clear understanding of the concept that Lean is not a new "thing," rather is a "completely new way" to do the things we need to do to survive and prosper. It explains how to get organized at the corporate as well as the value stream levels. After selecting our change agent and *sensei*, we make evaluations using the following tools:

- The Five Leading Indicators of Cultural Change, outlined in Chaps. 5 to 9.
- The evaluation of the present manufacturing system, including the commitment evaluations and Five Precursors to Lean.
- The educational evaluation of the workforce.
- Specific value stream evaluations, as detailed on our present state value stream map.

These evaluations, and the countermeasures, then create a huge list of *kaizen* activities that can be included in a Gantt chart or an appropriate project planning and tracking tool. We can then evaluate and determine completion dates for *kaizen* activities in the project, set specific goals for the value stream, prioritize the activities, and implement the *kaizen* activities. Our approach to implementation, be it for a single value stream or for the entire corporation is done in two parallel paths: growing the lean competency of the management team and learning by doing.

Appendix A—The Second Commitment Evaluation of Management Commitment

The Process of the Second Commitment Evaluation

The second commitment evaluation should be done in a facilitated session, which your *sensei* should be capable of facilitating. It is not unusual to take a full day for this evaluation and problem-solving session.

As preparation for this session, make sure each manager has a copy of The Five Tests of Management Commitment to a lean transformation. In addition, ask them to do a personal evaluation, and also an evaluation of the management group as a whole. When you ask them to do these evaluations, tell them the results from the tests are personal and will not be shared.

What you want to do is uncover the issues, frame the issues into problems, and begin the problem solving to eliminate the issues. Actually, most of the issues will be known, but most will be difficult for people to discuss and bring out in the open. We find that once the problems are on the table, 70 percent of them solve themselves. However, you will also find that some problems are more difficult to solve.

The agenda could be something like:

1. Group icebreaker on teamwork.
2. Break into small groups of five to seven and ask each group to answer the question, "What is commitment?". You and the *sensei* should move from group to group to make sure they are on track. Give them about 15 minutes for this exercise. This is for the small group's purpose only.
3. Follow this by bringing everyone together in a group discussion facilitated by the *sensei* about "How do we measure commitment?" This must be done in a "spin-around" brainstorming session using strict brainstorming rules. Document this discussion and all subsequent steps on flipcharts that you will post on the walls

of the meeting room. Beyond posting the flipcharts, nothing more needs to be done with this session. Very likely, you will want a break here.

4. Again facilitated by the *sensei*, do a brainstorm of, "What are our problems with commitment?", or tackle a similar question. Document all the comments, concerns, and issues on flipcharts and post them. Make no effort at this point to discuss, validate, or reduce the list in any form. Just let it sit there. I cannot emphasize those last two sentences enough. This step, to be effective, must be totally nonjudgmental.

5. A group exercise on values may be appropriate at this point. Many exist. I like the "Lifeboat Decision." In this story, ten people have clambered onto a lifeboat, but it will only hold seven. Thus, three must be thrown overboard and they will die. If you do not throw three overboard, all ten will die. For instance, in the boat are a young child, a blind man, a priest, a prostitute, a mother and her baby, a grandmother, a convicted killer (who is strong and muscular)…well, you get the picture. There are no right answers. It just forces the group to discuss values. It is a great exercise.

6. At this point, if there are many problems on the list you made in step 4, discuss each item in turn until the group is reacquainted with the list. The discussion must be "to a point of understanding"—we *still* do not want to pass judgment on the projects. We only want to understand the context of the problem.

7. If the list is large, we will need to select the most critical problems. It is likely that we have a number of very good projects, so almost any from the top of the list will be productive to solve. To find a few of the better projects, use nominal group technique or multivoting to select say the top four or five. Once this is done, we need to select just one of these projects to work on. Probably the *sensei* could say something like, "We will eventually solve all these problems, but for the first problem I would like to see the group solve the problem of…."

8. For the first problem, brainstorm. "What are the thoughts, concerns, issues, and so forth about this problem?" When the group has all the issues down on paper, do not reduce this list. Instead, proceed to the next step.

9. For this first problem, brainstorm again: "What are the possible solutions?" When the group has all the possible solutions down on paper, do nothing more with this list for the moment. Instead, proceed to the next step.

10. Conduct another brainstorming session on "What are the criteria we will use in our decision-making process?" Discuss this and reach a consensus. This second step of reaching consensus on the criteria is often difficult. If you are not familiar with consensus, it is not disagreement, nor need it be 100 percent agreement. It is the concept whereby everyone involved can say, "I may agree with the group decision or I may not agree with it, but regardless, I recognize it is in the group's best interest and I will give it my 100 percent support and commitment." Like I said, it is not easy, but to reach consensus on this step is crucial.

11. Select the best of the possible solutions, and then develop action plans with responsibilities and due dates. Very often after the consensus on the criteria is reached, the process moves very rapidly.

12. Return to step 7, select the next problem, and move on.

It is important that the group completely resolve at least one problem. If they can and want to take on more than one problem, that is even better. However, here the major benefit comes not from the problem solution itself. Instead, it comes from the process of solving the problem. The spin-around technique requires:

- Good listening
- Understanding problems from different perspectives
- Nonjudgmental discussion
- Introspection
- Patience and respect shown by all, to all

Frequently, these are behavioral traits that are not found in abundance in the typical manufacturing plant environment. Consequently, and nearly always, the results of the process are more both important and more lasting than the actual problem that was solved. Following this, facilitated spin-around—the managers will be better equipped to work together to solve their problems. In addition, they now have a behavioral model they can take back to their individual groups to use in resolving their own internal issues.

It has been my experience that this evaluation is a very sensitive one—everyone thinks they are committed, but this is simply not the truth. I wish I could give you a prescription on how to do this comfortably, but I can't. The best advice I can give you is to do it—but do it carefully.

Simply because there is a possible downside is no reason to avoid it—yet avoid it is precisely what most people do. Unfortunately, when either fear or denial sets in and begins to rule the culture, the progress stops and the end is in sight. There is no substitute for simply fighting through these two problems of fear and denial, because they will appear again and again. Many of these issues test the courage and the character of the lean transformation leadership. If they waver, the effort will suffer.

There is a wonderful quote from a movie where the protagonist, who is only 17, has thousands of dollars of video and sound equipment that he purchased with the profits from his marijuana sales. When he was asked by a friend if his father knows how he financed the purchases, he says, roughly, "My Dad thinks I can afford this on my minimum wage job," and then adds, "never underestimate the power of denial." This is true of denial, and the same maxim applies to fear as well.

Both fear and denial are two extremely powerful detractors that will rear their ugly heads time and again as you pursue this journey into Lean. The leadership has to be aware of these issues and must handle them in a professional and open fashion. This is necessary for the success of your lean effort…well, for any effort you might embark upon.

CHAPTER 17

Planning and Goals

Aim of this chapter ... to explain the power of doing planning well and why it is worth a great deal of management's time and effort. All too often I find it is only a superficial perfunctory exercise—some do it better; yet very few do it really well. We will explain how good planning is both a strong leading indicator of success as well as an activator for the intrinsic motivators. As we discuss the elements of goal creation and deployment, we will introduce the powerful technique of *Hoshin–Kanri* (H-K) planning.

The Inherent Leverage in Planning

First, Some Background

As a young engineering manager, I was assigned the job of managing a group of engineers whose task was to design and install numerous capital projects. In fact, at the refinery where I worked, there were several such groups. As luck would have it, one day a man who chose to be my mentor, out of the blue asked me about a project in his area: "What's your plan, man?". So I briskly took out our construction schedule to review it with him. At this, he snapped at me and told me in no uncertain terms that all construction schedules at this facility were crap. He was both more, uhm, "verbose" and more "graphic" in his terminology in relaying this to me. He went on and said, accurately so, that the primary purpose of our schedules was to publish something so we could later revise it. There was no real expectation by anyone, including the author, that the target dates would be met. Although no one had the courage to admit it, he was right, as usual. Yet great effort was expended to make these schedules. They were carefully and painstakingly produced and published.

At least now I knew what he did not like; however, I still did not know what he wanted. So, with a certain amount of trepidation, I asked him. With a distinct passion in his voice, he told me that the main problem the company had with schedules was that there were not enough *engineering leaders* who would put together plans and then stick with them. He referred to us and our managers using a term that was less than "manly." Our problem, he went on, "We bow to the god of company politics and it's the squeaky wheel that gets the grease not the wheel that needs to be greased." He continued, "The Division Production Superintendents look at the schedules, see their project, and complain that it takes too long. They then go to your bosses who crater immediately, and your bosses then come to you and you revise the schedule and republish it. Then the next superintendent complains, and the cycle starts all over again." I countered with, "But we get no guidance from them, only criticism, they are more like pressure transmitters than bosses." At which he said, "Your boss is not doing his job, and he never will. So quit whining. You need to do some things."

He told me to "put together a preliminary plan and highlight each commitment you will need from each superintendent. Then put together a serious plan that you *can* do, and then do it. Go directly to the superintendents, get those commitments from them, and don't let them off the hook. Badger those @#$%!s until you get the needed commitments. Make no mistake about it. They will turn on you like a dog when things go wrong, unless you get their fingerprints on your murder weapon." My mentor then went on, "Review it with all the superintendents and make them commit to the needed completion dates, and also commit to doing the work they need to do so your engineers can complete their projects."

As was normally the case, I took his advice and we did just that. I persisted, and persisted, and persisted, and we finally got all the needed commitments. We then put together a plan that included all the jobs of all the engineers, all on one schedule. It was reviewed and we put it in service. It worked, and very soon our group was significantly outperforming the other design groups.

The division goal was to have the cost of engineering and drafting less than 15 percent of the project total—a goal that was more often missed than met. After we got ourselves organized, we averaged 8.6 percent and led all groups by a large margin. In short order, this was noticed. I'm not sure how it got so quickly noticed, but I believe that this man who had adopted me as his mentee had done some "behind the scenes politicking." Not only did we spend less engineering time and money on each job, we routinely met the project startup dates, projects took less time from start to finish, and our schedule meant something. The schedules of others were still looked at skeptically, but ours now had credibility.

Some other unintended consequences arose from this effort. First, no good deed goes unpunished. I found that the size of my group grew with the resultant increase in responsibility. Most engineering groups had five to seven engineers; ours routinely had 11 to 15. Although this was more work, I took it as the supreme compliment: They trusted us to get things done. Second and most importantly, I learned the value and power of good planning. While others were busy revising schedules and making excuses, we were finishing projects at a record pace, gaining self-confidence, and earning respect—both of the latter borne from our success.

The lessons I was taught by my mentor—about leadership, working together, making and meeting commitments, and planning—I never forgot. In every management position I have held, we have used goal development and deployment, and it has offered immeasurable success in our ability to meet our objectives.

Yet I see so many businesses slight this powerful tool, which further baffles me because it's relatively easy to use. I find time and again that poor planning is a critical weakness in plant performance and plant improvement. Consequently, I have expanded this chapter in the hopes that it will not only help in the implementation of your lean transformation, but that it will carry over into the entire management of the business as well. So I ask you, as my mentor asked me: "What's your plan, man?"—and I have included some materials that should help you.

Good Planning Is a Leading Indicator to Cultural Change!!!!

In fact, as we dissect the Five Leading Indicators of Cultural Change, we find that planning is a major factor, enabling each to be successful. Regarding leadership, it is axiomatic that you cannot lead without a plan. When we speak of motivation, we speak of reaching some future state, and is that not simply the destination of our plan? Regarding problem solving, the first element is to recognize "the difference between what is and what should be." In the "should be state," there "should be" is nothing more than the goal level you are trying to achieve in your plan. And when we speak of engagement, planning is the embodiment of that element by determining what needs to be done and resource allocation is always an element of any plan? It is pretty hard for the organization's management and leadership to be successful in providing the Five Leading Indicators of Cultural Change unless they also are good at planning.

Good Planning Is a Leading Indicator to Individual and Group Motivation

Of all the tasks of management and leadership that I can think of, planning is one that take so little time to do well, yet is so powerful in terms of driving the plant to higher levels of performance. There is no real secret to this once see how good planning feeds each of the five intrinsic motivators.

Good planning is also a leading indicator for each of the intrinsic motivators as well. Think about a sense of control. If you are involved in HK planning as described later in this section, you will be directly involved via "catchball" and participate in "how things get done" in your area of interest. Beyond that, just having a good description of what is to be done and when it is to be done, the very output of a plan, gives you a sense of predictability and control. And then when your production-by-hour board shows you are meeting these goals, that fosters a strong sense of both contribution and accomplishment, and two more of the intrinsic motivators are fed. In addition, in today's environment nearly all of this is done in a team environment so the sense of community is further fed when you realize "we" completed all these tasks as planned. And finally when you can tell that your cell or value stream is performing well and you learn that the company is doing well, very much due to you and your team's efforts, it is very likely you will feel a strong sense of meaningfulness as you contributed to not only the well being of your employer, but your fellow employees and stockholders as well. Even beyond that, well thought-out goals also provide a future state, the Biblical equivalent of the promised land and in a word, this creates hope. Since good planning is literally a motivational force of itself, goal setting, deployment, and execution have become a passion of mine.

Herzberg on Planning

In his seminal book, *The Motivation to Work* (Transaction Publishers 1993) Herzberg regarding the importance of planning, states, "….the single most important goal in the progress of supervision is the development of new insights into the role of the supervisor *so that he may effectively plan and organize work.*" Furthermore, regarding individual involvement in the planning process he later says, "…………. it is certainly possible that the *ways in which these goals are to be reached (italics mine)* can be left to the judgment of the individuals." Here he is stating in simple format, the key concepts of "catchball" and doing so in 1959, long before *Hoshin–Kanri* was popularized. And quite frankly he is speaking to a huge opportunity we still have—well over 50 years later.

> ### Is Not the Importance of Planning Self-Evident??
> After spending years in teaching and using the PDCA cycle, it seems odd trying to emphasize the need to plan. Is it not self-evident in the first step of the Shewhart cycle, the very heart of a lean transformation?

Why Are Goals and Goal Deployment so Important?

What Is the Purpose of Goals?

To answer this question, first let's explore the purpose of goals. Let me be simple and straightforward. The primary purpose of goals is to guide behavior.

Often the most critical and important thing a leader or a manager must do is act with courage and conviction on the plan he has created. When he does this, he not only acts in the best interest of the facility but also shows, by example, the appropriate way to lead. If he does not act in this fashion, everything else he will do will be compromised. If he has developed a good plan for the facility, he then wants all those in the organization to act in consort with that plan, trying to reach those goals. This is his leverage. This then is his ability to get the work done through others and have confidence that the right work is being done, by the right people, in the right way, to reach the right objectives.

> **P**oint of Clarity   The only economic purpose of inventory is to protect sales.

- If his goals are not meaningful, then the organization will not get to the right place.
- If his goals are unclear, then the organization cannot proceed with confidence, undermining their ability to reach the goals.
- If the goals are not directed to—and understood by—the right people, then they fall on deaf ears and will not be reached.

> **"I**f we don't know where we are going, we might end up somewhere else. **"**
>
> Yogi Berra

The goals become the primary tool, used by the manager, so he can convey both the needs of, and the desired destination for, the facility. Good goals, well deployed, will not only leverage the manager's ability to get the right work done and improve the performance of the facility, they will actually be a primary motivating tool for the workforce. People will always act better and more decisively when they know where they are going. Weak, unclear, or poorly deployed goals will doom the facility to perform at an inferior level.

Think about It—Clarity of Purpose

Imagine an organization where:

- Every employee knows what he needs to do, when he needs to do it, and how to do it in order for the organization to run smoothly.
- Every employee manages by facts and knows how to analyze and correct problems.

- All the needed information flows smoothly and concisely to the people who need it.

- Managers can establish the few key goals to meet the needs of the customer and so best leverage the facility. And these managers will have the time to perform daily, weekly, and monthly management at the plant.

This is all possible, with good policy deployment.

Policy Deployment

The Manager's Task

Policy deployment, more than just goal setting, is management's way of:

- Communicating
- Guiding
- Following up on issues
- Changing the important aspects of plant operation

When the manager creates goals, he will:

- Convert business concepts into understandable performance metrics
- Be able to compare these performance metrics to a standard
- Implement just in time (JIT) corrective actions

Hoshin–Kanri Planning

Few things are as powerful for the manager as H-K planning. Any manager who has his pulse on the business can learn H-K planning and execute the basics of it rather quickly. It is not very time consuming to develop the plant goals, nor is it very difficult to review the goals monthly. Even if the manager is locked in his office and does not have any—or any real—contact with the floor (which is the real-time consuming step). H-K's goal formation, deployment, execution, and review can be powerful techniques.

I can think of no management technique that will leverage a manager more than goal creation, deployment, execution, and review, yet it is so seldom done well. It is one of the most powerful management tools, and one of the most humane, yet it is not difficult to do. It baffles me.

Goals or Metrics...Which Word to Use

I tend to use the terms interchangeably but regardless, they have a

- Conceptual name and often an acronym, for example, on-time delivery, OTD

- An operational definition to define and use in calculating the metric and, for example, OTD is to deliver per the customer's schedule the full load quantity of good product with no internal or external expediting

- A quantifiable level that must be attained, for example, 98.5 percent, as a minimum

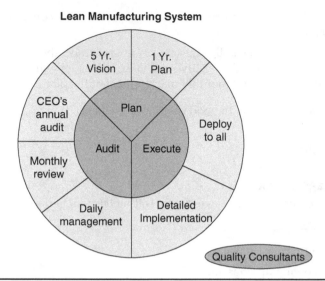

Lean Manufacturing System

FIGURE 17-1   The H-K planning model.

Many managers have found that H-K planning (see Fig. 17-1) has proven itself to be a superior model of policy deployment. I strongly support it and have found that it is a superior method to create both focus and alignment throughout the enterprise. It creates a horizontal and vertical integration of activity. It is a detailed method to achieve:

- The implementation of mission and vision
- The alignment of goals
- A self-diagnosis of progress
- The process management of a plant
- Targeted focus at all levels

Some Unique Strengths of H-K Planning

In this chapter, we will touch only briefly on H-K planning, but I strongly suggest you pick up a good book on it and learn it more fully. H-K planning has many strengths, but I wish to highlight three.

From The C-Suite to Floor Execution   First, H-K planning is an integrated continuous process that is graphically described in Fig. 17-1. It includes long-range planning such as the 5-year vision and the 1-year plan and also requires periodic reviews as well as daily management. Connecting the 5-year vision to daily management of the process is absolutely crucial. You see, for the managers in the C-Suite, all too often they have lost sight of the fact that regardless of the degree of sophistication and skill they use to create their strategic plans…*they are executed on the floor by the hourly worker.* This is in stark contrast to most planning efforts, which have a huge influx of effort and management time at only the goal development and maybe the monthly review. In leanspeak, H-K planning is not only a continuous, rather than a batch, planning process; but it integrates the management's highest level plans to actual floor activity.

Floor Execution Using "Leading Indicators" and Real-Time PDCA   Furthermore to execute at the floor level, a key driver in H-K process is the use of transparency. The key benefit of this is then the transparency becomes a set of "leading" indicators so problems can become surfaced and real-time problem solving can occur. As goals are deployed from the C-Suite to the workstation, key visual controls and monitors are put in place to be monitored and managed at every level for every time period, not just monthly, but daily, hourly, and sometimes each cycle. This form of visual control may be a posting of the key plant goals or even something as basic as a production by hour board. In H-K, the emphasis on daily execution is manifest in a variety of lean tools. First is the Leader Standard Work. With this tool, in addition to operator surveillance, the leaders at all levels are continually reviewing the key metrics whether they are at the plant, value stream, cell, or workstation level, and all critical goals get continual attention in a cascading layered methodology.

H-K planning has at its core PDCA with a number of check intervals: yearly, quarterly, and monthly for sure. But the problem with all these periodic reviews is that they are all lagging indicators and as such do not lead to rapid corrective actions. However, with the concept of transparency using real-time floor data and real-time floor management as a part of H-K planning, we connect those leading indicators to the vision. "Et voilà" we not only have the daily production plan being managed, we have the weekly, monthly, yearly, and 5-year vision managed all through daily management and transparency. This is truly a leveraged aspect as now every employee is not only focused but aligned to the strategic plans.

> **P**oint of Clarity ... the purpose of "catchball" is to align the "results" of the goals with the "means" to achieve them.

Catchball   The third major uniqueness in H-K planning is the extensive and intensive use of "catchball." This is both taught and most evident at the goal development phase. Just what is this thing called "catchball"?

In "catchball," goals are first created by top management based on the needs of the business. These goals are "what" must be controlled to be successful. The manager then introduces this goal to the next level and asks "how" do we do that? The next level replies as to "how" it must be done, including help that may be needed from top management. The next level turns their "how" into a "what" and asks their subordinates, "How will you do that?" and so forth... .

This creates a down-up-down-up-down process that looks a lot like a negotiation, except it is not a negotiation. A negotiation too often has winners and losers and its chief technique is compromised; that is not at all what "catchball" is all about. *Its purpose is to make sure the goals are properly aligned and the means exist to execute the goals so the results are achievable.* So contrasted to a negotiations, there are only winners and nothing is compromised. The end result is alignment and focus on meaningful and achievable goals.

For example, the plant manager decides, based on the specific needs of the business, that we must improve OEE by 15 percent to remain competitive. That is the "what" that must be accomplished. This goal then goes to the production manager, and the plant manager asks him "how" he will do it. The production manager says he will improve OEE by reducing machinery downtime, but he needs another engineer... . Maybe both agree, if so, the goal goes to the next level. Hence a down-up-down process that leads to agreement of the goals and the means to achieve the goals.

The process of "Catchball" has a number of extremely strong benefits, including:

- Goals are thoroughly deployed.
- Goals are mutually understood.
- Means are addressed.
- Ownership is clear.
- Measurement is clear.
- Priorities are clear.

H-K Planning Contrasted to Many Typical Planning Models   When I thoroughly scrutinize the planning in most organizations, I find the predominant method of planning is "top-down-autocratic" goal setting, usually trying to do some type of Management by Objectives (MBO) but being rather eclectic in the method used, ignoring much of what Peter Drucker proposed when he created MBO. In deference to Drucker, I call this Dysfunctional MBO or DMBO. Using DMBO, management passes the needed goals, "the whats," to the next level and expects them to be done. There is little iteration with the next level down and hence there is no real alignment of the desired results with the means to achieve the results. Normally in DMBO worker performance reviews are still tied to the objectives although they have had little input and almost without exception these reviews are performed in a ritualistic fashion at years end using such draconian techniques as 360 degree performance reviews. In addition, it is based on an underlying but incorrect belief that "only the boss knows best." All these factors lead to low levels of "buy-in," lots of morale problems and ultimately to very low levels of goal attainment. These deficiencies are cured by using the H-K model and executing it using the Six Skills of Lean Leadership.

H-K Planning Success Rate

All my experiences with H-K planning have been positive and this early up-front time investment has proven to be an outstanding investment, without exception. In the end, I have found that the accomplishment of goals created using H-K planning has been far more successful than conventional goal setting. Once managers start using H-K planning, I find they continue with it, because it is an effective management tool that allows the facility to perform at a much higher level.

Additional Information on H-K Planning

A great deal of good literature exits on H-K planning. I recommend the book by Bob King, *Hoshin Planning: The Developmental Approach* (Goal/QPC, 1989), and *implementing A Lean Management System* by Thomas Jackson (Productivity Press, 1996) (yes, the *i* in implementing is lowercase, which, by the way, is more about H-K planning than lean management), or the classic is, *Hoshin Kanri: Policy Deployment for Successful TQM* by Yoji Akao (Productivity Press, 1991).

Goal Development

At the heart of policy deployment is the development of goals that have:

- Purposes
- Characteristics
- Foundational concepts

- Deployment characteristics
- Owners

The Purposes of Goals

The purposes of goals are multiple. The primary purpose is to guide behavior. Let me say that again, **the primary purpose is to guide behavior**. Goal development and deployment also provides a reason and means for rewarding the correct behaviors. The two guiding rules we use in directing human behavior are given in the quotes here.

The secondary purposes of goals are psychological/sociological in nature. One purpose is to convey where the company is at this moment and convey what it must accomplish. By conveying where the company must go in terms of performance, the goals create a future state. This future state is the Biblical equivalent of the "Promised Land." In other words, once we get there (the future state), we will be in the land of milk and honey. Since the goals are, by defini-

> **"T**ell me how a man is measured and I will tell you how he will behave.**"**
>
> Unknown

> **"A**ny action you want repeated, first define, second support, third model, and finally reward.**"**
>
> Wilson

tion, attainable, this has the effect of creating hope. Hope that the company will become "the best," "more competitive," or whatever is the desired vision of the company. It is a psychological fact that there are few positive motivators that are as strong as the concept of hope.

In addition to creating hope, properly deployed goals will have owners and clear objectives. This then has the highly desirable effect of instilling both confidence and a sense of ownership in those who must execute the goals. They will proceed with more confidence when they know they are the ones who are both responsible and accountable for the goals. Furthermore, they know that the goals are clearly an extension of the plant goals: important to the success of the facility. For those who achieve their goals, all these factors combine to develop a strong sense of accomplishment with the attendant self-satisfaction attached to accomplishment. You can see that good goal development, deployment, and executing feed directly into the five intrinsic motivators.

Goal Characteristics

All goals must be:

- Written
- Challenging
- Believable
- Specific
- Measurable
- Have a deadline

Goal Deployment

Goals must be deployed in the right context, to the right owner, and with the right expectations. The context and expectation include the expected results in the expected time frame. In addition, good deployment will include the manager pointing out possible failure paths to avoid, and possible future conflicts that might naturally result from attaining the goal. It is also an aspect of good deployment to reach agreement on the available resources to allocate to the goal efforts and the consequences of achieving and failing to achieve the desired results. This is achieved using catchball.

The Owners of the Goals

Good deployment requires that the goals must have a clear owner who is responsible for the attainment of the goal. By responsible, I mean just that: "able to respond," and more than just being accountable, being "able to count." Hence, the owner must have:

- The awareness and the tools to determine if the process is performing properly. This is transparency.
- The imagination and values to determine what action is required.
- The desire to make a change when one is needed.
- The power to make it happen.
- The courage and character to accept the consequences of those actions.

Furthermore, the owner—and everyone for that matter—must recognize that we *cannot* live in a dependent world, and that total independence is neither real nor healthy in a society. The reality is that we live in a world of interdependence. Consequently, no one person can actually be totally responsible. In other words, we must work together and synergize for the common good; and therefore, the owner does not have total control, but he does have *functional control*—that is, he can make things happen so that progress toward the goal can proceed. This too is achievable via catchball.

Leadership in Goal Development, Execution, and Determining What "Should Be"

In Chap. 5, we enumerated the three requisite skills of leadership as:

- The ability to develop a plan
- The ability to articulate this plan and engage others
- The ability to act on the plan

Which Goals to Choose?

The first aspect of leadership is manifest when the goals are developed. The goals form the plans the manager will use. Consequently, the manager must have the skill to discern the few key metrics that will best guide the facility to success. I call these the "Plant Level" goals.

Plant level goals are almost always a subset of the three key customer needs of production:

- Quantities on time (on-time performance)
- High quality (usually something like first-time yield for internal quality and PPM customer returns for external quality)
- Fair priced (usually these are cost and productivity goals for the typical manufacturing plant)

There should be only a few—five to seven is ideal. For many of my clients who are tier 1 and 2 automobiles supplier, we typically use just five goals: Safety, Internal quality, External quality, On-time delivery, and Productivity. Too often, where there are goals, there are too many. I frequently find 30 or more. Who can remember 30 goals? Furthermore, with 30 goals, the focus is being lost. Even if they can remember 30 goals, who can focus on 30 different areas?

In addition, often the wrong goals are chosen for plant level goals. A goal I see very often is the goal to reduce the cost of expedited freight. Not that the cost of expedited freight should not be reduced—that is not the point. The point is that these are just not the key metrics that should be used to guide behavior in the facility.

In fact, there is hardly a better red flag to indicate that a facility is in trouble than when you see that one of the plant metrics is expedited freight. Think about it. That means they have problems with on-time delivery, which signals a whole foray of production problems. Also, it means that expediting costs are a significant part of the plant's operating expenses. The choice of this metrics sounds like a strong reason, in and of itself, to look into Lean.

Now, back to the development of metrics by the plant manager (PM).

Pick your metrics carefully because......

(1) Needs drive metrics

(2) Metric drive choices

(3) Choices create consequences

....hence those metrics will determine if you are successful or not.

Once the correct metrics are selected, the PM must determine which levels these metrics "should be" by year's end, for example. This creation of a "should be," as you recall, has now just created a "problem" for his staff. It is an oddity of management that one of their key roles is problem creation. They do this by creating the possible future state of the facility—what "should be" attained.

When the PM selects the specific goals and the levels that need to be achieved, he is starting to make a very certain and definitive commitment. Actually, there are two major commitments. The first is about the facility. Based on his experience and abilities, he is saying that the attainment of these goals is how the plant needs to be "best" or "competitive," for example. Second, he is making the commitment about his future actions, including what he will support, what he will model, and finally the rewards that will be given out for successful goal attainment. Both of these place a great deal of pressure on the PM. But make no mistake about it, he must do both. He must select the best goals to guide the best behaviors, and he must reward those behaviors he wants repeated. Recall:

If the correct metrics are not chosen with the appropriate performance levels, the policy deployment will not start properly. In short, the plan to improve the performance of the facility will not be a good plan. If the plan is not deployed well, it will not be understood and accepted by all, and execution of the plan will suffer. Finally, if the follow-up elements of H-K planning are not executed well, the leadership will not be acting upon the plan. All these issues are just symptoms of weak leadership, which:

- Causes poor goal creation, which in turn…
- Causes people to pay attention to the wrong metric, which in turn…
- Causes them to act in a way that is not in the best interests of the plant, which in turn…
- Causes the plant to be less robust, which in turn…
- Is exactly *what we do NOT want!*

Goals and How to Stress the System

You will recall that Step 8 of The Prescription is to "stress" the system. (See Chap. 16.) To this, there is no real mystery. It is so obvious yet most folks look for some magic pill in the toolbox of lean tricks. Well the magic pill is this; in business, if you want to guide people's behavior, create goals to give them direction and then create a set of recognition and rewards as confirmation. There is no mystery in this.

However, there is a trick, two to be exact. Trick one has to do with what goals you create and how you create them; trick two has to do with what you use for recognition and rewards …. what motivators you use. Each of these tricks has a simple yet elegant solution.

For trick one, which goals you create and how you create them? ….. the solution is the interactive process of (H-K) planning using catchball as outlined above.

For trick two, what to use for motivators?…. it is the intrinsic motivators coupled with the proper use of Herzberg's hygiene factors as outlined in Chap. 6.

So just how do you use these techniques to create goals? Exactly the same way you set them 25 years ago; you use all the data to your avail and select goals that will push the organization to higher levels of performance based on the needs of the organization, consistent with its growth strategy. (See the sidebar, Goals and the Ideal State). Well __almost__ exactly the same. When doing goal setting in a lean environment, you are able to set far more aggressive goals since the people are naturally more focused, more engaged, not to mention more competent. This allows you to ratchet up your goals to tell your team "what" you must do to survive and prosper. And if you are using catchball and have your team intrinsically motivated, they will tell you "how" to get there. This is an extremely powerful combination of tools to have at the disposal of a manager, and it allows you to set much more aggressive goals. So how can we "stress the system"— just set higher goals. However, now you have two new weapons which will assist you in achieving those higher goals. First, you have engaged, skilled workers who not only are capable of meeting those goals, they want to attain those goals. Second, you have the full toolbox of lean tools at your avail. Put together they are an awesome combination.

Stressing the system by setting tougher goals is not a new concept. The difference is that now you have at your disposal a group of workers that are focused and aligned as they were, via the process of catchball, critically involved in the creation of those goals. They are ready, willing, and able to attain those goals. You have at your disposal a completely different workforce. They are a highly motivated group of **all leaders,** who are learning their way to greater and greater levels of competence, and who are totally engaged in problem solving your organization toward the Ideal State. That group can accomplish amazing things and they only need challenging goals and an appropriate recognition and rewards system to test their mettle. In that regard, be aggressive, set them high. Let everyone know you have high regard for their abilities.

Goals and Leading into the "Unknown"

In most goal setting, I find that most organizations are very reluctant to set challenging goals, usually for fear of not meeting them. Simply put, they are reluctant to create a goal—unless they already know the solution. Quite frankly, if I review the activities of many management teams, I find they do so, often with good cause, and with lots of data to support their trepidation. Keep in mind that one of the leading indicators to cultural change is to create an environment of learning/teaching/experimenting (see Chap. 9). Inherent in this indicator is the concept of experimentation and a certain resilience for what many call failure … me I call it "an opportunity to learn." This cultural change leading indicator of learning/teaching/experimenting, when properly managed will transform your organization from having a culture of fear to a culture of challenge. Once this is achieved, you will find it possible to convince your workforce to "problem solve into the unknown." This is a lean principle in problem solving and a valuable concept in goal setting. Until you venture into the unknown, real progress is very slow. This is such a basic concept for Toyota, they included it in their "Toyota Way, 2001" booklet (Toyota Motor Corporation 2001). They state:

> "A Spirit of Challenge…
>
>> A spirit of challenge; a Drive for Progress…
>>
>>> We accept challenges with a creative spirit and the courage to realize our own dreams without losing drive or energy. We approach our work vigorously, with optimism and a sincere belief in the value of our contribution.
>>
>> A Sense of Self-respect and Self-reliance and the Acceptance of Responsibility…
>>
>>> We strive to decide our own fate. We act with self-reliance, trusting in our own abilities. We accept responsibility for our conduct and for maintaining and improving the skills that enable us to produce added value.
>>
>> The Acceptance of Competition….
>>
>>> We welcome competition, knowing that we will learn from the challenge and become stronger because of it. We demonstrate respect for our competitors and fairness in the competitive contest, even as we maintain our fighting spirit and our will to win … ."

These are some pretty strong words and form, in part, a culture that is willing to count very heavily on the organizational strength to take on the challenges to " …. produce added value." and "… to win."

Goals and the "Ideal State"

It is more than just interesting to note that the "Means to Lean" is to problem-solve our way to the Ideal State. Here is one very large concept that gets clarified. Specifically, the "destination" has been determined; the "should be" condition is defined. The goal no longer has some fuzzy end state…we are to achieve the "Ideal State." And our job is to continue to work, toward that end. So "what" we are to achieve has been defined. The questions that remain in goal setting is now some combination of "when" and "how."

More often than not, the Ideal State cannot be attained. Although this is interesting, it is not relevant. The relevance is that we will work "toward achieving the Ideal State"—regardless. If we are not, we are not making progress. So we can add concreteness to this concept, we develop "target conditions." These are measurable end results and positions that we wish to attain on our way to the Ideal State in the goal attainment period. This target condition is almost always somewhere in the "unknown territory."

The Ideal State

The Ideal State is a conceptual "end state" of your production system. It is the one that is:

- Safe
- Produces defect-free products
- One piece at a time
- with zero waste
- Can immediately respond

Chapter Summary

This chapter was added to first explain the strong connection between planning and the Five Leading Indicators of Cultural Change as well as the Five intrinsic motivators. In addition we have included information to and assist the management in overall goal development and deployment. H-K planning has proven itself to be a superior management tool for the development and deployment of plant-wide strategies and goals. Goals will not only guide the behaviors of the workforce, but goal development, deployment, and execution will create the hoped-for future state, and create confidence and a sense of accomplishment. Properly developed and managed goals, as outlined in the H-K methodology, can be a large motivational tool, and a strong weapon for the manager as he tries to leverage his power and maximize the potential of the plant.

CHAPTER 18

Constraint Management

Aim of this chapter … to show the power of understanding and managing the constraint, the "bottleneck" in a process.

The management of bottlenecks is often overlooked in lean applications because it is not a large part of the Toyota Production System (TPS). It, however, is often a strong tool since it can be used very powerfully for process improvements. In this book, I show numerous examples of bottleneck reduction which then led to huge process gains. Often, bottleneck reduction provided the "low-hanging fruit" for the early process gains. For this reason, it is included herein.

Bottleneck Theory

What Is a Bottleneck?

Bottlenecks are the limiting aspects of a process, much like how the neck of a wine bottle limits the flow of its contents as you pour. Often, instead of "bottleneck," we use the more sophisticated term "constraint." Every process has a constraint. There are constraints to our manufacturing processes, our engineering processes, and constraints to our business process. Every process that has an objective has a constraint, unless the objective is fully met. In a typical manufacturing process, the constraint is usually the step in the process that has the longest cycle time. If so, it can normally be identified with a standard time study, as described in Chap. 14.

Moving Constraints

In Lean Manufacturing, we are always striving to eliminate wastes. Often, this entails using the tool of line balancing (see Appendix C, Chap. 14). When we balance a line, we try to design the process steps so they all have the same cycle time. This brings about another phenomenon. When the cycle times are very similar, a small variation can cause a cycle-time increase, making it the constraint at that moment. The variation disappears and yet another process step incurs a minor variation that causes *it* to now be the long cycle-time step, and thus the new process bottleneck. In this manner, the constraint moves from one process step to another. We call this a moving constraint. (Refer to Chap. 24 for an example of this.) If the variation of the individual process steps is not too large, we are often not concerned with this moving constraint and do little about it. If, however, the variation of the individual process steps is significant, this variation will measurably affect the process performance and thus the process will not produce to the design cycle time. There is a cycle-time loss. In this case, there is a process loss. This loss can be quantified by the lean metric of Overall Equipment Effectiveness (OEE) discussed in Chap. 12, and once it is quantified we can decide if we wish to work on this constraint to improve the process.

Some Constraints That Are Often Not Called Constraints

For the moment, think about a product you make, produced by some process. What if the process is performing well? For example, the quality yield is 100 percent, as is OEE and on-time delivery. The process is producing to *takt*, making good margins, and meeting customer demand in every respect. Do we still have a constraint? Well, if your company is in business to make money and you do not have 100 percent of the market for that product, then the answer is a resounding yes. In this case, the constraint is likely your sales department. Why sales? That sounds odd! The logic goes like this:

1. The business objective is to make money.
2. There is more market share to be had and more money to be made.
3. What is our limit?
4. Answer: The constraint is probably sales.

The constraint is not always a step in your process. It can be any aspect that limits your ability to meet your objective. The constraint can be the process itself; it may be a raw materials supplier; it may be a machine, a workstation, or any other resource; or it may be another aspect of your business.

Policy Constraints

The most disturbing constraints are often policy constraints. This can occur when the company makes policies that turn out to limit the facility. Seldom are these policies designed to be limits. Rather, they are often created with the best of intentions but without a good understanding of the intended or unintended consequences of the policy. Two types of policies drive me absolutely crazy.

> **P**oint of Clarity All businesses have constraints, and these constraints limit the business's ability to make money!

- The first type is "We don't know what we are doing so we will create a policy to cover it." For example, one client explained to me that their policy on inventory was to have 30 days on hand. It was a corporate-wide policy. And no one could give even a rough explanation of why this 30-day policy existed.
- The second type is the "I don't trust you, so we will create a policy to limit your authority." An example is described next.

On one occasion, I was called in to make an evaluation of a production line. The plant manager needed to increase production by 38 percent and knew that the capacity constraint was their electrical testers. In an evaluation that took less than 1 hour, I was able to spot potential capacity increases of over 22 percent with no capital investments. The recommendations consisted of staffing the test station during breaks and lunch and moving one test station that had significant scrap. This test station was after the bottleneck, and by placing it in front of the bottleneck it would improve throughput instantly. I was feeling pretty good about myself and just figured he would jump at these ideas.

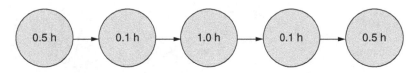

FIGURE 18-1   The 5-step process.

We met for lunch and although the plant manager was intrigued, he flatly rejected both ideas. These ideas would add about $200 per day in labor costs to the operating expenses but would increase revenues by over $26,000 per day. They were making about 22 percent on sales, so this was extremely profitable, to say the least. The plant manager was interested, but it was the policy of the company (that ugly word, *policy*) that they would not increase manpower for any reason, above their current levels. Hence, he no longer had the authority to add these people. Well, as you might expect, there was a lot more to this experience…but this was not the first, nor the last, time I encountered a business decision where "The Policy" was the system constraint. These constraints are usually very costly and frequently the management is blind to them and does not see them as constraints to the business.

The Economics of Constraints

The system constraint will limit the ability of the business to make money. However, in most cases, if the constraint is broken, the resultant incremental increase in production is often the most profitable production the company has. Let's look at a simple example of a 5-step process, shown in Fig. 18-1, which shows the process cycle times for each process step.

It is clear the 1.0-hour process step is the constraint and will limit production to 24 units per day and, as shown in Table 18-1, profits will be $20/unit.

Let's say we have added possible sales and we want to increase production. We could design and build a complete new line, just like our 5-step process, but someone notices that the line is not well balanced and suggests we break the system constraint. The constraint is the 1-hour cycle time at step 3. It is easy to see that if we wished to double production, we could duplicate the third step and place it in parallel with the third process step. We do this and have a new 5-step process, as shown in Fig. 18-2 .

The third step would now produce two units in 1 hour. We have broken the process constraint and the new process constraint would be any one of the 0.5-hour process steps—either steps 1, 3, or 5. Very likely, we could now produce up to 48 units per day, and on a good day our sales department could sell them and everyone would be happy. But just

Cost Category	$/Unit
Sales Price	200
Variable Costs	20
Fixed Costs	60
Raw Materials Cost	100
Profits	20

TABLE 18-1   Economic Profile, One-Hour Constraint

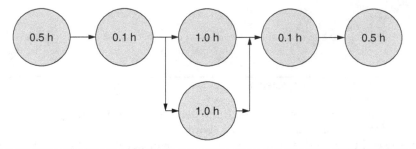

FIGURE 18-2   The new 5-step process, with added capacity at step 3.

how happy would they be? The economics of breaking the constraint are very important to understand. In the case with the 1.0-hour constraint, our profits were $20/unit.

How does breaking the constraint affect the profits? Our production unit will still need $100 of raw materials, and we will likely have most of the $20/unit variables costs as we use the consumables that it takes to make the unit. But the fixed costs are driven to almost zero. For example, our building and all the property taxes and all those fixed costs are already covered and will not be affected by the rate increase. Neither will such things as home office burden or staff and management costs, which are often very large. Simply put, the incremental fixed costs are very low, and for the sake of argument we will say they are zero. So let's look at Table 18-2 to see the economics for the incremental production volume after the constraint is broken.

In this case, with a $200 sales price, our profit per unit has risen from $20 for the first 24 units produced (up to the constraint) and $80 for all units produced after the first 24. Our profit margins have *quadrupled,* for all incremental production after the constraint was broken. If our sales department can sell, say 12 more units per day, our gross profits will go from $480 per day to $1440 ($20/unit × 24 units + $80/unit × 12) per day. With this you can see the power of "constraint economics."

Point of Clarity The most profitable production occurs after a system constraint has been broken.

System constraints need to be understood and aggressively attacked because this is one of the most powerful tools available for improving the economics of a business. This concept cannot be understated. When you read the stories of the Bravo Line in Chap. 21 and Larana Manufacturing in Chap. 22, pay particular attention to the

Cost Category	$/Unit
Sales Price	200
Variable Costs	20
Fixed Costs	0
Raw Materials Cost	100
Profits	80

TABLE 18-2   Economic Profile, One-Hour Constraint Broken

huge early gains that were achieved through breaking the system constraints. To a large extent, this was how they "harvested so much low-hanging fruit."

When Is It Best to Address Constraints?

Very often, during the review of the foundational issues, constraints will be found and highlighted. It is also, in this timeframe, that the constraints should be aggressively attacked and removed. First, it will make money for the facility and you will be able to accrue huge early gains. And second, it is more efficient to achieve the appropriate process flow rate, stabilize the flow, and then work on quantity control techniques like *kanban*. However, everyone should constantly have an eye out for constraints and should be attuned to removing them. After all, once the system constraint is broken, the system's profitability is vastly improved.

How Do We Spot the System Constraint?

This is the simplest of questions. Look for the inventory build-up. It will always be in front of the constraint.

What if there is inventory between all process steps? It will likely be the largest pile of inventory. What if there is inventory galore, as is so often the case? Then do a process study to calculate the time for each step. The step with the longest cycle time is usually the constraint. I say usually, not always. It is possible that a given process step may not have the longest average cycle time, but if it has huge variations in the cycle time, this step may be the system constraint.

Often when a multistep process has a lot of WIP (work in process), it may experience a broad array of problems. One of which is that it may operate at a cycle time that is well above its potential, thus producing at a rate well below its design. This is especially true if it is dominated by manual operations. In this case, unless there is some clear way to measure and control the cycle time, the system constraint may bounce from one station to the next and the whole line will under produce. This is commonplace (see Chap. 22). Consequently, at the end of the day when the production quota was not met, no one can point to any specific problems that the line experienced. Often, what follows is some of the most elaborate, but incorrect, rationales for why the production goal was not met.

The answer, of course, is first understand and reduce the variation and then get rid of the WIP inventory so you can "see" the process. This will allow you to find the constraint and work on it—if necessary.

Almost always, once the inventory is removed, the entire process, often every step, will speed up. To most people's amazement, the quality will improve as well. It is not logical to most that this should occur—that is, a simultaneous improvement in both quality and rate. However, it is a

> **P**oint of Clarity   When we rationalize, we generally just create "rational lies."

metaphysical truth, and also a very common occurrence, that production will speed up and quality will improve once the inventory is reduced.

Why Ohno Does Not Even Talk about Constraints?

If this constraint stuff is so important, why is there nothing of it in either Ohno's writings or those of Shingo? There are two explanations. First, Ohno and Shingo are so good at process design and management that achieving minimum cycle times is second nature to them, they take it for granted—we seldom fret about those things we take for granted. Second, their approach gets them to a minimum cycle time anyway. They break constraints by first removing the waste and then rebalancing the work.

Chapter Summary

Every process has a bottleneck, a constraint that limits its ability to produce. The constraint can be a specific step that has the longest cycle time, or if the steps have large variations in cycle time, the constraint can be a moving one. Generally speaking, the worst constraint is a policy constraint because they are often not recognized as things that can limit the system. The beauty of breaking through system constraints can be seen in the economics of constraint removal, because the incremental product produced, once the constraint is broken, is much more profitable. Constraints are generally easy to find since the inventory will accumulate in front of the constraint.

CHAPTER **19**

Cellular Manufacturing

Aim of this chapter … to show how cellular manufacturing is a key element of Lean Manufacturing. We will also explain a near-secret of cellular manufacturing; that is, it is a very powerful variation reduction tool in its own right. We will do this using a case study, the history of the Gamma line. This chapter is also designed so you can follow through the mathematics of cellular design. Grab a pencil and pocket calculator and follow me through the calculations as we redesign the line and *improve the production rate by 63 percent, with virtually no capital expenditures—and less manpower!* You will be amazed!

Cellular Manufacturing

The Definition of a Cell

A cell is a combination of people, equipment, and workstations organized in the order of process flow, to manufacture all or part of a production unit. I make little distinction between a cell and what is sometimes called a flow line. However, the implication of a cell is that it:

- Has one-piece, or very small lots which flow
- Is often used for a family of products
- Has equipment that is close-coupled and right-sized specifically for this cell
- Is usually arranged in a C or U shape so the incoming raw materials and outgoing finished goods are easily monitored
- Has cross-trained people for flexibility

The Advantages of Cells

Cells are an integral part of Lean Manufacturing. The use of cells is so basic that in the TPS (Toyota Production System) it is not even questioned. For Toyota, that works. Unfortunately, for others, cells may not work in quite the same way, so it is worth understanding the benefits and drawbacks of cells before we embark on a full-blown effort to convert everything to cells.

The primary purpose of a cell is to reduce wastes in the manufacturing system. The seven wastes again are:

- Transportation
- Waiting

- Overproduction
- Defects
- Inventory
- Movement
- Excess processing

It is easy to see that cells reduce transportation due to the close-coupled nature of the workstations. They do nothing directly to reduce waiting since that is a function of work balancing and variation. However, cells make the balancing much easier to manage. Cells, in and of themselves, also do nothing directly to prevent overproduction or defects. They do minimize the inventory when one-piece flow is achieved, which is their basic design. As for movement, they actually promote movement, more efficient movement, depending upon the cell design and do nothing to reduce excess processing.

Thus, cellular production is designed to reduce the wastes of transportation and inventory. Consequently, it is also designed to speed up the process and make it flow. Cells almost always do this, resulting in production advantages such as:

- The reduction of first-piece lead time
- The reduction of lot lead time

So with the reduced lead times, we have greater flexibility and responsiveness.

However, both of these benefits can be achieved with a flow line. A flow line is a linear, rather than a U- or C-shaped, closed arrangement. To achieve these benefits in a flow line, there needs to be the same low inventory approach as well as making sure the stations are close coupled as in a cell. So why are cells so popular? There are other benefits to cells over flow lines that may be a bit harder to quantify, but are possibly even more powerful reasons to choose cells over flow lines in many manufacturing circumstances.

Cells or Flow Lines?

Just what are these other benefits in choosing cells over flow lines?

The first and often the largest reason to select cells over flow lines is the production rate flexibility possible with cells. For example, let's say we have a typical balanced cell staffed with six work stations and six workers. If we use three workers instead of six, we can have each worker do the work of two stations and this will double the cycle time or halve the production rate. If the workers are designed to move from station to station, it is possible to use either one, two, three, four, five, or six workers and get 16, 33, 50, 67, 83, or 100 percent of capacity with no increase in labor unit costs. This allows the cell to operate at different rates as the customer's demand changes. This rate modulation is much more effective than starting up, running the cell at full rate, and then shutting down when the month's demand has been met. The cycling up and shutting down is nothing more than creating batches. The TPS is a batch destruction system, not a batch creation system. This rate modulation by modifying staffing is not practical to do with a flow line.

When cells are arranged in a C or U shape, worker communication is facilitated. For example, all workers are in proximity to one another, so worker interaction is encouraged. Worker interaction to assist in cross training and worker interaction to assist in problem solving are just two such benefits. This proximity just makes communication much simpler. They are also not only able to communicate better but assist each other as well. The positive impact of this ease of communication should not be underestimated.

The typical U-cell situates the first and last work stations near each other. This makes cell supervision much easier and gives everyone a better sense of work completion. Also, in the typical U-cell, workers usually sit or walk in the center of the cell. This frees up the exterior of the cell to supply materials to the cell more easily.

Two Hidden Benefits of Cells

All the items listed earlier are benefits, but in my experience I have found there are two major benefits of cellular manufacturing that are seldom mentioned, but that are very real, very positive, and very powerful.

First, the very nature of a cell creates a team with a sense of flow and synchronization not seen in flow lines. In the flow line, you have two neighbors; in the cell, everyone is in close proximity. The personal dynamics are changed considerably, leading to a feeling of a group, of a team. The team concept is very powerful and there is a real sense of assisting each other. In the cell, since the process is all around the worker, there is a sense of flow and a sense of synchronization that is not present in the flow line. We have documented cases that show this sense of flow and synchronization actually creating a faster pace in the cell with reduced cycle times. We have found that it is not uncommon for cells to reduce cycle time by 10 and even 20 percent as they mature. I have often witnessed this cycle-time improvement in cells, yet I hear many engineers attribute it to training and worker maturity. These same engineers, however, cannot explain why we do not see the same benefits in a flow line as it matures.

Nevertheless, the greatest benefit of cells is a well-kept secret. Cells are a tremendous tool to assist in reduction of variation.

I will later describe a case study of a high-volume production flow line that was converted to cellular production. The plant achieved the rate modulating benefits earlier mentioned, but in addition—with less staff and the same equipment—the cellular option, compared to the flow line, was able to improve production by over 63 percent.

The Gamma Line Redesign to Cellular Manufacturing

The Background

The Gamma Line was a 21-station flow line with a 16-second cycle time. The first 16 stations were all manual assembly, followed by two tests and three packaging stations, which utilized some expensive test and packaging equipment. Material was delivered to one side of the 200′ long assembly line, the same side on which the workers were stationed. Each station was staffed with one worker. The first 16 stations had almost no automation; the most sophisticated tools were some ergonomic screwdriver stations. It was a highly labor intensive production line.

Since most of the skills were very basic, the operators would learn all 21 stations in less than six months with little effort or inconvenience to the work schedule. They had an aggressive operator cross-training program.

For this corporation, new production lines were designed by the home office engineering staff and then sent to the facility to debug them and bring everything up to speed. Consequently, this facility did not have a full engineering staff.

This production line was a straight flow line with a conveyor. The conveyor would advance and then remain stopped for the 14-second work cycle time; the transportation time between work stations was two seconds. The appliances were mounted on rotating tables to facilitate construction, and the tables were fixed to the conveyor. There was no new technology in this line, but the design cycle time was considerably shorter than prior designs. It stood out in the facility as being different and placed significant pressure on everyone, especially the materials delivery staff. But the demand for this product was high and the management wanted to only invest in one set of equipment—hence, one high-velocity line.

Earlier, we had done significant training in statistical problem solving at the facility and conducted two waves of Greenbelt training, as well as two waves of Blackbelt training. Following this, Greg, the general manager asked us to review the operation of the Gamma Line. He said he was in a hurry and wanted us to evaluate why the labor efficiency was so low. Labor efficiency was one of their most important plant metrics.

Greg explained the situation. The line had been placed in service over three months before and had never achieved design rates. As the demand ramped up, they had to schedule Saturdays and even some Sundays to meet demand. That was why the labor efficiency was low. In addition, the union was becoming a significant obstacle. From the beginning, the union was against the design. The shop steward claimed the short cycle times placed too much pressure on the workers. Greg had the same concern.

We were not familiar with the metric of labor efficiency, so we asked about this metric and the 0.85 standard. It was explained thusly: It is the allotted labor, which is based on the cycle time and design line staffing, divided by the actual hours worked for all the hourly workers. Each line is calculated based on only their direct labor headcount for those on the line, including those not working. The 0.85 factor was to account for labor that was scheduled and working but not producing due to machinery failures or anything that would cause production to be reduced. We asked if the production losses included quality dropout, stock outs, machine downtime, and not performing to cycle time. They said that was the concept and the minimum standard was 0.85. Anything over 0.85 (or 85 percent) was gravy. On this line, the best they had done was just recently when they struggled to get to 70 percent.

Without knowing it, their labor efficiency was a type of OEE. Not as powerful and not as usable as OEE, but a very similar concept. The unfortunate part of this metric was twofold. First, their objective was to reach 85 percent. If they got there, they would be happy. There was no strategy to go beyond the 85 percent. Second, this labor efficiency was affected by all the aspects that affect OEE, such as quality, machine availability, material supply, and cycle-time stability—to name the major issues. Yet they framed it as a labor issue. Quite frankly, it was everything *but* a labor issue.

A problem exists when a facility is managed as only a cost center. Clearly their only effort was to make sure they did not use too much labor. But as we will see, labor was not their problem—they had neither begun to understand, nor begun to attack their real problems. It shall be shown that their true problems were waste and variation—which is just waste by another name.

Greg went on to explain some more history of the line. He said that upon startup, unlike prior startups, tremendous scrap was generated, over 30 percent. When they slowed down the line for a while, scrap dropped, but when the line was sped up again, scrap increased to unacceptable levels. The workers claimed they did not have enough time to complete the tasks and that the short cycle times caused a lot of stress.

To remove the stress from the workers, they tried the Red-Tag Procedure. The company told the workers that if they could not finish a unit, they should leave it on the line and simply place a red tag on the unit. When the unit progressed down the production line, no one would work on it, and just prior to the test station all red-tagged units would be removed and sent to the rework station for completion. Unfortunately, this did not work at all. The first rework station quickly became overloaded and when the second station became overloaded also, the company decided a change was necessary. To make product, they again slowed down the line while they worked on a solution.

After some thought and interaction with the union, it was decided to give the operators a little control over the line. At each work station, they placed a delay button so the operator could delay the advance of the conveyor and hold it for another five seconds. If the operator was afraid, he/she would not finish the unit in the allotted 14 seconds; the operator would hit the delay. Greg said it had been the plan to institute line stoppages for quality issues, their attempt at *jidoka*, but since the line was not yet stable, they had not implemented this feature. The company thought this might be a good transition into *jidoka*, as well as solve the current production problem.

Immediately, the number of red-tagged units dropped to practically zero and the line produced with only a minor quality issue. Their data showed there was 1.2 percent rejected product at functional test and electrical test combined. This could all be reworked. In addition, there was another 2.9 percent rejected from final inspection, mostly cosmetic defects associated with handling. This, too, could be reworked. Unfortunately, the line rate, since installing the operator-controlled line delays, had not achieved rate. They struggled merely to reach 70 percent. (The design rate, with no losses should be 225 units per hour, [$60 \times 60/16$] but were only able to produce 155 units per hour, or about 70 percent).

Greg mentioned he was impressed by our training and asked us to review the line operation and see if a redesign was in order. We had previously spoken with him about some of the problems inherent in a long flow line—especially one with very short cycle times. He asked for our recommendations and requested we look at converting the flow line to cellular manufacturing with the caveat that there was very little capital available for new equipment. His guidance was to maximize the production rate. Although customer demand was 1,000,000 units per year, demand changes were common, and if the production could be increased and costs reduced, the company would consider dropping the price in an effort to increase market share.

Strategy 1: Synchronizing Supply to the Customer, Externally

Customer demand was produced by running 24 hours per day, five days per week, less 11 holidays. They had 250 scheduled days per year to produce the 1,000,000 units, hence daily production should be 4000 salable units. Each of the three operating shifts had 50 minutes for lunch, along with their breaks, so the available time was 21.5 hours per day with a design rate of 186 units needed per hour. Hence, *takt* should be about 19.4 seconds ($21.5 \times 3600/4000$), and with an OEE of 0.85 the necessary cycle time should be around 16.5 seconds. Their design of 16 seconds should have worked nicely... but it didn't. In fact, it wasn't even close. But why wasn't it?

Our first step was to look at 24 hours of production. The data culled from this are shown in Table 19-1.

The average for this day was 163 units per hour (3516/21.5), which was well below the design rate of 186 units per hour, although our data were slightly above the recent average of 155 units per hour. In addition to the low rate, there was large hour-to-hour

Shift	Hour	Production	Cumulative per Shift	Cum Total
Shift No. 1	7 am	156	156	156
	8 am	180	336	336
	9 am*	132	468	468
	10 am	180	648	648
	11 am**	84	732	732
	12 n	192	924	924
	1 pm*	132	1056	1056
	2 pm	204	1260	1260
Shift No. 2	3 pm	156	156	1416
	4 pm	168	324	1584
	5 pm*	132	456	1716
	6 pm	156	612	1872
	7 pm**	72	684	1944
	8 pm	156	840	2100
	9 pm*	120	960	2220
	10 pm	156	1116	2376
Shift No. 3	11 pm	144	144	2520
	12 m	156	300	2676
	1 am*	132	432	2808
	2 am	180	612	2988
	3 am**	84	696	3072
	4 am	156	852	3228
	5 am*	132	984	3360
	6 am	156	1140	3516
	Total	3516	3516	3516

*30-min lunch.
**10-min break.

TABLE 19-1   Gamma Line Production, Base Case

variation. Hourly production rates varied from a high of 204 to a low of 156 (ignoring the first hour of each shift, which is naturally erratic due to the shift change). This range of 48 units is plus or minus 15 percent of the average. No one could give any rational explanation for the rate variation beyond the obvious explanation that the operators were hitting the delay button. At first glance, the production numbers do not look too bad until we recall that the design of 186 was based on an OEE of 85 percent, or 15 percent losses due to quality, availability, and cycle-time losses. The data we analyzed showed:

- Quality losses to be 3.1 percent
- Zero availability losses
- All other losses were cycle-time losses

So what were the cycle-time losses? The design hourly rate at 100 percent efficiency was 225 units but we produced 163 units so losses were 62 units or 27.6 percent. If quality and availability losses total 3.1 percent, this means cycle time losses total 27.6 – 3.1 percent or 24.5 percent … which is huge! With all these losses the effective cycle time is 22 seconds ($21.5 \times 3600/3516$).

Our design is synchronized externally since we need a 16.4-second cycle time and have a design of 16 seconds, which supposedly should work nicely. It does not, however, and the reason it does not work is because the production cycle time is not producing anywhere near *takt* × OEE.

So, just what can we do about it?

Well, we need to redesign the production process so it will produce to *takt*, and to do that, we will proceed to Strategy 2: Synchronize Production, Internally.

Strategy 2: Synchronize Production, Internally

A New Approach, Multiple Work Cells
We will attack this problem with cellular design, which will directly reduce the variation and better balance the flow, using multiple cells in our new design.

The questions we must first answer are:

- How many cells?
- How many persons per cell?
- What is the cell layout?

The Time Study and Line Balance Review
A time study was performed and a line balance chart was constructed (see Fig. 19-1).

As we evaluate the line balance chart, we find the following.

The first evaluation to make on a balancing chart (see Appendix C in Chap. 14) is to evaluate the waiting time. Recall that the waiting time is the distance from design time to the station cycle time, plus the time for transportation. In this case, the waiting time for station 1 is two seconds, plus two seconds for transportation for a total of four seconds, station 2 waiting time is three seconds plus two for transportation for a total of

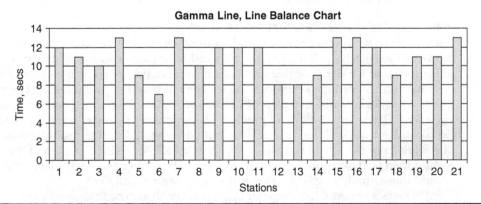

FIGURE 19-1   Gamma Line balance chart, base case.

five … and so on, rendering the total wait time as 108 seconds per unit produced. This is the *design* waiting time, not the actual waiting time. Quite frankly, that is both huge and amazing. The actual work time is 228 seconds per unit for a total labor time of 336 seconds per unit, which means the designers planned for labor losses, due to waiting alone, of 32 percent $[100 \times (108/336)]$. However, since the process is not able to perform as designed, our labor used is 462 seconds per unit (22 seconds $\times$ 21 stations). The *actual* waiting losses are 234 seconds per unit $[(22 \times 21) - 228]$. And now it is obvious that the time losses exceed the actual work time. To put this in simple terms, "we need a lot more time and a lot more labor than we planned for." So, in summary:

- We have 228 seconds of planned work in each unit.
- We are paying for 462 seconds of labor per unit.
- Our actual wastes due to waiting are 234 seconds per unit.
- The planned waste was 108 seconds per unit.

These represent huge losses that we may be able to exploit in the redesign.

Second, we evaluate the balance, qualitatively. In this case, the balance is very bad. There are five stations with a 13-second cycle time, and seven with cycle times of ten or less. That is a 30 percent variation, which then translates into a lot of wasted labor. Clearly, rebalancing will benefit this process.

Third, we find the bottleneck. In this case, we only evaluated to the nearest second, so it appears there are five bottlenecks. There is, by definition, one bottleneck in any system.

Although we do not know which station is the bottleneck, the evaluation is extremely revealing. First, it shows the average times are all below *takt*, yet we know the process cannot sustain this average rate. The graph also shows that there are five stations right near the 14-second maximum time. Consequently, if a slight variation in cycle time occurs, the operator would likely hit the delay button and this would delay the entire process for five seconds. Since on every unit of production this opportunity is presented seven times, I was actually amazed we were able to produce as much as we did. Recall that our real cycle time was 22 seconds. If we add the designed cycle time of 16 seconds (14 seconds of work plus 2 seconds of transportation) to the time of one delay, 5 seconds, and compare this total of 21 seconds to the actual cycle time of 22 seconds, it means, that on average the delay button is struck once per production unit. It is clear that a redesign is needed.

How Many Cells? How Many Stations per Cell?

With the time study and balancing study completed and evaluated, we needed to come up with a reasonable cell design. As you recall, little capital was available to spend on this project, so we were not able to change the end of the line, which had some expensive machinery. Stations 17 through 21 consisted of two tests and three packaging stations, which culminated in pallets of 12 units being shrink-wrapped and ready to load on a semi. These five stations all produced at cycle times below *takt* and had a total of 56 seconds of work. Consequently, we decided we would design multiple cells using the work from the first 16 stations, (which totaled 172 seconds of work) and these cells would then feed into our final cell, which would be testing and packaging. Now, what would those multiple cells that feed test and packaging look like?

We knew we wanted to increase the cycle time. The short cycle times were a root cause for much of the waiting time, and the increase in cycle time was causing the low production.

We have found that, for repetitive small work of this nature, cycle times of 30 to 90 seconds seem to work best. So we did some rough calculations.

Each cell will now have 172 seconds of work, if we use four work stations and can balance perfectly, each station will have 172/4 = 43 seconds of work plus 2 seconds for transportation. With three cells, operating in parallel, that means we would produce 3 units per 45 seconds, or a theoretical cycle time for overall production of 15 seconds compared to our design cycle time of 16.5 seconds. On the surface that looks good, but we know we would not be able to balance the work stations perfectly and since we were not really worried about overproduction, we decided upon four cells. At this point, Greg chimed in and mentioned they had already approved some design improvements in final packaging and test, which would reduce the required work on the fifth cell.

So our final design was four cells of four stations, each working in parallel, and feeding a fifth cell. We decided to lay out all four new cells in the U design layout and leave cell 5 in a straight flow line with five work stations.

We now need to calculate the approximate cycle time. We have 43 seconds of work for the cell, and if we allow two seconds to pass the unit to the next station in the cell, we would have a cycle time of 45 seconds. If we use four cells in parallel, we could produce four units in 45 seconds, or about an 11-second cycle time. We are well below *takt*, were producing more with the same staff and so we knew we were on the right track.

Wow! Were we happy with that! And we had just begun!

Why Not More Cells?

First, when the restraint of no available capital was placed on us that set the lower limit of the cycle time at about 12 seconds. This is the current cycle time for the two expensive testers at the end of the line, hence it made them the de facto bottleneck and the limit on our rate. Now, in all designs of this nature—single station in series—the theoretical number of work stations will be the work time divided by the cycle time, or in this case 172/12 = 14. If we add a little for OEE losses, it is easy to see we need 15 or 16 stations.

Next, at this point it becomes a matter of style if we want 3 five-station cells, 4 four-station cells or 5 three-station cells. This calculation can be refined in additional ways, but we did not feel our data were accurate enough to draw these conclusions. Somewhat arbitrarily, we selected 4 four-station cells, which allows production modulation in 25 percent increments yet is easier for material supply than five cells of three each. Quite frankly, any of the three would have worked well in the beginning.

Balancing the Work within a Cell

Now, just how do we balance the work in the four stations of any cell? In our new cell, we will have four work stations, which we will call stations A, B, C, and D. As a first pass, we will try to combine the work from the existing stations 1 through 4 into the new workstation A in each of the four new cells. Likewise, existing work stations 5 through 8 were combined into new work station B; 9 through 12 into C; and 13 through 16 into D, respectively.

Using the data from our time study, we now have:

- New station A is 12 + 11 + 10 + 13 = 46 seconds
- New station B is 9 + 7 + 13 + 10 = 39 seconds
- New station C is 12 + 12 + 12 + 8 = 44 seconds

Work Station	Work in Secs	Trans in Secs	Total in Secs
A	46	2	48
B	39	2	41
C	44	2	46
D	43	2	45

TABLE 19-2   Gamma Line Balance Chart, New Cellular Design

- New station D is $8 + 9 + 13 + 13 = 43$ seconds
- Flow Line E is $12 + 9 + 11 + 11 + 13 = 56$ seconds, less 3 seconds from the modifications

We will then need to look at the balance. The results are shown in Table 19-2.

In checking the balance for workstations A thru D, we find about seven seconds difference, or about 15 percent. Although we would not normally consider this good enough, we accepted it for now. We had made so many changes we wanted to get the cells in service and check it out. In addition, we were under severe time pressures. The risk here is that any changes we now make in the cells will need to be done four times. Even considering this, we were comfortable and decided to proceed with this design, in addition, this would make for a neat transition. For example, we could reuse almost all the work procedures.

We Plan the Modifications and the Test Run

With this design in hand, and plans to implement these changes, Greg placed yet another restriction on us. We would need to prove this design before he was willing to convert the entire line. Although we were very confident, his stipulation was reasonable, so we agreed to construct one cell (as if we had any choice at all) and direct its product to station 17, while the original line continued to produce normally. We would operate this cell for one or two shifts per day, measure the performance of the cell and then proceed from there. Meanwhile, we implemented the changes on the final test. They were simple and successful and reduced three seconds of work.

So far we have designed a cell, operating at a lower cycle time (remember that *takt* is not really of interest to them, they want more, faster), and balanced so it will flow nicely with one-piece flow. Our practical limit was the inability to add capital. Practically, this meant we wanted to increase the flow rate up to the ability of stations 17 to 21, which was about a 12-second cycle time.

The Test Run

Over the weekend, during the construction of the first cell, we had problems. We planned to convert the automatic conveyor to a track so the platforms could be manually advanced. The conveyor track would not fit in the U cell as designed. We either needed to enlarge the size of the cell or change its shape. We immediately converted this cell to an L-shaped cell and Sunday night tested it. It worked just fine.

Monday morning came and we trained the operators assigned to our new cell, arranged for a new materials handler and by morning break we had the cell in production, although nowhere near the design cycle time. Materials delivery was a problem,

Work Station	A	B	C	D
Design—original work, secs	46	39	44	43
Design—trans time, secs	2	2	2	2
Design—total, secs	48	41	46	45
Current cycle time, including transportation time	45	40	44	42

TABLE **19-3**   Cycle Times for Work Stations, Cellular Design

but that was quickly ironed out. By the end of the shift, the bugs had been worked out and the cell was producing, with no quality losses, at a 55-second cycle time. Although we had hoped for 45 seconds, we were still pleased. The workers were pleased as well. They responded extremely well to the longer cycle times. They simply said that they felt more comfortable and it wasn't as stressful. The looks on their faces were unmistakable—they were grateful. We made some minor changes and prepared for the next day.

Tuesday was a better day. Many of the efficiencies we had hoped for were realized. The cell increased in speed and the cycle times for each station had improved per Table 19-3.

We were really pleased, but let me tell you: Greg was ecstatic. Not only was our experiment clearly showing success, but for the first time in a very long time the shop steward visited him with good news. The steward had gotten unsolicited comments from the cell operators. Uniformly, they liked the new cellular arrangement. They particularly liked the longer cycle times. Their stress was reduced significantly since they could now advance the unit when their work was done. They no longer had to worry about the conveyor taking their work away before they were ready.

We spent the rest of that week using only one cell while we planned for installation of the other three cells. Meanwhile, the original line was in operation as usual. The cell gave the line-added production and this gave the plant a chance to catch up on production. While this was underway, each day we would rotate a new crew into the L cell, and after a little training they would begin production. By the end of the week, nearly all operators had been trained on the new work layout. It was received very well.

Over the weekend, we dismantled much of the old line and constructed the three additional cells, per Fig. 19-2. The construction was simple and did not take long except for the raw materials supply which needed to be arranged for each cell. On Monday morning we started up and there were a few problems, but by the end of the third day the cells were stable and producing at a record rate with all four L cells producing. We completed a time study and the new cycle time was 12.5 seconds.

The Results

Actual production was as shown in Table 19-4.

- Production had increased from 163 units per hour to 266 units per hour!
- Cycle time had reduced from 22 seconds to 12.5 seconds!
- This was a 63 percent increase in production using the same people, on the same machines, with the same raw materials ... and all we changed was the work environment by the conversion to cellular manufacturing.

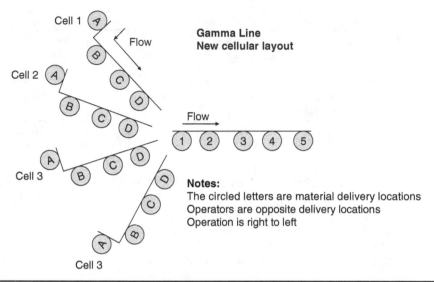

Cell 1

Flow

Gamma Line
New cellular layout

Cell 2

Flow

1 2 3 4 5

Cell 3

Notes:
The circled letters are material delivery locations
Operators are opposite delivery locations
Operation is right to left

Cell 3

FIGURE 19-2   Gamma line, new cellular layout.

Where All These Production Gains Came From

Just to refresh, see Chap. 13 again and let's review the seven ways we can reduce lead time and improve flow:

1. Reduce production time.
2. Reduce piece wait time.
3. Reduce lot wait time.
4. Reduce process delays.
5. Manage the process to absorb deviations, solve problems.
6. Reduce transportation delays.
7. Reduce changeover times.

We did nothing to reduce the production—that is, work time—and since it was already one-piece flow with standard WIP of one, lot wait time was already zero. We had no process delays and there were no changeovers, so from this list we see that three of the tactics were employed. They were:

1. Reduce piece wait time
2. Manage the process to absorb deviations, solve problems
3. Reduce transportation delays

Recall that the design had 42 seconds of *transportation time* per unit. We effectively reduced that to 18 seconds (2 seconds per station × 9 stations) by the use of cells. That directly translates into a faster cycle time by about 1.5 seconds [(42 – 18)/16]. This does not sound like much, but on a 16-second work cycle, it is a 9 percent improvement in production rate. Or viewed from a cost context, we just converted 9 percent more raw

Shift	Hour	Production	Cumulative per Shift	Cum. Total
Shift No. 1	7 am	264	264	264
	8 am	276	540	540
	9 am*	216	756	756
	10 am	288	1044	1044
	11 am*	144	1188	1188
	12 n	264	1452	1452
	1 pm*	204	1656	1656
	2 pm	276	1932	1932
Shift No. 2	3 pm	252	252	2184
	4 pm	276	528	2460
	5 pm*	204	732	2664
	6 pm	276	1008	2940
	7 pm**	144	1152	3084
	8 pm	276	1428	3360
	9 pm*	204	1632	3564
	10 pm	264	1896	3828
Shift No. 3	11 pm	240	240	4068
	12 m	276	516	4344
	1 am*	216	732	4560
	2 am	264	996	4824
	3 am*	132	1128	4956
	4 am	264	1392	5220
	5 am*	216	1608	5436
	6 am	276	1884	5712
	Total	5712	5712	5712

*10-min break.
**30-min lunch.

TABLE 19-4   Gamma Line Production, New Cellular Design

materials into finished goods at no increase in operating expense. (See Chap. 18 for more on this.)

Next, we eliminated the *wait time* of the workers. For the first 16 stations, there were 52 seconds of wait time—we turned this into productive time. That accounted for about three equivalent seconds of cycle time (52/16 = 3.25).

Turning those wastes—the waste of transportation and the waste of waiting; actually, in this case it was all waiting—into productive time effectively reduced our cycle time by over four seconds.

But, hey, not so quick! That doesn't fully explain all the gains.

If the line had been actually producing at a true cycle time of 16 seconds originally, even with all the waste of waiting, it would have been making 225 units per hour (3600/16 = 225) at 100 percent OEE. The real OEE was not 100 percent, rather the line had 4 percent scrap and less than 1 percent availability losses, so OEE was really about 95 percent. Consequently, we should have had about 214 units per hour (225 × 0.95 = 214). We did not have 214 units per hour; rather, we had only 163 units per hour.

So how do we account for this missing 51 unitsper hou (214 − 163 = 51)? The answer has to do with the fact that the line could not perform at the design cycle time of 16 seconds.

And why was that?

You got it! The answer is *variation and dependent events*! (See Chap. 24 for more information on this.)

Yes, this effect is huge, and in this case it is easy to understand. Whether the process is in lock-step when the process is fully synchronized with a conveyor, as this one was, or if there is no inventory, the effect is the same. Any time one station performs at a time above the cycle time, the effect is felt in all stations. This effect accounted for a huge loss of production, 51 units per hour, over 20 percent of the design rate of 225 units per hour! Most people find this interaction of variation and dependent events amazing! Well, amazing it is, but it is also true, and it is also often an overlooked phenomenon.

Think about this concept of variation and dependent events for just a second. Since all 21 process steps were synchronized by a conveyor in this case, any time one station would perform at a time above the design cycle time, there was a time loss for the whole line—for all 21 stations. It was exacerbated by the short cycle times, but the basic problem was that 21 people had to be totally synchronized to make this work. In this case, each of the 21 work stations were dependent upon the other 20, otherwise no station could maintain its cycle time. That is the nature of variation and dependent events and it must be understood. Conversely, in the four-person cells there are now only three levels of dependency, so even if one person slows down, they now only affect three others, not 20, and with 4 cells that one person only slows down 25 percent of the production, not all of it.

> **P**oint of Clarity There is always a loss associated with the variation in the system… Always!!

In this example is explained one of the beauties of cellular design. They are a natural variation reduction device and they help us to execute the flow improvement tactic of, "managing the process to absorb deviations" (see Chap. 13).

Summary of Results

So, summarizing, just how did the conversion to cells improve our production?

- With minor losses, the 21-station flow line could only produce 163 units per hour.
- By using cellular design we reduced the waste of waiting, caused by the interaction of variation and dependent events and we should have been able to produce about 214 units per hour.
- However, we produced 266 units per hour, and this increase above the 214 units per hour was due to the reduction of the wastes of waiting and transportation, which was also made possible by the cellular design concept.

Make no mistake about it. Properly designed cells are:

- a variation reduction device
- a waste reduction device
- a productivity improvement tool!

> **P**oint of Clarity   Cells are a natural variation reduction device.

Chapter Summary

So what happened to Greg's labor efficiency? Well, it skyrocketed and is now well over 100 percent, which of course cannot be. Needless to say, he was very very happy.

The Gamma Line is a story often repeated in manufacturing and differs from the situation that Ohno and Toyota had to deal with in the automobile business. The Gamma Line produced a household appliance where the life of any given product is seldom three years. In addition, the monthly demand for household appliances has huge swings due to the sales and promotions by their customers. In short, the concept of *takt* does not really apply to these products, and overproduction is handled by running the line in less time—seldom is the production line slowed down. Nonetheless, we were able to apply the concepts of Lean and achieve huge process gains. To achieve these gains, exactly which tools did we utilize?

First, we looked at the waste in the system; our focus was waste removal. Then, utilizing cells, we were able to eliminate most of the waste. They already had a pull production system in place, using one-piece flow, but we utilized line balancing and conversion to cellular manufacturing. This allowed us to reduce the wastes of waiting and transportation and more effectively utilize our manpower. Cycle-time reductions were a large part of our gains, as was an understanding of how variation and dependent events were affecting our production rate. Changing the design to cellular production allowed longer cycle times so that small variations in techniques did not affect production nearly as much as with the shorter cycle times.

> **"E**veryone is somewhere on the journey to become Lean; no one has yet arrived. **"**
> Lonnie Wilson

These were the lean techniques employed to achieve the 63 percent increase in the hourly production, as well as achieve a more level hourly production.

However, these gains were not achieved in a vacuum. A number of lean techniques had already been implemented and were functioning. First, to achieve flow, the original line already had one-piece flow. Consequently, their first-piece lead time and lot lead time were relatively good. To supply the line, they used a two-container *kanban* system for the small items, which represented 75 percent of all parts. The large parts were delivered on a scheduled route by the materials handlers. In addition, they obviously had multiskilled workers, which made this conversion possible. All of this helped as we embarked on an effort to improve "labor efficiency."

CHAPTER 20

The Story of the Alpha Line

Aim of this chapter … to show, by example, how the management of Bueno Electronics stepped up to the challenge, recognized their responsibilities, and provided the leadership to guide the company through the needed cultural changes. In so doing, they managed the five cultural change leading indicators very well, practically textbook in nature. (See Chaps. 5–9.)

How We Got Involved

The Alpha Line was the first of many production lines at this Mexican maquiladora, Bueno Electronics. It was the first plant in Mexico for this European-based manufacturer, who moved to Mexico to take advantage of low-cost labor and the proximity to the U.S. auto industry. Since this was their initial interaction with the U.S. auto industry, and didn't know how to deal with their customer, nor had the required skills to meet their customer's demands, they retained us to assist them.

At Bueno Electronics they started up the Alpha Line, and after their initial customer audit, in which they achieved a failing score of 41 (75 was the minimum acceptable), we were contacted to assist them. Their primary weakness was in statistical techniques, where they scored 0 out of a possible 10 in four different areas of statistical applications. Of particular concern to them was the need to implement SPC (statistical process control) for all critical product and process characteristics—a skill they were completely lacking. Here is where their story of leadership and management commitment unfolds.

For several years, as Bueno Electronics grew, we helped them implement a number of lean systems, including pull production systems, cellular manufacturing, and an OEE transformation, to name but a few. In addition, they made great strides in improving both their flexibility and responsiveness through lead-time reductions. However, the best part of the Bueno Electronics story occurred when we were initially asked to assist them on the Alpha Line. Of particular importance was how their management responded when they first started up this production line. That set the stage for all the success that followed, including their journey to becoming a lean facility. This is that story.

Initial Efforts to Implement Cultural Change

We were retained by the plant manager, Kalista and reported directly to her. Our first assignment was to teach the required skills of SPCs, measurement system analysis (MSA), correlation and regression (C&R), and designs of experiments (DOE).

Of particular significance was the way in which the training was done. This sent a clear message to the facility and set the stage for later successes.

The training started with top management in the various departments, including production, engineering, purchasing, maintenance, and human resources—all of which attended every class. The initial training was an SPC class: 36 hours of training that covered SPC and focused on the topics of attribute and variables control charting techniques. In addition to the classroom work, they needed to complete a project. At the initial class, Kalista addressed the class and "set the tone" very well. She stated, "… management has a distinct role in the success of this and there are no shortcuts. They all must be directly involved …" More importantly, she attended the entire training, and like all the others—managers and non-managers alike—she completed the required project and passed the final exam. Kalista set an unmistakable example. Following this management group, other supervisors, engineers, and technicians were trained in SPC. Following the SPC training, other techniques such as MSA, C&R, and DOE were taught. In each case, the management team was the first group trained. In addition, classes were given in Kepner-Tregoe problem solving, and later we were retained to assist in the implementation and support so often needed in these efforts.

> **"P**ut everyone in the company to work to accomplish the transformation. The transformation is everyone's job.**"**
> (Point 14) W. E. Deming
>
> **"A**dopt a new philosophy. We are in a new economic age. Western management must awaken to the challenge. They must learn their responsibilities and take on Leadership for Change.**"**
> (Point 2) W. E. Deming

It was not surprising that the entire effort had great traction and very rapidly the process improvements became obvious.

Around this time, Kalista, the plant manager, approached me with a specific concern. She had been given an appropriations request to construct the second rework facility. It disturbed her. She had been assured that all work stations (they had a 28-station line) were at 98 percent effectiveness or higher, except for one that was struggling at 88 percent. Rework for this product was permitted, but she immediately knew something was amiss. We did a quick analysis of the line and found that its first-time yield (FTY) was less than 50 percent. This means that less than 50 percent of the total product went through the production process the first time, with no rework. Over 50 percent of the product needed to be reworked at least once, with some units getting reworked more than once. When this was explained, she was amazed but immediately approved the rework station. Being the good manager she was, she called together the managers of engineering and production, and with us also in attendance, issued the following instructions:

- Start using the metric of FTY as the plant's measure of internal quality. She wanted it to be calculated and posted by next Monday.

- Develop and execute a training course in FTY, that not only taught engineers and managerial personnel the concept of the Poisson distribution, but also its detailed calculations. Until this was done, the production manager would calculate and post the FTY daily.

- Create a specific plan to improve the FTY.
- As part of the plan to improve the FTY, one of the action items would need to be the dismantling of the second rework station.

The metric of FTY became the facility's measure of internal process quality, and the key tool to improving the FTY became the use of SPC.

Kalista, had shown excellent leadership in many ways and certainly a strong point was how she modeled the behaviors she expected from others. For example, she could always be found with her ESD strap and gloves while on the floor and her office clearly met the same standards of 5S they required on the floor.

And her leadership in this effort was no different. First, she had required that the entire management team know SPC, and then she had created the key metric for the plant in FTY. She and all her managers were the first wave of SPC students taught and each one had to complete their project as well as pass the final exam. All did. Next, she set about supplying all the needed training to the rest of the organization. Operators were taught how to take data, make Xbar-R chart calculations, and plot the data. They were also taught how to read the charts for special causes (1 point beyond +/−3 sigma, runs and trends, for example) and then the operators would solicit assistance for out-of-control conditions from their supervisor. Supervisors and engineers were taught the necessary problem-solving skills, plus how and when to recalculate limits and general chart management. In addition, one engineer and one full-time assistant were dedicated to this effort as the number of charts skyrocketed (see Fig. 20-1) with the success of the effort (see Fig. 20-2). She created the support mechanism so the entire production line was capable of Rapid Response PDCA (Plan, Do, Check, Act). They were using line personnel to solve the problems in real time. This was a classic example of lean leadership in action.

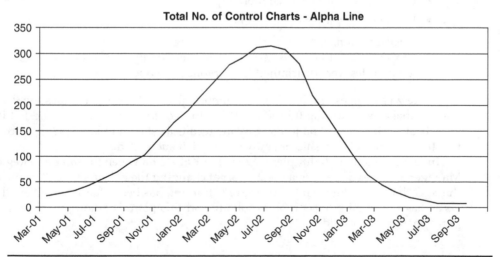

FIGURE 20-1   Total number of control charts: Alpha line.

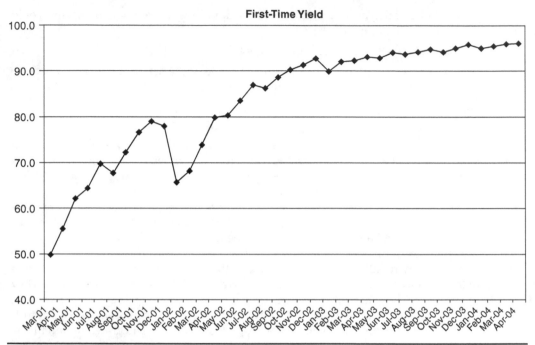

FIGURE 20-2   Alpha line, FTY.

The leadership shown by the plant manager went well beyond the training. She made sure that:

- The metric of FTY was clearly posted at the front of all production lines.
- FTY was discussed at each morning meeting.
- FTY was a key topic in the weekly production planning meeting.
- Once a month, chaired by the plant manager, there was a meeting to address FTY and FTY alone. This meeting was attended by all plant management, and the minutes were distributed to everyone in a supervisory position.

There was no question that FTY was important, and there was no question that the plant manager was leading this effort. This effort could be heard, seen, and felt in all aspects of the business. And judging by the results as shown in Fig. 20-2, it is obvious that this effort and leadership not only persisted, it succeeded.

In fact I worked with this plant, even after Kalista had been promoted to Regional Manager. She had done such a great effort at changing the culture, that 10 years later, the SPC charts are still used today, the SPC training has been standardized at all levels as has the small group problem solving and this plant is among the world class performers in their field.

Some of the Results

As you can see in Fig. 20-2, yield climbed significantly and the entire effort was a huge success. Though it was a success because the yield climbed, the effort's success went beyond that.

As it turned out, that very year, demand for the plant increased dramatically, largely due to the process improvement efforts. The customer of the plant had three alternate suppliers (each of the four now had 25 percent of the demand) of this product, but the customer's stated desire was to reduce that to only two suppliers. One would be the primary supplier; the other a backup supplier. The purpose was clear: The customer wanted to reduce his variation—that meant fewer suppliers for each product, which meant the plant needed to be the best supplier or lose some or all of the business.

> **"C**reate constancy of purpose toward improvement of product and service with the aim to become competitive and to stay in business and to provide jobs. **"**
>
> (Point 1) W. E. Deming

Because of its efforts, by the end of the first year this plant had outperformed all others. At the annual "Give me your quality improvement ideas" meeting called by the customer, we were able to finesse the following information from the customer. This plant was the only plant of the four that was meeting the customer's goal of less than 500 ppm field failures—this plant was at 220 ppm. The other three suppliers were at 700, 1100, and 3300, respectively. It was no surprise that by the end of year 1, the plant now had 45 percent of the demand, while the fourth supplier had been eliminated entirely. The remaining two backup suppliers each had 27.5 percent of the business. In subsequent years, the Alpha Line increased in capacity, and after three years it was supplying 70 percent of the demand, with one backup supplier taking on the remaining 30 percent. A look at the progress made by the plant, as shown in the Quality History table in Table 20-1, shows how well the plant performed.

Continuous Improvement, as It Should Be

There is yet another interesting aspect of this story. Take a look at the FTY graph in Fig. 20-2. It started at about 50 percent and rose almost steadily to over 96 percent, over a three-year period. It should be noted that this was an old design with lots of manual operations, and with 28 stations that meant each station on average was over 99.8 percent effective. During this period of continual process improvement, the key tools used were the statistical tools of MSA, DOE, and SPC. In addition, small group problem solving was implemented, and other than basic good work practices, nothing of note was implemented—yet the line continued to prosper. It prospered month after month and year after year. This was a classic, almost textbook, example of good leadership, good process management, and process improvement. It was continuous improvement as it should be.

	Mar-91	Jan-92	Jan-93	Jan-94	Apr-94
First-time yield (%)	49.9	65.7	89.9	95	96.1
Assembly line returns (PPM)	1200	780	320	155	62
Field failures (PPM)	220	120	47	22	8
Cost of scrap (K$/Mo.)	$150	$120	$46	$4.2	$0.5
Production rate (units/day)	2800	3300	5200	8300	10,500

TABLE 20-1   Alpha Line, Quality History

The Cool Story of SPC: SPC Done Right!

There is another fabulous subplot in the story of the Alpha Line. It is the story of SPC and how to implement it.

In our SPC trainings, we tell our students that one of the purposes of SPC is "to cease doing SPC." This sounds odd, but recall that SPC is not a value-added process. It is a waste. Maybe a necessary waste, but a waste nonetheless. So just how do we get out of the business of SPC once it is started? That was the beauty of this effort. The Alpha Line did it right and this is how it was done.

The primary purpose of SPC is to gain intimate process knowledge. Then, using this knowledge, it is possible to stabilize and improve the process. Once the process is improved to a certain level and process parameters have been standardized, the SPC should show that the process is stable and capable [a high Cpk (Process Capability Index), for example). Once this is accomplished, there may be no need for the SPC. Let's discuss just one specific example on the Alpha Line.

In this process line, there was a solder "touch-up" station—actually, the plant had several, including the rework stations. Here, an operator would make minor repairs using a solder iron. This station had a high fallout rate. To better understand the process, a DOE was performed, and it was found that the solder quality and solder iron temperature were the two critical process characteristics. The proper solder was purchased and the solder iron temperature was monitored and placed on SPC. An Xbar-R chart was used. Subgroups of three were gathered every hour and plotted. The chart quickly showed the appearance of a special cause after about 20 hours of operation. A quick analysis determined that flux and residue had built up on the iron and additional mechanical cleaning was instituted. It was made part of the standard operating procedure, and at the start of each shift the operators all cleaned their solder irons. Also, at this time the sampling frequency was changed from hourly to every 2 hours.

Some minor problems were encountered, but FTY increased by about 4 percent at this station. (Please recall that one station was at 88 percent FTY—this was it.) Over the next 2 months, sampling was changed to every 4 hours and yield climbed another 2 percent, though we are not exactly sure why—I have always suspected it was due to the Hawthorne Effect.

The Hawthorne Effect

At the Western Electric plant in Cicero, Illinois, from 1927 to 1932, studies were done on workplace conditions and worker performance. In one series of tests, it was hypothesized that worker productivity would improve if workplace lighting was improved. They improved the lighting and, not surprisingly, worker productivity rose. Someone questioned the validity of this testing, so they performed another test. In this one, they told the workers that the test was such a success that they were going to increase the lighting even further. They made their changes and worker productivity rose once again. However, instead of raising the lighting levels, they actually had lowered them; leading the researchers to conclude that there were other dynamics to consider. This and other experiments led them to develop what has become known as the "Hawthorne Effect." Here is one definition of the Hawthorne Effect.

People singled out for a study of any kind may improve their performance or behavior, not because of any specific condition being tested, but simply because of all the attention they receive.

But then we noticed a common occurrence on many of the SPC charts. Somewhere between 45 and 90 days of operation, the process would go out of control. Upon investigation, we found that the solder irons just wore out. It seemed the tips would change metallurgically and not be able to hold the temperature. So we implemented a program to change out the tips. This was done to all solder irons during the monthly plant preventive maintenance (PM). The tips were only a few dollars each and even though we changed them out prematurely, we did not want to scrap even one electronic control unit (ECU) since each was worth about $65. Shortly after this process change, the solder iron SPC changed sampling to once per day and we used only one data point, transferring the information to an XmR chart for individuals. This continued for several months, and the process showed remarkable stability, with Cpks exceeding 3.0. At this point, we discontinued the SPC entirely and the process continued free of defects at the solder iron stations.

> "Cease dependence on inspection to achieve quality. Eliminate the need for inspection on a mass basis by building quality into the product in the first place."
>
> (Point 3) W. E. Deming

So just what happened? Exactly how did we execute this purpose of SPC, which is to cease doing SPC?

Well, we used the information from SPC to make the process more robust. First, we implemented a simple form of maintenance: tip cleaning. Since the process was stable, we reduced the sampling frequency to reduce the cost of the SPC. Then, through careful monitoring of the SPC, we gained additional information about the process. We used this information to make process changes—in this case, a monthly PM, which increased the robustness of the process. All the while, we were reducing the cost of sampling and analysis until we were able to do away with the SPC entirely.

So you see, here we were successful. By implementing SPC, we were able to cease doing SPC and in the interim we made the process robust. Neat, huh?

Just How Committed Was Management of the Alpha Line?

The Commitment Test

Let us test their commitment by grading it against the following:

The Five Tests of Management Commitment to Lean Manufacturing

1. Are you actively studying about and working at making your facility leaner and hence more flexible, more responsive, and more competitive? (All must continue to learn and must be actively engaged; no spectators allowed!)

2. Are you willing to listen to critiques of your facility and then understand and change the areas, in your facility, which are not Lean? (We must be intellectually open.)

3. Do you honestly and accurately assess your responsiveness and competitiveness.... on a global basis? (We must be intellectually honest.)

4. Are you totally engaged in the lean transition with your:
   - Time
   - Presence
   - Management attention
   - Support, (including manpower, capital, and emotional support)
     (We must be doing it; we must be on the floor, observing talking to people, and imagining how to do it better; lean implementation is not a spectator sport.)

5. Are you willing to ask, answer to, and act on, "How can I make this facility more flexible, more responsive, and more competitive?" (We must be inquisitive, willing to listen to all including peers, superiors, and subordinates alike, no matter how painful it may be and then be willing and able to make the needed changes.)

The evaluation below gives a test-by-test critique of how the management performed as they made the huge progress on the Alpha Line.

How Did They Score?

1. By attending all the training—and completing a project, too—they showed their commitment to learning. So on this point, they receive the top score.

2. As for point two, they showed intellectual openness throughout. First, when we recommended they be the first ones trained, Kalista, the plant manager readily agreed; they modeled the correct behavior. When the request for the second repair station was made and we gave them our recommendation, they listened. In addition, not only did they make a clear assessment of their skills—admittedly, this was highlighted by the low score on their initial customer audit—but as problems arose, they listened to others, accepted their responsibilities, and provided the leadership to succeed.

3. Regarding intellectual honesty, it was their hallmark, so they scored well in this category, too.

4. Their firm was the poster child for management engagement. I am not sure how they could have become more engaged as managers. They did the easy managerial tasks of hiring the right people, creating the goals, and finding the finances to pay for the efforts. They also did the more managerially difficult tasks of attending trainings, leading meetings, and spending time on the floor and with the people. These last three items cut deeply into a manager's schedule and are so often slighted—but in this instance, no such thing occurred, even though they were already working for long hours.

5. They were inquisitive and open to changing whatever was needed to attain the goals they had established.

Quite frankly, on all Five Tests, they scored extremely well, which is a sign of why they were able to make such huge gains over a long and sustained period. Results like this do not come about without a committed management team.

How Did the Alpha Line Management Team Handle the Fundamentals of Cultural Change?

Let's look at the actions of Alpha Line management in relation to the implementation of any cultural change transformation (see Chaps. 5–9). How did the plant manager handle the five cultural change leading indicators?

Did They Have the Necessary Leadership?

So did they have the necessary leadership? The answer is a resounding, "Yes!" The company's leadership was evident in:

- The direct involvement and commitment of management
- Their open and honest evaluation
- Their acquisition of the necessary resources, especially of people
- Their implementation of a plan to train
- Their plan to improve and all their actions to support the plan
- Their creation of a metric: FTY
- Their creation of a whole series of "problems" by establishing goals for FTY
- How clearly management communicated all of this to the entire workforce. Everyone knew the objectives and the direction of the quality transformation.
- The fact that every time there was a problem, management acted—doing so thoughtfully, quickly, and decisively.

Leadership was especially evident in the actions of the plant manager who spent time on all that was required to make this effort a success. At every step, she was not only involved, she was committed. She set an excellent example. Actions by the plant manager were swift and effective. Thus, no one at the plant ever doubted the plant's direction and its goals.

Did They Have the Motivation to Make It Work?

Absolutely, they were staring at an audit score of 41 and the thought of losing business if they did not improve. This was communicated to and understood by everyone in the plant. In addition, to management this was a grassroots effort at getting a foothold in the American automobile business as well as at moving into the low-cost Mexican labor market. No one wanted this to fail. Everyone worked long and hard to make it succeed. Overtime and weekend work was commonplace.

To further motivate the employees, plant metrics were posted and discussed often. This was not done to create a punitive or intimidating attitude, but to clearly and simply communicate to the employees in hopes of succeeding. Since the necessary work was done, the effort was a success. This early success fueled even greater motivation. The management team had created an environment of success—an environment that clearly said, "We will do what is necessary to succeed and we expect the best from you." Everyone responded accordingly, as people normally do.

Did They Have the Necessary Problem Solvers in Place?

At first, the necessary problem solvers were not in place. It was key that they recognized this and responded openly and honestly to it. In my experience, this honest evaluation and admission is not common. Most try to, instead, "just make do," and usually fail. However, the plant manager showed the commitment to not only start at the management level with the training, but as the effort progressed, frequent assessments were made and the necessary training and staffing was supplied. This included the addition of both engineering and support staff, along with the addition of a consulting firm—ours—to provide ongoing support. Within 3 months, all the necessary problem solvers with the necessary skills were in place.

Did They Have the Whole Facility Engaged?

Recall that in the assessment of management commitment we said, "Their firm was the poster child for management engagement. I am not sure how they could have become more engaged as managers." First, the management team got engaged. They learned what to do, how to do it, and acquired the resources. Next, they made sure the rest of the staff knew, what to do, how to do it, and had the resources to accomplish the necessary work. Finally, as the need for SPC support grew, they supplied the training and additional resources to keep it engaged. Like I say, they were the poster child for "how to properly engage."

Did They Foster a Learning/Teaching/Experimenting Culture?

Point of Clarity A good leader will make it so clear through his/her actions that he/she need not say much.

Again, I am not quite sure how they could have done this better. The training attendance was virtually 100 percent. Managers then taught their subordinates down to the factory floor. Small group problem solving augmented the information gained from SPC. They learned, they taught, and they prospered.

And the results were obvious, as we have seen!

Chapter Summary

As you can see, the story of the Alpha Line is one of management commitment, leadership, and support above all else. They managed all five of the cultural change leading indicators well and in turn created a culture that was producing a better product for their customer. The customer satisfaction was manifest in additional business. Not only did the actions of management allow them to better satisfy their customers, but they simultaneously made the plant a better money-making machine, with increased employment and job security for their employees.

It is not just a wonderful story of cultural change managed well; it is a wonderful success story as well.

CHAPTER 21

The Story of the Bravo Line: A Tale of Reduced Lead Times and Lots of Early Gains

Aim of this chapter … to show in dramatic fashion, using a typical plant scenario, how process improvement, the elimination of waste and lead-time reductions can be accomplished. This was done on the Bravo line and is described herein. This case study will also show just how important *jidoka* is to these process improvements. Finally, we explain how the seven techniques to reduce lead time and improve flow were applied at the Bravo Line and how these changes resulted in a reduction of first-piece lead time by 97 percent, a reduction of lot lead time by 81 percent, and a greater than 60 percent reduction in manpower consumption.

Background Information

My company was called into "use lean technology to improve the performance" of the Bravo Line. The problem was stated as: "The line cannot meet the demand of 10,000 units per week. It is well laid out, has a cycle time that should easily meet demand in a 5-day work week, but we consistently need to work overtime. In fact, we are working on Saturday and Sunday even though the production plan says we should not have to. It has gotten so bad that we have even had to put the utility line into service on occasion, effectively working an eighth day, just to meet regular demand. Quality is extremely good, a process with Six Sigma yield, but we just cannot meet demand."

We inquired about the OEE (Overall Equipment Effectiveness) for this line, and they did not know, but we were able to discover the following information:

- The production process consisted of two cells, operating in series.
- Nearly all steps were simple manual assembly, with no major equipment until final test.
- They believed that line availability was well over 95 percent, close to 100 percent.
- Operators were highly cross trained.
- They had not had a stock out in over a month.

- They had recently conducted line time studies, and had both line balancing charts and standard combination work tables posted at the line for the product.

- The entire plant had begun a lean initiative with some withdrawal *kanbans* and had begun utilizing U-shaped manufacturing cells.

- On this line, which produced 11 different models, they had also made an effort to go to small lot production.

- Within each cell they had a pull system in place, using *kanban* spaces, and the operators would not forward the small lots unless the *kanban* spaces were empty.

- All models were very similar, with over 90 percent of the component parts being the same in all 11 models.

- They described a small lot as 50 units, which was a tray.

- Four trays were stacked in a small box and five small boxes were packed in a larger box, for a total of 1000 units per box, and a typical shipping lot was 2000 units, or two large boxes on one pallet.

- The customer demand was 10,000 units per week of a model mix, but since changeover time was minimal, product mix was not an issue. Their problems were threefold:
  - They could not produce to schedule and frequently missed shipments
  - It took nearly twice the time
  - It took nearly 70 percent extra labor to make a batch

Implementing the Prescription

We did some preliminary calculations and could easily understand some of the problems with production times and missed shipments. We could not yet explain the magnitude of the extra labor required, however. Nor could we explain the need to work 8 days to produce 5 days of product.

Our approach was simple: We used the prescription outlined in Chap. 14 and we decided to:

- Synchronize supply to the customer, externally

- Synchronize production, internally

- Create flow, including
  - Establishing *jidoka*
  - Working to destroy batches

- Establish a pull-demand system

- Standardize and sustain

Synchronizing Supply to the Customer, Externally

The *Takt* Calculation

Demand was 10,000 units per week and the normal workweek was 5 days. Available time was 24 hours per day, less a 30-minute lunch and two 10-minute breaks during each of the three shifts, so we needed to produce 2000 units in 21.5 hours or generate a 39-second *takt* ($21.5 \times 3600/2000 = 38.7$).

We checked the standard work combination table and it listed the cycle time as 28 seconds, but the line balancing studies appeared to be balanced to 25 seconds, so we were baffled. First, why have two different cycle times? And second, if the cycle time design is way less than *takt*, where are their problems? None of these questions could be answered by the production supervisor or the process engineer—or anyone, for that matter.

Synchronizing Production, Internally

The Basic Time and Balancing Studies

Even though they had done time studies, we redid them and found the following (as shown in Figs. 21-1 and 21-2):

It is obvious they had made an effort to balance the cycle times to 25 seconds. We could not find a basis for the 25 seconds, which was also the basis used in their

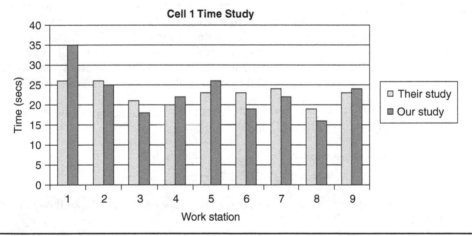

FIGURE 21-1   Cell 1 balance chart, base case.

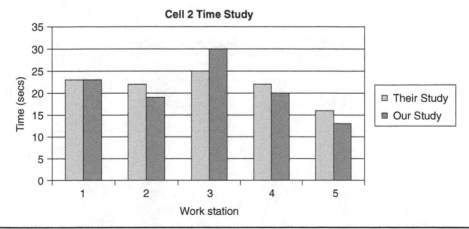

FIGURE 21-2   Cell 2, balance chart, base case.

planning program. Ever since the original balancing was done 6 months earlier, problems had occurred at both cells: at Station 1 in Cell 1, and Station 3 in Cell 2. As a result of these problems, the work procedures had been modified but the balance chart and planning numbers had never been updated. This, of course, was part of their problem, but only a very small part.

We completed the three-part evaluation—waiting time, station-to-station balance check, and bottleneck analysis—of the balancing chart, but it was not at all revealing as to why the system could not meet demand. *Takt* was 39 seconds, well above the line bottleneck of 35 seconds. Yet they say it takes twice as long to produce an order, which means, they have an effective 78-second cycle time. It is clear that the design cycle time was not an issue. This evaluation was not particularly revealing, but it did show that the line balance was simply terrible. Cycle times varied from 35 to 13 seconds—this needed to be corrected, but as I said, it does not explain our problems. At some level, that explains why they had extra staffing. But why did the jobs routinely take the extra lead time?

To unravel this question, we needed to do a line balance chart with our data. First, we needed to find a reasonable cycle time. Normally, this is the *takt* time multiplied by the OEE (as a decimal fraction). In this case, we chose to take management at their word. Recall, they said, that quality losses are less than 3.4 PPM, and availability losses are practically zero. In that case, all the OEE losses were due to cycle-time variations, and so the OEE calculation basically lost its worth. We now knew where to direct our attention. Consequently, we calculated the rest of the redesign based on an OEE of 1.00, so *takt* time effectively became cycle time.

The work (from the time study) for Cell 1 was 207 seconds and the work for Cell 2 was 105 seconds. Consequently, for a 39-second *takt*:

- At Cell 1 we needed 207/39 = 5.3, or six stations
- At Cell 2 we needed 105/39 = 2.7, or three stations

The line balance charts for the redesigned flow now looked like those in Figs. 21-3 and 21-4.

We now needed to redesign the work stations. It turned out to be remarkably simple. We created a "Time Stack of Work Elements" (see Chap. 22 for an example) and were able to redistribute the work very nicely at both cells. This was easy to do since the work

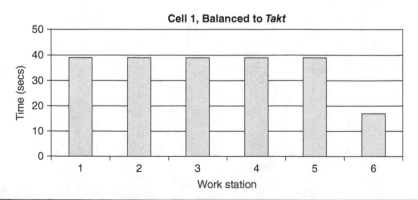

Figure 21-3   Cell 1, balance chart, redesigned flow.

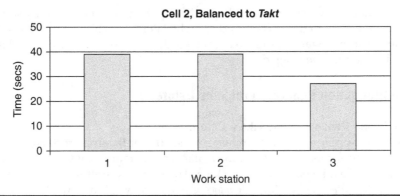

FIGURE 21-4 Cell 2, balance chart, redesigned flow.

elements were almost all manual and almost all short-duration process steps. In addition, each work station had several process steps and the staff was highly cross-trained.

Creating Flow, Including

- Working to destroy batches
- Establishing *jidoka*

Working to Destroy Batches

To destroy batches, we wanted to implement one-piece flow in place of the 50 unit batches. Although the workstation design did not allow close coupling of the workstations into a tight cell, this was relatively simple since the line previously had a conveyor and all we needed to do was move it into place to connect the processing stations. The conveyor eliminated the need to pass 50 unit trays from station to station, and in this case, no *kanban* system would have been better. The conveyor had gates to divert the product to the work stations. Since we knew little about the process stability, we modified the *kanban* spaces to hold a maximum of five pieces.

To better connect the cells, we decided to use the *kanban* cart operator to transfer the small boxes from Cell 1 to a makeshift storehouse we created in front of Cell 2. We changed the transfer lot from 1000 units to 50 units.

Establishing *Jidoka*

We made one other agreement with management, which we knew would be critical. The agreement was that if we had a quality problem, we would stop the line and not start it up until the problem was fixed. We knew it would be crucial to have this type of *jidoka* concept in place. At first, they balked at this proposal. We did not even try to explain logically how important this was, but when we reminded them that they had characterized the line as having few quality problems, it was hard for them to object. They reluctantly agreed.

Establishing Pull-Demand Systems

Our *kanban* system provided a good pull system within each cell, but we had no pull signal from the storehouse. We had enough information to design a good *heijunka* board with a make-to-stock system. We inquired about implementing a *heijunka* board with a production *kanban* system but, like modifying the work stations, they were opposed to taking the time to do it. They said "Maybe later," which usually means, "No." It was clear what they wanted: to increase the production rate and reduce the labor it took to produce a unit. We obliged.

The Production Run, with Problems Galore

Our *Jidoka* Concept Works Like a Charm

With these preliminary steps in place, we set up for the first production run. It took less than 1 hour for us to see one of the major issues. Station 6 of Cell 1 was now only used for final visual inspection and packing. The very first unit it received had an incorrect O-ring installed. Following our *jidoka* guidelines, we shut down the line, gathered the cell workers, and began to investigate. It was simple: The changeover had missed that this model needed both a different shaft and a different O-ring. The shafts had been changed but the O-rings had not. Fortunately, due to the *kanban* system, we had less than 25 units to rework. We reworked the work-in-progress, modified the error in the changeover procedure, acquired the correct O-rings, and started again. On this run, we had to stop four more times for quality problems in the first 4 hours of production alone. It was clear at this point that process stability was nonexistent. They needed a good dose of lean foundational issues—that is, quality control (see Chap. 24).

Back to the test run and the finding of defective products. In each case, with small lot production, the problem was easily corrected and we only had a few units to rework. If we had maintained the lots of 50 as before, when the incorrect O-ring was found we would have had 50 units per work station to rework. That would have been over 400 units on the old nine-station arrangement for each of the five quality problems we encountered. Combined, that would have been 2000 units reworked with the old cell design in the first 4 hours alone—recall that daily production was only 2000 units! It was becoming obvious to everyone how important the small lot and *jidoka* concepts were. And it was equally clear that we had found at least one major source of the extra time it took to make an order. The rework, in the old arrangement, could easily have taken as much time and labor as the design workload.

Suspicious that quality might be a larger problem than previously portrayed, we did a little questioning and found that almost every run had one or two, or sometimes more "start-up burps," as they called them, and rework like this was common. In fact, that was one of the reasons they had everyone so cross-trained: so they could do the rework. There were several, huge, absolutely unmistakable messages in what we had just uncovered.

- Their concerns about quality were focused on the product only. They had a very low interest in, and understanding of, process capability and process stability.

- Management never once mentioned that rework was a concern. They literally had to be blind not to see it. It was a culturally acceptable norm to overlook rework. Once we started the process, and observed its operation, the defects stood out. Our *jidoka* concept made it procedurally unacceptable to proceed without fixing the problems.

- Their view of downtime was odd, to say the least. They did not consider the line to be out of service while all this rework was going on, even though they were able to produce precisely nothing. That was "just how we do things around here," they said.

And the list could go on. At this point, we "began to see" that the required rework both extended the time to produce a batch and also consumed much more labor than was designed.

Production Improves Even Further

With these problems behind us, the line speed picked up and right before the shift change we increased the conveyor speed so there were only two pieces between stations. Production went along without a hitch. After about 16 hours, we completed the first batch of 1000 units in front of Cell 2. There was already a queue of 46 hours in front of Cell 2. The next day, we returned and were able to break into Cell 2's production schedule, which had operated flawlessly with only three people. When our model got there, it went through without event. In the ensuing days, we were able to work off the materials in the Cell 2 queue and get it more in balance with Cell 1. After one week, we had the flows essentially balanced with a small buffer (2–10 hours) in front of Cell 2. We again tried to push the concept of a *heijunka* board. Had we done that, we felt we could have effectively eliminated the queue in front of Cell 2. They again decided not to change and continued to plan the two cells independently.

At this time, something very interesting, although not uncommon, happened. By the end of the first week, the whole line production had sped up—some work stations were under 30 seconds, and it looked like we could rebalance and thus need only five people in Cell 1 and only two in Cell 2. No one could really explain it, but this phenomenon is known as the "Hawthorne Effect" (see the explanation in Chap. 20). After a week, the line was producing smoothly, making over 2000 units per day and meeting the model mix. Furthermore, they were doing it with less people, less space, and most importantly, with less chaos.

We Try to Implement Step 5 of the Path to Lean, Standardize and Sustain

At various times we tried to get some standardization implemented and all the issues, even things as simple as standard combination work tables were met with resistance. They said they did not want to waste our time on something they, themselves could do. I am not sure if they ever completed that step.

The Results

Let's now look at the impact of our efforts on the performance of the line. We will review several aspects, including:

- The impact on lead time, which was our original intent
- Other typical gains achieved in a lean initiative
- How the gains were achieved
- What should be next for the Bravo Line improvements

The Lead-Time Results

As a result of the *jidoka* and the batch destruction techniques, the process improved dramatically and the lead-time results in Table 21-1 display just some of the benefits achieved.

Cell No.		Base Case—14 Work Stations	Redesigned Flow—9 Work Stations	% Reduction
1	Lead time, first piece	3.9 h[1]	6.5 min[2]	97%
1	Lead time 1000 unit box	13.6 h[3]	—	
	Queue time for Cell 2	46–52 h[4]	2–10 h[4]	88%
2	Lead time, first piece	1.7 h[5]	5.2 min[6]	99.5%
2	Lead time, 1000 units box	10.0 h[7]		
	Total shipment lead time	145 h[8]	28.4 h[9]	81%

[1]$([35*50]/60)*8/60 = 3.9$ h (the line fill time for eight stations, first piece appears at Station 9 for inspection; total lead time would be over nine stations)

[2]$(39*2*5)/60 = 6.5$ min (at two-piece flow fill time for the five stations, appears at No. 6)

[3]$([([35*50]/60)*9] + [([35*50]/60)*19])/60 = 13.6$ h (the line fill time for all nine stations, plus the time to produce 19 more trays to make the 1000 unit transfer batch)

[4]By observation

[5]$([30*50]/60)*4/60 = 1.7$ h (line fill time for four stations; Station 5 also was an inspection station)

[6]$(39*4*2)/60 = 5.2$ min (line fill time for four stations; Station 5 also was an inspection station)

[7]$([([30*50]/60)*5] + [([30*50]/60)*19])/60 = 10.0$ h (line fill time for five stations, plus the time to produce 19 more trays)

[8]$(13.6*2) + 46 + 52 + (10*2) = 145.2$ h, or 6.05 days (lead time to make two 1000-unit batches in Cell 1, plus two queue times, plus the lead time to make two batches in Cell 2)

[9]$(6.5 + [(39*50)/60] + 360 + 5.2 + [(2000*39)/60])/60 = 28.4$ h (this is the first piece lead time in Cell 1, plus the time to fill one tray, the transfer batch; plus average queue time, plus Cell 2 first piece time, plus the time to finish the batch

TABLE 21-1   Lead-Time Results

Other Benefits Achieved

When we started redesigning the Bravo Line, they had already begun a lean initiative. But, as you can see, they had not made much progress. In addition to the process lead-time improvements in Table 21-1, other benefits are summarized in Table 21-2. All these changes were accomplished with virtually no capital investment.

Other Impacts	Original Case	After Leaning Out the Process	% Reduction
Space Impacts	8 line-days per week, plus 600 sq. ft. of buffer inventory	Less than 5 line-days per week, with less than 100 sq. ft. of buffer inventory	~70%
Manpower Impacts	14-person line running 3 shifts for 8 days per week = 336 md/ 10,000 units	9-person line running 3 shifts for 5 days per week = 135 md/10,000 units	60%

TABLE 21-2   Other Benefits

Some Points Worth Repeating

The following are some points worth repeating, a few with a slightly different twist:

- The most important quantity control tactic for process improvement is often first-piece lead-time reduction. It is accomplished mainly by a reduction in lot size, ideally going to one-piece flow. Once achieved, this will allow the problem-solving process to be responsive, and in this case, rework can be reduced on both a batch and permanent basis.

- By implementing *jidoka* as we did, using line shutdowns, we literally equated quality with production. If quality was bad, production went to zero until quality was reestablished as good. This is the beginning of a huge cultural change from an attitude of "production trumps all" to an attitude of "we will only allow the production of product that is satisfactory to our customer."

- Overall lead time is a very important metric. It is the sign of both flexibility and responsiveness. The reduction of overall lead time is basically done by reducing lot wait times and process delays, or "queue time" as some call it. In this case, process delays, due to the imbalance situation also contributed greatly to the overall lead time.

- By converting to one-piece flow (actually two in this case), we allowed the quality problems to be found and solved. Finding and eliminating quality problems are not basic lean elements. Finding and eliminating quality problems is not *quantity* control, it is *quality* control. It is a Precursor to Lean. This allowed us to reduce the manpower needs from 336 to 210 md per 10,000 units. That is a huge savings, or better yet a huge loss reduction. (See Table 21-2: 210 md is calculated by 14 persons, over 5 days for three shifts to produce the 10,000 units—in other words, their planned staffing.)

- The second reduction, from 210 to 135 md per 10,000 units came about by balancing line production to *takt*. This is not a new lean technique. It is an age-old industrial engineering tool known to every manufacturing firm that has significant labor costs.

- After the line was balanced and flowing, it improved itself! Recall that production rates increased even after we had completed our changes.

In Summary

- The company was now able to produce to schedule, and missed shipments dropped to zero.

- First-piece lead time was reduced from 3.9 hours to 6.5 minutes.

- They had now reduced the lot lead time from over 6 days to less than 30 hours.

- They found they could achieve these numbers with much less manpower. They reduced from 336 md/10,000 units … to the plan of 210 md/10,000 units … and then to a new low of 135 md/10,000 units. All with virtually no extra investment. And they were just getting started!

How the Gains Were Achieved

The story of the Bravo Line is like many lean stories. It was not just an effort at making a process Lean through quantity control techniques. It was both an effort at taking care of the foundational issues of quality and utilizing quantity control tools to make further waste reduction improvements. Note that the majority of the huge gains came from the implementation of the foundational issues.

The Five Leading Indicators of Cultural Change

Just how did the company score on the five fundamental issues of cultural change? A good argument can be made that the reason they got into these problems was due to an inherent weakness in their culture.

However, for the time being let's focus on the culture that surrounded this change in the process. They affected a successful change in the process because, at least for this project, they had good leadership. A plan existed that was understood by those involved and they acted upon that plan. Second, they clearly had the motivation to proceed. The escalating labor costs for the line were making the product unprofitable, and the situation was deteriorating. Third, they recognized they did not have the necessary problem solvers in place, so they acquired them. It remains to be seen if they will institutionalize these cultural issues, but for the moment they properly addressed the first three of the five fundamental issues of cultural change. Regarding the engagement factor, they were clearly lacking. We found numerous examples where operators had huge cycle-time variations because they literally did not know how to do their jobs. The most telling sign was the low level of engagement by the management team. It was obvious the moment we embarked on the project as we found that much of what management conveyed to us was incorrect. We readily and frequently found flaws in the information with only a cursory evaluation. Their lack of understanding of process fundamentals was obvious from the beginning. Finally this clearly was not a learning/teaching environment. The only reason we were commissioned to solve this problem was their inability to meet short-term financial measures. Virtually everything we encountered was driven by short-term financials from the outline of the project, to the reluctance to spent money to modify workstations and create *heijunka* board to the lack of interest in standardization.

Foundational Issues Addressed

Before the *quantity*-control issues could be implemented, foundational *quality*-control issues needed to be implemented. Review the House of Lean and notice all the foundational issues that were addressed, including:

- Problem solving by all
- Understanding variation
- Process stability (Cp, Cpk)
- Cycle-times reductions
- Standard work
- Availability

Quantity Control Techniques Applied

While these were being addressed, we were able to also implement the *quantity* control techniques, including:

- *Jidoka*
- One-piece flow, small lot flow
- Balancing to *takt*
- Minimizing lot sizes
- Reducing both WIP and lead times, which caused a significant improvement in the flow

The quantity control tools would have been much less effective if the quality control issues had not been addressed. Taken all together, these tools then improved the performance and the following occurred:

- Quality was improved
- Lead times were reduced
- The cost to produce was reduced

These three are the key objectives of Lean…better, faster, and cheaper…. (again, refer to Chap. 24).

Application of "The Seven Techniques to Improve Flow"

We can employ seven basic techniques to reduce lead time and improve flow (see Chap. 13). The following describes how they were applied to the Bravo Line.

1. **Reduce processing time** Here the major gain was achieved through the reduction of defects and especially the reduction of rework, which had become a part of the normal process. All the rework was simply non-value-added time. Regarding the necessary work, here we did nothing to intentionally reduce the processing time, although the time decreased as the operators became more engaged in the process.

2. **Reduce piece wait time** This is the time a single piece is waiting to be processed. Here the wait time is reduced by balancing, so the flow is synchronized. In the original case, the cycle time was controlled by station one at 35 seconds, so it took 35 seconds per station, times nine stations in Cell 1, *to* go through the line, or 315 seconds, but only 207 seconds of work were performed. There were 108 seconds of wait time per piece caused by poor synchronization at Cell 1 alone. Again, in this case, the wait time was reduced, but it did not have a large effect on lead time; however, it certainly made the process flow better. Following the rebalancing of the work the gains achieved by reducing piece wait time were found in manpower reductions.

3. **Reduce lot wait time** This is the time that a piece, within a lot, is waiting to be processed. In this case, it is substantial and adds to the overall lead time, but more importantly it adds to the first-piece lead time. This time is often overlooked

but is incredibly important. There will always be quality and production issues and these issues must be uncovered quickly so they can be solved. If the lot wait time is like the original case of the Bravo Line, it takes 3.9 hours for the first pieces to get to final inspection. If a problem is found at final inspection, we now have over 400 pieces to inspect and rework—for each problem we find. If the lead time is shorter, the problems surface more quickly and we can react more quickly. In our first run, we stopped five times in the first batch to correct problems. This impact alone could have accounted for the huge excess in labor expenses to run this line in the base case. To reduce lot wait times, shrink lot sizes as we did here. The goal of minimum lot sizes is one-piece flow.

4. **Reduce process delays**   This is the time an entire lot is waiting to be processed. Often it is called queue time. Here we were able to reduce it from an average of 49 hours to 6 hours. This is caused by lack of synchronization and also by transportation delays. Close-coupling processes, or making them in the same cell, mitigates this problem. *Kanban* can help this, but *kanban* is only a way to make the best of a lousy situation. Better ways to reduce processing delays are to synchronize operations, close-couple processes, and production rate leveling for both rate and the model-mix.

5. **Manage the process to absorb deviations, solve problems**   For the Bravo Line, we did not specifically highlight any applications of this technique. There are many sources of deviation that increase production lead times such as machinery breakdowns, stock outs, and stoppages for quality problems, to name just a few. All these deviations cause inventories to rise, and inventories are the nemesis in Lean Manufacturing—we want to go to zero inventory if we can. So whenever we have variation in the system, do not add inventory … instead, attack the variation.

6. **Reduce transportation delays**   One-piece flow, synchronization, and product leveling all place emphasis on transportation, which is also a waste. To reduce this waste, several strategies can be employed:

   * *Kanban* is the first thing most people think of, but *kanban* has inventory, and kanban creates a second delay, the delay of information transfer, so it is a double waste in itself. Try to avoid *kanban* systems by instead using close-coupled operations such as those in a cell and the use conveyors, as we did here.

7. **Reduce changeover times; employ SMED (Single Minute Exchange of Dies) technology**   The Bravo Line had very short changeover times, so we made no applications of SMED technology here .

This is clearly a Lean success story!

What Could Be Next for the Bravo Line?

Further Analysis

Now it is time to do another time study and rebalance the line. We did paper *kaizen* and determined we could:

* Standardize the improvements to date

* Make minor processing changes

- Rebalance the workload
- Create better flow by moving Cell 2 to be next to Cell 1 and eliminate the queue, then convert to a single seven-station U-cell arrangement
- Eliminate the conveyors, redesign the workstations, and go to one-piece flow

Potential Results

The paper *kaizen* improvements showed:

- The company could reduce space by another 55 percent
- The company could operate at a cycle time slightly less than *takt* and run both cells with only 6.4 people—really 7
- The labor consumed would be less than 105 man days per 10,000 units

And the Story Goes On

In this case, we should make these changes and then proceed to implement the prescription and utilize even more *quantity* control techniques. Each improvement brings about a new present state that will have another set of opportunities—and the story goes on, forever.

Chapter Summary

The story of the Bravo Line starts with a simple request. The plant manager wants to make more product, faster, using less labor. Something all managers want. To reach that end, earlier, the company embarked on a lean initiative, but it can now be seen that they had some serious flaws in their approach. First, it was apparent they wanted more to *appear* to be Lean, rather than actually *be* Lean. However, the biggest flaw in their approach was that they had not evaluated the foundational issues very well, if at all. Consequently, once we backtracked and took care of the quality control issues, progress was possible. After implementing the quantity control techniques on this stronger foundation, even greater gains were achieved. In all, we were able to reduce first-piece lead time by 97 percent, reduce lot lead time by 81 percent, and in the end reduce the required manpower by 60 percent. As a result, not only was the Bravo Line leaner and its process more robust, but the gains made were huge and achieved early.

An Initiative???

Many of you will notice that to describes the efforts of the Bravo line I used the term "lean initiative" rather than the more commonly used term of "lean transformation." Here is an example of an "initiative" rather than a "transformation." A lot was "initiated" much less was transformed, and it is pretty clear it was done "to the people" not "for the people." The Bravo line had what I sarcastically refer to as a "Seems to be Lean" system. Nonetheless, we helped them and showed them new ways to manufacture.

CHAPTER 22

Using the Prescription— Three Case Studies

Aim of this chapter … to show the technique explained herein by the use of three case studies that follow the prescriptions. In each case, the prescriptions are applied somewhat differently. These examples are included so you might see the variety of applications for the prescriptions. It is not, "my way or the highway," nor is it "all or nothing." Nevertheless, the prescriptions will guide you to the huge early gains that most find so beneficial.

Why These Case Studies?

In my experience, especially since starting my consulting practice in 1990, I have found that *applying* techniques, principles, and theories for most people is very difficult—more difficult, for example, than *understanding* the techniques, principles, and theories. I have found there is an "applications gap" in industry. In addition, especially with Lean, I find there is frequently a mentality that to do Lean is an "all or nothing" proposition. I do not ascribe to that principle at all. I do, however, believe it must be done well or not at all. It is very practical to implement just parts of the entire House of Lean without danger of regressing as long as those parts are done well. Of course, the lean principles work together, and to get the full benefit, you must apply the entire complement of strategies, tactics, and skills, but you can apply *some* of them and get *some* of the benefits. And almost always, some of the benefits are better than none of the benefits.

These three examples will show that there are different ways to apply these prescriptive methods and still achieve significant results. These examples will also show that there are different degrees or depths of application and you can still achieve huge benefits.

Larana Manufacturing

The first study, which concerns Larana Manufacturing, is the story of a facility that wanted to implement a lean transformation. This company is a good example of the second prescription of How to Implement Lean—the Prescription for the Lean Project, without really applying the first prescription, the Five Strategies to Becoming Lean. In fact, they had so many foundational issues (quality control issues) to work on that their transformation did not get very deep into real quantity control efforts. In addition, this facility was under severe financial pressures, and although the plant management team was very talented and motivated, they were relatively inexperienced and had myriad obstacles to overcome. Nonetheless, they applied the second prescription very well, took on the foundational issues, and made monstrous early gains in spite of the obstacles. It is an amazing and energizing success story.

The Zeta Cell

Next is an application to a single processing cell, the story of the zeta cell. I have included it since they were able to apply the first prescription, The Five Strategies to Becoming Lean, and make huge early gains with it alone. I have also included this since it is often the type of project undertaken as part of a lean effort. That is, conversion of a single process or a single cell to lean Manufacturing. Again, here, the gains made in this production cell were truly outstanding. They, too, were made under some very adverse circumstances.

QED Motors

Included also is an example of a much more complicated processing arrangement for a larger value stream in the case of QED Motors. In this case, they applied the entire second prescription: How to Implement Lean. They addressed the fundamental issues to cultural change, completed an entire systemwide evaluation and thoroughly addressed the first prescription—the Five Strategies to Becoming Lean for one of their production lines, the Motors Line. So they addressed the quality control and the quantity control issues including doing a full value stream analysis. Like the previous two examples, their gains were also huge and rapid.

Maybe You Can't Fix It All ... But

My hope is that this book and these techniques will reach a broad range of people who want to make their facilities leaner, making those facilities: better money-making machines, which are a more secure workplace for all; and the supplier of choice to their customers.

> **"D**o not let those things you cannot do, prevent you from doing those things you can ...**"**
> John Wooden

I often find an engineer for a value stream who would like to implement lean techniques but tells me, "You know I would like to apply lean principles but the facility is just not interested." That may prove to be an obstacle, but don't let that stop you. The following real-life examples show why you must not stop:

- That was almost the case at Larana Manufacturing. Some of the local management wanted to change but there was literally no support at the corporate level.
- That is precisely what we did in the case of the Zeta cell. In fact, the plant owners DID NOT WANT to change and we found a way to make them change.

Just because we cannot change the whole plant does not mean we cannot change our value stream. And just because we may not be able to change the entire value stream does not mean we cannot change things in our area of influence. Which is the basis for our lean initiative mantra of:

- Start where you are
- Use what you have
- Do what you can

Lean Preparation Done Well: The Story of Larana Manufacturing

Background Information

The story of Larana Manufacturing is a story of real success, not only of a lean implementation but how to work through the Precursors to Lean and achieve success all the way. It is the story of getting huge early gains on the way to becoming Lean.

- Larana Manufacturing makes electrical parts. They are a tier-2 supplier to the automobile industry.

- The facility was hampered by a large list of what most people would call weaknesses. First the plant was old. Only three of the nine production lines were less than 3 years old and none of the lines had the newest technologies to produce their products. In addition, the process flow was poor. Most lines were poorly balanced, set up in islands, and saddled with mismatched production capacity. Larana Manufacturing was burdened with a dangerously small technical staff and had only the basic quality skills in place with no lean experience at all. The one staff person who was earmarked to be the lean implementation manager quit, taking another position outside the company. Their employee turnover was about 8 percent per month at the line operator level. To make matters even worse, the plant was under severe financial pressures. I will spare you the details, but the bottom line was this: If the plant was to survive, they had to reduce operating costs. If that was not enough, their home office exhibited a rather heavy hand in the business, sending representatives there almost constantly. With all this visibility, the "home office help" was, more often than not, the "home office burden."

- To top all of this off, the most severe problem was that senior management (not the local plant management) was using old-world paradigms and was not lean thinking at all. Unfortunately, they actually thought they were lean thinkers. That proved to be a significant burden to everyone as we embarked on this transformation.

Even with all these issues and obstacles, Larana Manufacturing did an exceptional job of working on the Precursors to Lean, generating huge financial gains for the facility. It is truly a story of "Lean done well … under some very grim conditions."

How They Handled the Five Fundamental Issues of Cultural Change?

Leadership

First, the movement toward cultural change was led by the plant manager, Kermit, who was in his first major manufacturing leadership position. Kermit had a degree in engineering and was one of those natural problem-solvers that all companies look for but few can find. Even though he was quite young, 30 at the time, he had a maturity beyond his years. This maturity was manifest both in his excellent judgment and his wonderful rapport with the entire workforce. He turned out to be an outstanding leader and directed the effort extremely well.

Together, we created a plan. He made sure all the training was delivered on time, that all follow-up was done, and he was actively involved in all aspects of the transformation. Leadership was a strong point for this effort, and it was seen in the results.

Motivation

The motivation of the local workforce was not an issue. While I was at this facility, I can say with certainty that the people were not only engaged in the effort, they put in extra time to *make* it work. With all the distractions mentioned earlier, I was both surprised and pleased with the level of employee motivation.

> **"W**e are told that talent creates its own opportunities. But it sometimes seems that intense desire creates not only its own opportunities, but its own talents. **"**
>
> Eric Hoffer

Adequate Problem Solvers

As for adequate problem solvers, that was an issue both in numbers and in the quality of their problem-solving skills. To enhance the abilities of the problem solvers, we provided training, which helped immeasurably. Regarding the number, we were always understaffed in that regard. Consequently, we made some changes to augment the problem-solving staff.

Whole Facility Engagement

This was an initial problem. Mostly the line operations and support staff, although highly motivated really were not doing the correct things as you will see throughout the story. They were "moving and shaking" just not doing what the facility needed and it was incumbent upon management to supply them the guidance and resources they needed. Management did just that, virtually without exception, thus showing their engagement

A Learning/Teaching/Experimenting Environment

This was something that Larana Mfg. grew into and did so quickly. To begin with, before we arrived, Kermit, had created an Information Center on the floor, so all data could be managed in real time and was available shortly thereafter. When I saw this, I knew we had a great chance to be successful. And then at every step of the way we spent the time to teach the supervisors, engineers, technician as well as the line workers themselves as we made changes. Without exception the teaching preceded the work in order. They did very well on this cultural change leading indicator and it certainly showed in the results.

The System-wide Analysis of the Present State

The Commitment Evaluation

We discussed the commitment evaluation briefly and at the senior management level there were serious problems. The evaluation of the local management was very good, however. Nonetheless, we decided to proceed, knowing full well, that at some point we would run into some serious obstacles. Our hope was that we would have huge early gains, which we did, and thereby get a greater commitment from senior management, which unfortunately we did not.

The Evaluation of the Five Precursors to Become Lean

Kermit and I performed independent evaluations, the results of which are shown in Table 22-1.

The Five Precursors	Kermit's Score	My Score
High levels of stability and quality in both the product and the processes	0.5	0
Excellent machine availability	1.0	0.5
Talented problem solvers, with a deep understanding of variation	0.5	0.5
Mature continuous improvement philosophy	0	0
Strong proven techniques to standardize	0	0
Total	2.5	1.0

TABLE 22-1   The Evaluation of the Five Precursors

Process Maturity

We completed a review of the production lines for process maturity and none reached level 2. Stability was a problem in every case.

Both rate and quality levels were unstable when evaluated on a Shewhart control chart. This was true of all three lines we evaluated. Based on this, we decided we needed to do more foundational work before we advanced to a quantity control transformation. Consequently, we then proceeded to implement "An Overall Equipment Effectiveness (OEE) Transformation." We also decided that Line 9 was the most stable and felt we could improve it rapidly, so we choose it to implement some appropriate lean strategies, tactics, and skills—that is, quantity control techniques on this line only. We would address quantity control on the other lines after they had taken care of the issues uncovered in the systemwide evaluation.

The Educational Evaluation

The educational needs were quite extensive and we began immediately to train. Training the engineers and supervisors on SPC (statistical process control) and problem solving was our first task. We taught basic problem solving to all and the Kepner-Tregoe methodology to the supervisor and process engineer on each production line. Later you will see we spent considerable time on TPM (Total Productive Maintenance) as well as MSA (Measurement System Analysis) to a key group.

Introductory Training

An introduction to Lean Manufacturing was given to everyone by human resources as part of a 2-hour awareness training, which also included a review of the implementation plan.

Management Training

Management was given training over a 1-month period, with each session lasting roughly 4 hours. Topics included:

- The role of management in Lean
- The House of Lean
- Introduction to OEE

Other Training

Specific skills training was given to those involved, primarily engineers and supervisors, in the areas of:

- The House of Lean
- Introduction to OEE
- OEE calculations for engineers, OEE strategy
- SPC, focus on control charting
- MSA
- TPM
- *Kanban* design and calculation
- Cycle, buffer, and safety stock calculations
- Flowing to takt, production leveling and *heijunka* board design

Just How Was the "Rest of the Prescription" Managed

I have summarized the second prescription here for your reference. It is an eight-step prescription for the Lean Project, which includes the following measures:

1. Select and prepare the value stream change agent.
2. Assess the status of the five cultural change leading indicators.
3. Create True North metrics and a balanced scoreboard.
4. Perform the systemwide evaluations.
5. Document the current value stream condition.
6. Redesign to reduce waste.
7. Document and implement the *kaizen* activities.
8. Do it all over again.

The OEE Transformation

It can be seen that Larana Manufacturing had completed steps 1, 2, and 4. Our evaluations, especially the evaluation of the Five Precursors, told us clearly we were far from being able to get deeply involved in true quantity control. Consequently, we chose to implement an "OEE Transformation" because we had to work on improving the foundational issues of quality control. This OEE Transformation effectively replaced steps 5 and 6. We then set OEE goals, step 4, and evaluated each line for the appropriate *kaizen* activities. For each line, we developed an 8-week improvement plan, and these plans were executed by the process engineers and line supervisors. To provide support, each week Kermit would review the progress with each engineer, and once a month we had a formal OEE review, which took about 90 minutes. In the formal review, progress was checked, priorities were assessed, and any needed resources were addressed. A summary of 8-week status report is shown in Table 22-2, which is updated for week 4 information.

The transformation took off, everyone knew their role, and progress was made almost instantly. This goal-setting, implementation of *kaizen* activities, and follow-up

OEE Improvement Plans—Gains ... Line 3				
	First 8-Week Plan			
	Period	Weeks 12–19		
	Week 1 Actual	Week 4 Actual	Goal for End of 8 Weeks	Deviation,* Week 4 Actual - Goal
Production Information				
Total units	20,289	20,289	21,100	−811
Total losses	4,027	2,928	2,932	4
Total salable units	16,262	17,362	18,178	−816
Cycle-time improvements	5.20	5.20	5.00	−0.20
OEE performance	81.17%	85.84%	86.81%	−0.97%
Availability	87.38%	87.76%	91.60%	−3.84%
C/T performance	94.74%	99.65%	96.30%	+3.35%
Quality	98.04%	98.16%	98.63%	−0.47%
OEE Losses/Units				
Availability	2,563	2,483	1,866	−617
C/T Performance	1,067	71	777	+706
Quality	398	373	286	−87
Summary gains		This 8 weeks	Total to date	
Total units/shift		0	0	
Total losses/shift		−1,100	−1,100	
Total salable units/ shift		1,100	1,100	

*All positive deviations mean the goals were exceeded.

TABLE 22-2   Typical OEE Summary Status Report

were steps 4, 7, and 8 of the prescription. In short order, we had made a great deal of progress. Tables 22-3 and 22-4 show our results after the completion of the first and third 8-week plans.

Comments on the Results

So just how did they improve? Well, immediately upon implementing the OEE transformation, process variation began to show improvements. Not only did quality yield rise, but line cycle times improved. The assigned process engineers had individual work plans and executed them with energy. The engineers grew with the implementation. Problem solving became a way of life with root causes found and eliminated. We developed and defined a structured continuous improvement methodology and it was employed throughout the plant. The processes were showing improvements and the short-term gains were sustained. It was a joy to be a small part of this effort.

	Line 7	Line 9	Line 3
Total production	+9.2%	+18.8%	+6.3%
Quality yield improvements	+2.0%	+2.6%	+0.0%
Cycle-time reductions	5.5%	15.8%	3.9%
Capital investment	$0	$0	$0

TABLE 22-3   Progress after the First 8-Week Plan

	Line 7	Line 9	Line 3
Total production	+27.01%	+42.8%	+11.7%
Quality yield improvements	+8.1%	+5.73%	+0.63%
Cycle-time reductions	16.4%	23.5%	3.9%
Capital investment	$0	$0	$0

TABLE 22-4   Progress after the Third 8-Week Plan

This means that:

- With the same staff, the same machinery, and the same raw materials, these three lines could now produce, on average, 24 percent more product. The per-unit cost dropped dramatically. Most of the extra product which they made, they could sell—where they had no incremental demand, they were now able to run the lines fewer hours per week.

- On line 9, their monthly demand was 1,000,000 units when we started, which could not be met—even when they worked 24/7. Now the demand could be met working only 5 days per week, with a resultant huge savings in manpower and overhead.

- With 5 percent less raw materials, they could produce more product, and quality leakage to the customer was also reduced. Raw materials costs dropped dramatically.

- The financial performance of each line was increased by an average of over 20 percent, with all factors taken into account. Line 9, which, when we started, was about 30 percent in the red, had turned the corner and was now in the black, making money.

The Financial Results Put in Perspective

The financial results are even more impressive when put into perspective. Prior to the transformation, the process engineers were working almost solely on solving quality problems. If they were able to improve the quality yield by 1 or 2 percent on a yearly basis, that was considered good progress. So, figure it out. At about $0.50 per unit, making 30,000 units per day, a 2 percent increase meant an incremental gain of over $100,000 per year to the bottom line. That was their definition of "good progress."

We had completely changed the paradigm of achievable progress. Now we were making progress at not 2 percent per year, but 20 percent, and in less than 6 months. This was 10 times as much in less than half the time. *Our rate of improvement was now 20 times larger!*

How Did the Quantity Control Activities Work Out?

Besides working on OEE, during this period we carefully implemented some other lean techniques on Line 9. For example, on Line 9, which had a particularly problematic process flow, several changes were made:

- Several work stations were modified and two were eliminated by combining with other work stations. This involved process simplification and improved flow.

- *Kanban* were installed to connect the "island processing scheme." This resulted in improved flow, using pull systems with vastly improved inventory controls.

- We slashed the size of the *kanban* containers (transfer lot size). Here, the lean technique was that the minimum lot size was reduced to accelerate flow, this reduced the manufacturing lead time from 38 to 4.5 hours.

- The final goods inventory was recalculated, set up with cycle, buffer, and safety stocks. A *heijunka* board using production *kanban* was installed to complete the pull system.

- There had been a materials delivery position that was staffed during only part of the shift. We made it a full-time position and added the *kanban* movement to his/her duties. We formalized both a materials delivery route and a schedule.

- In addition, we prepared a present state and a future state VSM for Line 9.

Were They Now Ready for Quantity Control?

Following our third 8-week plan, we completed another assessment on the Five Precursors. These three lines received scores of 12 to 13.5, with no precursor scoring less than 2. We weren't yet done, but we had already made huge progress. Each line still had a huge upside, just in the precursors alone. Look at the chart in Table 22-5 to see the huge remaining opportunities—and all this without any major capital expenditures. Now Lines 3 and 7 are also ready for quantity control activities. It was time to apply the Five Strategies to Becoming Lean (described in detail in Chap. 14), modify the plan, and have some more fun.

So What Is "The Rest of the Story" Behind This Success?

How did they achieve these huge gains despite the plant's weaknesses?

You have seen the results. To me, these results were not surprising—pleasant for sure, but not surprising. I was not surprised because, in the first case, we had properly addressed the five cultural change leading indicators. The plant manager provided excellent leadership, in spite of his inexperience and his lack of support from his senior leadership. The staff was and remained highly motivated and engaged throughout. With the information center, the training focus and the management driven teaching they were able to create a culture of teaching/learning/experimenting and this was supported by a nucleus of problem solvers who were solving problems in a JIT (just in time) fashion. With those basics in place, there is always a decent chance of success.

Next, the implementation went well as we followed "The Prescription," as outlined in Chap. 16. There was a good leader and a *sensei*, we made an assessment using the assessment tools from Chap. 24, and utilized the assessment to create our plan. Next, we deployed the plan and turned it into "worker level" goals. This, coupled with some JIT training, fueled the effort and the results speak for themselves.

	Line 7	Line 9	Line 3
Operational Improvements			
Cycle time—week 12 (seconds)	6.10	6.65	5.20
Cycle time—week 35 (seconds)	5.10	5.09	5.00
Achievable cycle time (seconds)*	4.40	4.20	4.50
OEE week 12	79.00%	76.24%	81.17%
OEE week 35	89.83%	84.31%	88.68%
Achievable OEE**	96.00%	96.00%	96.00%
Production gains (units/month)*			
Monthly salable units—week 12	812,930	724,559	1,274,778
Monthly salable units—week 35	1,032,475	1,034,748	1,423,667
Monthly achievable salable units	1,354,579	1,419,083	1,765,970
Gains—first 24 weeks (wk 12–35)	219,544	310,189	148,889
Additional achievable gains**	322,105	384,335	342,303

*Based on 6 days per week.
**Achievable without any major capital investments.

TABLE 22-5   Gains and Achievable Gains

What about the obstacles and plant weaknesses we mentioned earlier, with all this going against it, just how did the plant achieve all these gains?

The answer to that is simple. Excellent leadership with a good *sensei* working together is the key to success. I cannot say enough about the leadership that Kermit supplied. He showed not only the intelligence and honesty to grasp the situation and see what could be done, he also showed the courage and character required in these stressful situations. He is an impressive young man. Couple that with a good plan, excellent problem solving, and some motivated staff, that combination will trump all these other so-called problems.

More specifically, if none of the problems stated earlier existed: If the plant was new and was equipped with the latest technologies and with well-designed process flows, and if the plant had great home office support, with plenty of experienced talented staff, and financial solvency, it might have succeeded much more quickly and easily. But if the leadership or the *sensei* were mediocre, or the plant personnel lacked motivation or engagement, or there was a lousy plan with no good problem solving, such circumstances would doom a transformation to failure. Simply put, those plant weaknesses mentioned earlier are simply not all that relevant.

Do not underestimate the need to have:

- Good leadership with an experience-based *sensei*
- Motivated and engaged staff and workers
- Good problem solvers

With this formula, you can attack nearly any lean issue with a strong probability of success. That is the message of Larana Manufacturing and their success.

The Zeta Cell: A Great Example of Applying the Four Strategies to Reduce Waste and Achieve Huge Early Gains

The story of the Zeta cell is a "back to basics" story. Sometimes when new tools come out, we become so enamored with them that we forget the basic foundation on which these tools were built. The Zeta cell example shows, in glowing detail, that there are huge early gains to be made, sometimes from the basics alone.

Awareness

One of the better books I have read on Lean Manufacturing (besides Ohno and Shingo's works) is *Learning to See* by Mike Rother and John Shook. As I mentioned in an earlier chapter, the key focus of their book is to "learn how to see," to become more aware of how your production processes are performing. They helped us "see" using the performance metrics of:

- Percentage value-added work
- Production lead time

Value Stream Mapping Is Sexy, but ...

Another contribution Rother and Shook made to the lean movement was popularizing the technique now called *value stream mapping* (VSM). Prior to the publication of their book, VSM was known to a small group within Toyota, and only a few outside of Toyota. Within Toyota it was not called a VSM but was called the Material and Information Flow Analysis. VSM is a wonderful tool to use in a lean transformation. It is also a new and it is certainly, "in vogue," and like many good things, when they hit the market, such items oftentimes are overused due to their novelty.

There's More to Reducing Waste than Value Stream Mapping

Today, value stream mapping is being treated by lean practitioners the way a new diet fad is treated by those wishing to lose weight.

Very often, to those seeking weight loss, they put all their faith in the diet and plan to do nothing more than "stick to the diet." Their hope is that in the diet, and in the diet alone, they will reach their goal of permanent weight loss and improved health. If the dieter has the ability to stick to the diet, the results are predictable. Lots of early weight loss. That's not too difficult, but can they sustain the weight loss and improve their health as well? That usually depends on what else they have done. Have they changed their sleeping habits? Have they changed their exercise habits? If the diet is all they did, with certainty the weigh will come back and with that ugly weight will come the emotional baggage of failure and guilt.

In this light, many managers view value stream mapping like the latest diet craze. They hope VSM is the "key" to their success and they think they need to do nothing more to analyze their situation than prepare a few value stream maps, from which they can fully analyze and fully improve their plant. If that is the only analytical tool they use, and they rely on it alone, the results, just like the dieters, are predictable. There will be short-term gains, encouraging gains, but again like the dieter's situation, over time these gains will likely regress and with this will come the emotional baggage of dissatisfaction and disillusionment.

In fact, putting all your eggs in the diet basket, or in the VSM basket, is just another version of the "silver bullet" or the "quick fix" mentality—and history has shown time and again it simply does not work in any field of endeavor.

Should We Ignore Value Stream Mapping?

So does that mean we should ignore value stream mapping? Of course not—no more than the dieter should ignore the diet. Value stream mapping is a powerful tool that allows you to see the production operation from a different perspective, and a powerful perspective—from a distance. It allows you to see how the various parts of the value stream are connected and how these pieces add up; how they increase the lead time and increase the waste accumulation in the value stream. Its major advantages are threefold. First, it focuses your attention on the total value stream and assists you in avoiding the problem of point optimization at the expense of system optimization. Second, it is the most direct and powerful way I have seen to focus on overall lead time, the key metric of whether a facility is truly Lean. Third, almost without exception the major contributor to the magnitude of the overall lead time is the inventory that is held in front of and after work stations, normally called WIP (work in process). So consequently, VSM is very good tool to focus on the removal of inventory to accelerate process flow and reduce lead time. This is its major strength.

But just what are its shortcomings as an analytical tool? Value stream mapping is a "macro tool" in that it has an overview aspect and does not get into many of the details that must be understood if you want to make your plant Lean. The biggest weakness is in analyzing cell performance. Cell performance is absolutely crucial to overall performance. So we need other tools to analyze cell performance, and if we do not properly analyze the cell performance, we will end up with good connections between cells that are individually performing quite poorly.

I find it is best to prioritize your improvement efforts, inside-out, so to speak. First, start at the cell, make it efficient and effective. Once this is done, connect the cells—this is where the VSM is powerful. There is nothing wrong with improving the cells and improving the cell-to-cell flow, simultaneously. But it is simply incorrect to ignore the cell performance and it is highly inefficient to do it after the value stream mapping effort.

Although that sounds trivial, it is missed by many and with the power of a value stream map driving a project, a lot of progress will be made. In short order, however, the poor performance of the cells will create problems, and these issues will need to be addressed. In the end, the macro work uncovered by the VSM will often need to be redone.

As I have mentioned repeatedly, the healthiest place to start a lean transformation is to evaluate the "Foundational Issues." Following that, I have found that the place to start is a detailed analysis of the work elements. My mantra is "You gotta know the work!" From there, you can create VSMs, make a spaghetti diagram, create flow, design pull production, do some more analysis, and begin the improvement activities.

> **P**oint of Clarity There is no substitute for "knowing the work."

You Gotta Know the Work!

In their book, *Creating Continuous Flow*, authors Mike Rother and Rick Harris spend some time on value stream mapping. However, the focus of the book is on improving "flow." We often use *Creating Continuous Flow* in our training since it is an excellent book on the subject.

In addition to Ohno's wisdom in the insert, we have found that establishing the flow, as Rother and Harris explain, coupled with line balancing, are two powerful manpower reduction techniques.

Part II of their book is entitled, "What Is the Work?". When we introduce this to the typical manager, usually they just yawn. It seems this is too beneath them to study. And we usually find the same response from the industrial engineers as well. And why not? That's how their boss views it. Ultimately, we find very few firms can really define the work down to the element level, and until they can, their efforts into Lean are inadequate.

> **"E**stablishing the flow is the basic condition …. Unless one completely grasps this method of doing work so that things will flow, it is impossible to go right into the *kanban* system when the time comes."
>
> Taiichi Ohno

An Example of "Not Knowing the Work"

For example, while working to Lean out a very new but large and complex production line at an international tier-1 supplier, we encountered this problem. Specifically, they had a detailed present and future VSM, as well as detailed flow with timing studies. These studies had been done on their standard format which was Excel based. Data could be entered and the output was a series of charts, including a line balance chart, a standard work combination table and several other important-sounding documents. The Excel program was filled with lookup tables, dynamic data interchange, data validation, and myriad techniques that, quite frankly, sounded very impressive.

There were two problems, however. First, the work elements, the basic input to the Excel spreadsheet, could not be described by the engineer, the supervisor, or the business unit manager. In fact, no one could. Worse, as I watched the process perform, although you could recognize from the documents that we were watching the correct process, a number of work elements were missing. In addition, the time study of the work was not correct at all. Waiting and work was all mixed up. All in all, the study, although it was presented in an impressive display of charts tables and graphs, was basically useless. Well, "you gotta know the work" … and they didn't.

I find this to be a frequent scenario—that is, much effort on the presentation, but much less on the substance. This always detracts from a facility's ability to attack waste.

They Were Also "Not Very Aware"

In this specific case, the OEE of this line was 61 percent, which was woefully low, all things considered. However, as so often was the case at this facility with its lean-appearing facade, they had an Information Center from which we were able to quickly evaluate only part of the problem. The telltale sign was that, for this process with an OEE of 61 percent; quality losses were 0.95 percent and availability losses were 9.5 percent. What they failed to recognize was that the cycle-time losses were nearly 30 percent! Ouch!

How did we know they failed to recognize the largest problem?

Well, the Continuous Improvement Activities board, at their Information Center, showed a number of projects to improve quality, and two or three to work on availability, but not one single job was focused on cycle time. We were not there to work on that issue but could not help but notice and point it out. They, very promptly, did exactly nothing. For the three months we worked with them, we made good progress on quality and availability, yet the huge losses due to poor cycle-time performance were never addressed.

"To Know the Work": The History of the Zeta Cell

Background Information

An example of not focusing on the work was also seen in the Zeta cell. It is an odd story in that the company we helped the most was not our client, rather it was the supplier to our client that really got the gains from our efforts. Our client was a robot manufacturer and the supplier provided a sophisticated controller to guide and control the robot.

The background to this is that we were hired for our problem-solving abilities—in this case, to solve a problem with the controller that was produced on the Zeta cell. The controller was used to guide a robot and would occasionally stick in the "full speed ahead mode," causing the robot to consequently crash into a wall or the production line. This was not only undesirable, it was dangerous.

We analyzed the data using Kepner-Tregoe techniques and found the root cause of the problem. We then assisted them in modifying the design for our client and the production process for the supplier. We implemented the change and began production with the redesigned process immediately. Meanwhile, we monitored both production and field operations while the company completed the necessary environmental and reliability testing.

The really interesting part of this experience was that during this same time period, our supplier embarked on what they called a "full-blown implementation into Lean." It was at this time that we discovered that this 900-person plant was extremely unprofitable and was up for sale. Their subsequent efforts all looked like a company in crisis. The actions were quick, direct, and their "full-blown implementation into Lean" was done in a very dictatorial fashion. Employee, supervisory and even mid-level management input was virtually nonexistent. This was classic top-down command and control management and it was obvious from the beginning.

Previous to our arrival, they had hired a new general manager (GM) and his objective was clear. On one occasion, he told me they had 12 months to make the place profitable or it would be sold. Their costs to produce were just too high and must be brought in line. He fired the current plant manager and replaced him with a man he knew. His "full-blown implementation into Lean" effort was "front-end" focused. He started at the front of the process, the raw materials supply, and worked through the process in the direction of flow.

A huge initial effort was focused on the raw materials warehouse and supplying all materials to the line via *kanbans*, replacing the current practice of kitting. He also started a lean implementation office with three engineers. They did some basic training regarding the transformation, but there was little if any individual skills training supplied. The three engineers spent the majority of their time working with the raw materials supply, and one thing they did extremely thoroughly was have a Plan for Every Part (PFEP). This is an extremely time-consuming activity, but it is also a good one. It is one I have found best to leave for the later stages of lean implementation, at least waiting until good flow is established. In addition to having a PFEP, they standardized on a specific work table design and an external U cell and began to change the layouts of all cells to match this. Doing so, they were able to save a great deal of space. Due to their efforts, in-plant supply lines shortened, materials delivery improved, and some clear lean benefits were achieved.

After 6 months into the effort for some reason, the GM called me into his office and gave me his 6-month update. It was a PowerPoint presentation that described their effort and showed the gains made. Many were paper gains, such as the space savings achieved and reduced lead times for raw materials delivery, and quite frankly the plant looked a lot better. The layout was improved and raw materials flowed more smoothly. He was rather proud that through the materials-handling effort they had been able to shorten delivery times, reduce stock outs, and eliminate three delivery positions. For a three-shift operation that meant a reduction in headcount of nine. The other gain was that on-time delivery had improved from a meager 68 to 94 percent, and it was still rising. Most of the delivery gains were achieved by outsourcing a large portion of the machining that had been done internally, effectively purchasing capacity. Considering the effort expended, with the delivery performance excluded, I was not impressed with the results. Since their basic problem was financial in nature, they had made very little progress on improving the operational efficiencies.

Some Specifics on the Zeta Cell

It was about this time that I was working with the production engineer to improve the cell that provided the controller to my client. The line was clearly under-producing and to the trained eye, the line was a "lean-opportunity-waiting-to-happen." I was buoyed by the energy of their lean implementation and thought this would be like shooting fish in a barrel.

At this cell, the workers, who were grossly underworked, would leave the cell without warning. Inventory would build up in front of their station and then the operator would return, concentrate on the work for a while, and the inventory would move further down the process. At times, if inventory buildups were too large—that is, they ran out of space—a worker might leave his station to assist in the work-off of the inventory at his colleague's workstation. In short, it was a herky-jerky inefficient operation typical of non-lean production facilities. No one found it odd. In fact, they were proud of the teamwork and the level of cross-training, which made all this manpower movement possible. Nonetheless, being a lean practitioner, I proceeded to Lean it out—on paper, that is.

The basic time study and balancing calculations for the Zeta cell are shown in detail in Apps. B and C of Chap. 14.

Applying the Five Strategies to Reduce Waste

Synchronizing the Supply to the Customer, Externally

First, we did a *takt* calculation. Since they have a 9.5-hour shift with 50 minutes for lunch and breaks, the *takt* was 39 seconds to produce the 800-unit weekly shipment for my client. (As you read on, you will find that the actual cycle time to produce an 800-unit batch was well over 80 seconds!)

Synchronizing Production, Internally

First, we completed a time study and a balancing study, which are shown in Figs. 22-1 and 22-2.

The time study showed we had 157 seconds of work. At a *takt* of 39 seconds, that would be 4.02 operators at 100 percent OEE. No one knew the actual OEE, but they believed it was over 90 percent. With five operators, we would average 31 seconds of work per position. This would be a reasonable place to start and give us a reasonable OEE target of 0.79. Even though one station, on paper, took 35 seconds (Fig. 22-3), I still

| Process to Monitor | | | Rayco 43-27 | | | Date | 3/9/2005, 2 shift | | | | | | | |
| Station: | | Zeta Cell | Done by: | | | J. O. Bengineer | | | | | | | | |
Step No.	FC Id.	Work Element	Cycle 1	Cycle 2	Cycle 3	Cycle 5	Cycle 6	Cycle 7	Cycle 8	High	Low	Range	Average	Final
1	10	Cut bracket	3	4	3	2	5	11	3	11	2	9	4.4	3
2	20	Assy bushing (3)	11	10	13	12	13	19	12	19	10	9	12.9	12
3	30	Install o-ring and clip	9	6	6	8	7	8	7	9	6	3	7.3	7
4	40	Place in jig, glue	7	8	9	11	10	10	9	11	7	4	9.1	9
5	50	Press in magnets (2)	4	5	6	5	4	7	17	17	4	13	6.9	6
6	60	Insert o-rings, cap, grease	14	12	12	13	19	13	14	19	12	7	13.9	13
7	70	Install support	7	8	7	8	8	9	7	9	7	2	7.7	8
8	80	Install o-ring and clip (2)	6	7	8	9	23	7	8	23	6	17	9.7	8
9	90	Apply epoxy, 3 locations	12	13	15	14	14	14	13	15	12	3	13.6	14
10	100	Install control capacitor	7	8	9	9	8	8	7	9	7	2	8.0	8
11	110	Apply epoxy, topside	7	6	5	9	6	5	5	9	5	4	6.1	6
12	120	Install retainer ring	9	8	9	8	9	7	8	9	7	2	8.3	8
13	130	Install cover cap	6	7	7	8	6	7	7	8	6	2	6.9	7
14	140	Unload/load machine (2)	2	3	3	2	12	3	3	12	2	10	4.0	3
15	150	Apply final sealant (1)	22	14	15	28	14	15	16	28	14	14	17.7	15
16	160	Final test, wrap leads	16	19	17	18	22	17	18	22	16	6	18.1	18
17	200	Package	12	10	28	12	13	11	12	28	10	18	14.0	12
18														
19														
20		Total	154	148	172	176	193	171	166	193	148	45	168.6	157

Notes 1 Gun required unplugging hence long times, place on PM program
2 Long cycle time was due to dropped parts, attention to details
3 Long cycle times were due to dropped parts, operator needs surgical gloves
4 Hard to do study, so much inventory and lots of movement plus lots of wait times
5 Numerous units dropped on the floor
6 Transportation times not taken
7
8

FIGURE 22-1   Zeta cell time study.

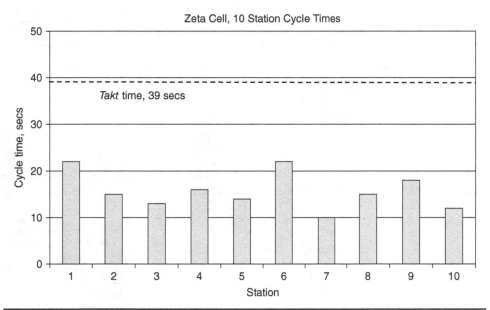

Figure 22-2 Zeta cell balancing graph.

felt it would work. Fig. 22-3 shows in tabular format and Fig. 22-4 shows in graphical format how the work times were allocated with the original ten person layout and how they were reallocated with the newly proposed five person cell. Fig. 22-5 shows the new balancing for the five person cell and compares it to *takt*. In addition to rebalancing the work, I would have liked to perform paper *kaizen*, to further reduce work time, but I knew that going from the current staffing of 10 people to 5 would be a huge cultural change for the group. I thought that would be enough for an initial change. I wanted to keep it simple. Consequently, we did not modify any of the work elements.

An Effort to Get This Approved

I presented these analyses to management and thought I would be embraced as a lean hero. Quite the contrary … no one in management wanted to implement the new plan; no one from quality, no one from production, no one from engineering; not one management person. As far as any logical reasons for them to resist the change: They had none. It fit with their lean transformation, the people were already trained, and we only needed to move one machine about 18 inches and reanchor it and then rearrange the work tables. It was a huge money maker, and it was as simple as it comes—but the unified reaction was that it should not be done now. "Maybe later," they said. I have heard this "maybe later" phrase many times, and it usually means, "not if I have anything to say about it." So I dropped the issue for the time being.

They Had Other Problems, Which Created Opportunity

Concurrently, we had been investigating some warranty returns that we thought could be related to the earlier problems. The failure analysis was not clear but it pointed to the root cause possibly being that the units were dropped during production, creating some hidden internal damage, and the hidden damage did not show up until after about 50 hours of operation, hence the warranty issues.

| | | | Zeta Cell, Original Design | | | | New Balanced Proposal | |
| | | | 10 Operators | | | | 5 Operators | |
Step No.	FC Id.	Operator No.	Work Element	Final Time	Time per Operator	New Design, Operator No.	Time per Operator	
1	10	1	Cut bracket	3		1		
2	20	1	Assy bushing (3)	12	22	1	31	
3	30	1	Install o-ring and clip	7		1		
4	40	2	Place in jig, glue	9	15	1		
5	50	2	Press in magnets (2)	6		2		
6	60	3	Insert o-rings, cap, grease	13	13	2	35	
7	70	4	Install support	8	16	2		
8	80	4	Install o-ring and clip (2)	8		2		
9	90	5	Apply epoxy, 3 locations	14	14	3		
10	100	6	Install control capacitor	8		3	28	
11	110	6	Apply epoxy, topside	6	22	3		
12	120	6	Install retainer ring	8		4		
13	130	7	Install cover cap	7	10	4	33	
14	140	7	Unload/load machine (2)	3		4		
15	150	8	Apply final sealant (1)	15	15	4		
16	160	9	Final test, wrap leads	18	18	5	30	
17	200	10	Package	12	12	5		
			Total	157	157		157	

FIGURE 22-3   Zeta cell rebalancing calculations.

That was all I needed. With the failure analysis report in hand, I explained that the 10-person, large work cell with all the excessive handlings, routinely allowed a lot of dropped product to the floor, which was accurate. They had two options. Scrap all dropped product or close up the cell and eliminate the opportunity to drop the units. Suddenly they were interested in redesigning the cell—so we did.

Creating Flow and Establishing Pull-Demand Systems

We cut the cell size dramatically. Using their new table design, we created a cell using less than 40 percent of the space of the 10-person cell. We moved the press and anchored it. The supervisor was not convinced we could go to one-piece flow, so they set up four-piece space *kanbans* at each work station. Workers were trained to stop producing if the *kanban* location was full. Work instructions were modified to match the work stations, and we were ready to start, which we did the next morning.

We Start Production

Work started and went exceedingly smooth. Quite frankly, since they had been forced into doing this, I expected resistance, but surprisingly none occurred—instead, the whole group was cooperative. With only four pieces at each station, the line filled quickly and production was moving smoothly at about 40 seconds cycle time. Station 4 was clearly the bottleneck. It was extremely awkward for the operator to transfer her part to station 5. The supervisor recognized the problem, had a small sheet metal slide

Time Stack of Work Elements, Zeta Cell

Time in seconds	Activity	Orig Plan	New Plan	Time in seconds	Activity	Orig Plan	New Plan
80 79 78 77 76 75 74 73 72 71 70 69 68 67 66	Apply epoxy 14 secs	Operator 5	Operator 3	157 156 155 154 153 152 151 150 149 148 147 146	Package 12 secs	Operator 10	Operator 5
65 64 63 62 61 60 59	Install O-ring and clip 8 secs	Operator 4	Operator 2	145 144 143 142 141 140 139 138 137 136 135 134 133 132 131 130 129 128	Final test, wrap leads 18 secs	Operator 9	Operator 5
58 57 56 55 54 53 52 51	Install support 8 secs						
50 49 48 47 46 45 44 43 42 41 40 39 38	Insert O-ring, grease insert, place in cap. 13 secs	Operator 3		127 126 125 124 123 122 121 120 119 118 117	Apply final sealant 15 secs	Operator 8	Operator 4
37 36 35 34 33 32	Press in magnets 6 secs	Operator 2		116 115 114 113 112	Unload-load machine 3 secs		
31 30 29 28 27 26 25 24 23	Place in jig, glue 6 places 9 secs		Operator 1	111 110 109 108 107 106 105 104 103	Install cover cap 7 secs	Operator 7	
22 21 20 19 18 17 16 15	Install O-ring 7 secs	Operator 1		102 101 100 99 98 97 96 95	Install capacitor retainer 8 secs		
14 13 12 11 10 9 8 7 6 5 4	Assembly bushing to bracket, install cap. 12 secs			94 93 92 91 90 89 88	Apply epoxy 6 secs	Operator 6	Operator 3
3 2 1	Cut Bracket 3 secs			87 86 85 84 83 82 81	Install control capacitor 8 secs		

FIGURE 22-4   Time stack of work elements.

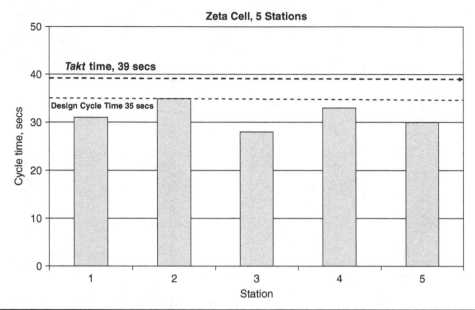

Figure 22-5   Zeta cell balancing graph.

made, and installed it. By noon, the line was producing at less than *takt*. In fact, by the end of the run, the line was producing smoothly and comfortably at a 28-second cycle time and the 800-unit batch was, for the first time, completed in less than one shift. We had only one major problem. The materials supply and *kanbans* were set up to supply at the older slower rate, so the materials handler had to make many emergency runs to keep up with the cell. Recall that they had started at the supply end with detailed plans for every part, so when we improved the cell, all their part plans needed to be redone.

Quick Early Gains Are Huge

In summary, an 800-unit shipment that had previously taken a 10-person cell over two complete shifts to prepare, was now being done in less than one shift with only five people. Hallelujah! (See Table 22-6.) It does not get much better than this when you deal in process improvements. All this without any capital expenditures, and no quality or other downside issues at all! I was pleased. Very pleased.

There also was a large unintended consequence that resulted from this process improvement. As it turns out, this controller was one of a family of controllers that were all made on this work cell, hence the new layout could be used for the entire family.

Metric	Original Case	Leaned Process	% Improvement
First-piece lead time	4.5 h	9 min	97% reduction
Batch lead time	20 h	8.5 h	58% reduction
Space utilization	425 ft^2	160 ft^2	62% reduction
Operators per cell	10	5	50% reduction
Labor costs/unit	15 min/unit	3.19 min/unit	79% reduction

Table 22-6   Gains Summary, Zeta Cell

The 10-person cell would normally work 24 hours, 7 days a week to make the demand for the entire family. Now they could produce the demand in less than 4 days with only five operators. The savings was a net reduction of 30 man days of work per day, forever.

No "Lean Hero of the Month" Here

Wow, think about this. The GM was so proud of his reduction of materials handlers by nine persons, I figured now I would surely be hailed as the Lean Hero of the Month. Instead, everyone just yawned and didn't even say "thanks." In fact, the lead industrial engineer, who resisted the suggested improvements in the first place, now said, "We had planned on doing that all along." So I guess my entry into the "Lean Hero of the Month Club" would have to wait a while. But the cell was certainly setting new performance standards and was ready for a new time study, and also ready to be rebalanced to achieve further gains.

Lessons Learned

So just what is the point of the Zeta line experience? It clearly points out that:

- You have to know the work.
- If you want to make materials flow in the basic work cell, you have to deal with the work at the element level.
- "Inside out" is almost always the best way to begin eliminating non-value-added work activities, something that can easily begin during the paper *kaizen* phase.
- There are huge early gains everywhere.

The Case of the QED Motors Company: Another Great Example of Huge Early Gains on an Entire Value Stream

Background

The Prescription—Revisited

The second prescription, How to Implement Lean—the Prescription for the Lean Project, is well detailed in Chap. 16. The story of QED Motors shows how all eight steps of the prescription are applied. As is normally the case, the prescription cannot always be adhered to in a linear straight-line fashion, but if you follow it through, you will see that all eight steps are well addressed by the team, which improved this process. As a refresher, the second prescription is summarized here:

1. Select and prepare the value stream change agent.
2. Assess the status of the five cultural change leading indicators.
3. Create True North metrics and a balanced scoreboard.
4. Perform the systemwide evaluations.
5. Document the current value stream condition.
6. Redesign to reduce waste.
7. Document and implement the *kaizen* activities.
8. Do it all over again.

Background Information

The QED Motors Company had been making motors for over 30 years. They had a plant in California and had just constructed this new facility in Mexico to take advantage of the low-cost Mexican labor. Their plan had been to run the two plants in parallel for 3 to 6 months while the Mexican plant came up to speed, and then shut down the California facility. It had now been over 15 months and the Mexican plant had just that month achieved the design capacity of 3500 motors per month. However, to meet this demand the plant was working 7 days versus a business plan of 5 days, and still had a number of production problems. The largest problems were:

- Low on-time delivery, only 76 percent
- 14 percent scrap rate
- 7 days production lead time

The New Mexican facility had not demonstrated sufficient capacity or sufficient stability to allow the plant in California to shut down. In addition, at this facility, they had serious management turnover. Since starting up, they were now on their third plant manager, second quality manager, and second purchasing manager.

QED had embarked on a corporate-wide lean transformation about 5 years earlier. The effort was being directed from the home office in Minneapolis. It was obvious they had a number of lean tools in place since *Hoshin* planning matrices were posted and radar charts of all kinds decorated their bulletin boards. But the most encouraging thing was that some of the training had been very effective because many employees understood the basics of Lean.

Nevertheless, this facility was anything but Lean. Due to all the inventory and recycling on the line, it was impossible to follow the flow even though all work stations were labeled and 5S had obviously been attempted. Scrap was high (with over 100 motors in-process at the rework station), pallets of inventory were all over the place, the CNC lathe had huge inventories in three different locations, and flow was virtually nonexistent. In addition, the process flow path was unnecessarily convoluted, material segregation was a disaster, and rejected parts were mixed with normal production.

Their basic design was to produce motors to a finished goods inventory and have a small stock of motors on hand to account for demand and production variations. Their current finished goods inventory was zero. They currently shipped what they could make. This created frequent and disruptive daily changes in the production plan.

We were asked to assist them to further implement their lean system, which they called the QED Production System (QPS). Luis, the QPS manager, who had been there since the startup, was our contact person. He reported directly to the home office and his job was to guide the plant into QPS maturity. Once that was achieved, he was to take over as plant manager and the plant manager would be promoted to a job in Minneapolis. Luis was very knowledgeable in both the motors manufacturing business and Lean Manufacturing, but he had not been dynamic enough to provide the necessary leadership to force this situation.

Specifically, we were asked to:

- Improve line capacity so the plant could provide demand on a 5-day basis.
- Make a 50 percent reduction in lead time.

- Implement a make-to-stock, pull system, operating at *takt*.
- Reduce line rejects by 50 percent.
- Increase on-time delivery to over 95 percent.

We were given some specific restrictions. First, no large capital outlays were available. Second, we had 60 days to achieve our gains. Third, and most restrictive, we could not shut down any facilities at all if it caused weekly production to fall below current levels.

The Process Description

The process description is as follows (Fig. 22-6).

Stators were prepared by welding and grinding, and then were placed at the front of four coil insertion cells. The insertion cells had a build schedule and the supervisor hung markers so the stator prep station knew which stator was required. Three models of stators were available: one large model and two smaller ones. Wire was coiled by the coiling machines in a "coiling island" and then mounted on rolling coil trees for transport to the coil insertion cells. The line operators would leave the cell and go get the wire they needed. Forty trees were queued up at the "coiling island." The trees were color-coded for different coil arrangements required on different motors. Four coil insertion cells were available, each staffed with four operators. All four cells were capable of making all 10 models.

From insertion, the wound stators went to lacing and press where a pool of nine operators would complete these tasks. The work load varied at this station, due to model mix, so these operators were also used to transport motors in the production line. From lace and press the motors were transported to Hipot, an electrical stress test.

After Hipot, the stators accumulated to form a batch prior to going to the preheat oven. Due to oven size, maximum batches would be 12 to 24 motors, depending upon the large/small ratio. After preheat, they were moved to a batch varnish operation, which varied in batch size from 6 to 12, and after varnish, the motors were moved to the curing oven where batches of 24 to 48 were cured.

Following the cure, stators were first polished and then moved to the CNC lathe where they were trimmed and sent to final assembly, which consisted of two final assembly cells, operating in parallel. Each cell was staffed by two operators. Following this, the motors underwent a series of tests, including visual inspection, and then were passed on to packing.

Some comments on the planning. There was a 3-month forecast and a monthly plan, both of which were largely ignored at production. A weekly update would come from Minneapolis on Friday, for the following week's production. The planning staff turned this update into a daily production plan and sent it to shipping along with a copy to the insertion cell and also the coil winding island. Several times each week the production schedule would change. Local planning would send updates to all three planning locations on the production line. Many of the required changes were due to internal production or parts concerns. This changing of the schedule created huge variation in the process, it was a major problem.

Some More Relevant Information

The production demand of the line was that shown in Table 22-7.

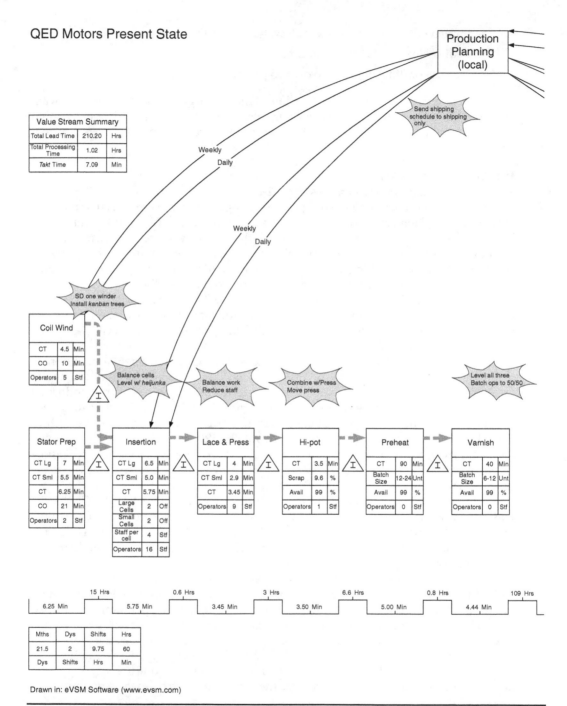

FIGURE 22-6 Present state value stream map: QED motors.

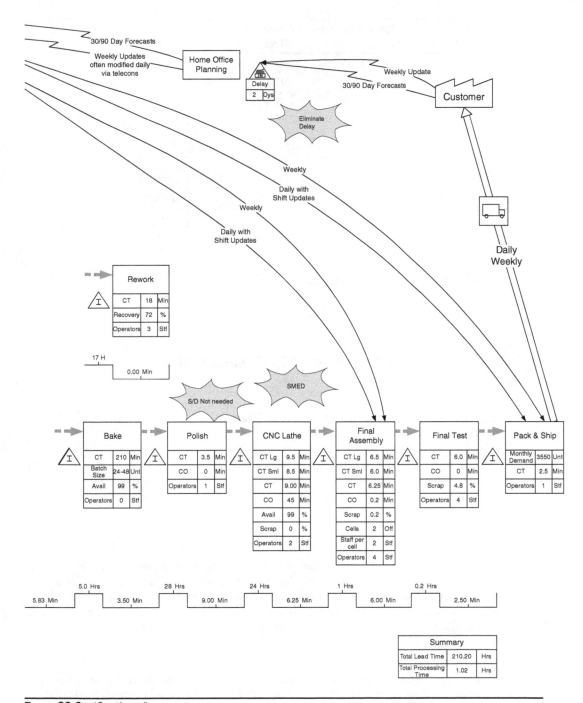

FIGURE 22-6   (Continued)

Model	Monthly Demand (Avg.)	Pickup Frequency	Stator Size	Wire Coil Pattern
Ia	1130	Daily	Large	A
Ib	720	Daily	Large	B
II	950	3/wk, M,W, F	Small-1	C
III	475	1/wk	Small-2	D
IV	175	1/wk	Small-2	D
V	90	1/wk	Small-2	D
VI	50	1/wk	Small-2	D
VII	20	1/wk	Small-2	E
VIII	20	1/wk	Small-2	E
IX	10	1/wk	Small-2	E
Total	3550			

TABLE 22-7   QED Motors, Model Mix Demand

Applying the Second Prescription at QED Motors—How to Implement Lean

The Five Fundamental Issues of Cultural Change

Leadership

Luis and I reviewed the level of leadership, for this short project. I was certain that with both of us working together we could provide the necessary leadership. However, the lack of leadership at the plant manager level and from the home office was the root cause of the current critical situation.

Motivation

As for the motivation, in spite of the problems, the plant line personnel were highly motivated and morale was very high. In most activities, they just didn't really know how to apply the lean principles in their situation. They knew what to do, just not how to do it considering the many obstacles they faced. However, individual motivation, which is often a serious obstacle, was a strong asset at this company.

Problem Solvers

Regarding problem solvers, the company did not have enough, but with Luis and I working full time on the effort, that temporarily met the requirements. With that in mind, we put together a plan and felt that inside of 3 weeks we could increase the production so a 5-day workweek was practical, and we could get on-time delivery to reach over 98 percent. We were uncomfortable about making scrap reduction projections. At least that was the plan we submitted to the management team.

Whole Facility Engagement

Workers knew what to do and how to do it from a technical standpoint, the basic problem was not skill based. At the worker and supervisory level, this was an asset. However, the same could not be said of the management. In many cases management and leadership were lacking and most of the problem stemmed from a highly flawed process design exacerbated by a chaotic work environment.

A Learning/Teaching/Experimenting Environment

They actually had the makings of a very good learning/teaching/experimenting environment. They were part of a larger company which was respected for its lean applications and many vestiges could be seen here. They were just poorly managed and poorly lead so the education really did not have the environment to take hold. We helped immensely in that regard.

The System Assessment

The Commitment Assessment

The commitment assessment was done by Luis and I (behind closed doors) and we concluded that a few problems existed but nothing that we wanted to work on now. We might do this more formally later. We were convinced that for the next 2 months we had the necessary commitments.

The Five Precursors to Lean

Even though our mission was clear, Luis and I began an assessment of the Precursors to Lean. I wanted to get a clear picture of where the facility was on Lean, and I thought it might be an eye opener for Luis. It was!! The facility graded out as shown in Table 22-8.

They achieved a score of 7.5 out of a possible 25. Quite frankly, for a company that is five years into a lean transformation, that is not very impressive, even taking into account the relative age of the facility. What was particularly disappointing was the poor layout, the absolute lack of flow, and the very low levels of process stability. Luis was extremely disappointed with the condition of the processes but agreed with the evaluation we had jointly compiled. As I hoped, it was an eye-opener for him.

Process Maturity

As for the specific assessment of process maturity, we did not do one. In this case, we knew that by this point we were going to make major process changes, so we felt it would be best to complete these changes first. However, after the project was completed,

The Five Precursors	Score
High levels of stability and quality in both the product and the processes	1.5
Excellent machine availability	2.0
Talented problem solvers, with a deep understanding of variation	1.5
Mature continuous improvement philosophy	1.0
Strong proven techniques to standardize	1.5
Total	7.5

TABLE 22-8 The Five Precursors, Summary Evaluation

Luis had this document turned into a plant standard. On subsequent projects, he later told me, all work stations needed to achieve level 3 status, by all measures, before start of production.

Educational Evaluation

Luis and I performed an evaluation and, since they had done a great deal of training earlier, we didn't really need to do anything at the outset. The entire facility had a reasonable grasp on what Lean was really supposed to be. Their problem was not understanding, it was application. However, as we got into the project, we needed to do some education on quick changeovers, *kanban* calculations, inventory calculations, and *heijunka* board design. When these issues surfaced, we trained only the necessary persons and trained them just in time.

The Implementation of the Five Strategies at QED Motors

Synchronize the Supply to the Customer, Externally

This step has three processes. First is the *takt* calculation; next, we must level production; and finally since we have a make-to-stock system, we need to calculate the cycle, safety, and buffer stock inventories.

- The *takt* calculation was: $Takt = (22 - 2.5)*60/165 = 7.09$ minutes. It was based on two shifts each lasting 11 hours. Each shift included 1.25 hours of meal and rest breaks. Production was 3550 units per month—with 21.5 days/month = 165 motors/day.

- The processing time for the 10 models was relatively straightforward to calculate and it made model-mix leveling easy. All models of small stators take one particular time, while all models of large stators take a longer time. Since large stator motors comprise 52 percent of the total, we decided to level with an initial model mix of 50/50, large and small stators. We were not comfortable designing a perfectly leveled system, rather we would level production based on stator size and work in the specific model planning with the *heijunka* board. For our design, at this point, that was all we needed to do.

- To size the buffer and safety stocks, we did not have good information, so we chose to do it rather arbitrarily. For buffer and safety stock combined, we decided to have 2 weeks for all weekly pickups and 3 days for all daily pickups. For cycle stock we used the pickup volume, as a starting point. Table 22-9 shows the outcome of the plan for the inventories.

Synchronizing the Production, Internally

Defining the Work

This was very difficult, and when we were doing it, I was convinced it would need to be done again because the processes were very unstable and cycle times had large variations. Nonetheless, we needed to start somewhere, so we elected to use the time for overall process steps, not element times. (For example, in the process to Hipot, the motor needed several steps, or work elements, performed. These work elements included burning the leads, straightening the leads, cleaning the leads, connecting to the Hipot machine, cycling the machine, disconnecting from the machine, and passing

Model	Monthly Demand	Pickup Frequency	Daily Prod.	Weekly Prod.	Cycle	Safety and Buffer	Total FG Inv.
Ia	1130	Daily	53		53	159	212
Ib	720	Daily	34		34	102	136
II	950	3/wk M, W, F	44		44	132	176
III	475	1/wk		110	110	220	330
IV	175	1/wk		41	41	82	123
V	90	1/wk		21	21	42	63
VI	50	1/wk		12	12	24	36
VII	20	1/wk		5	5	10	15
VIII	20	1/wk		5	5	10	15
IX	10	1/wk		2	2	4	6
Total	3550						

TABLE 22-9   Inventory Plans

to the next station.) Since most of the work elements could not really be transferred to another work station, this was not a good tool for us as we tried to internally synchronize the flow. (Recall that this was a key element in balancing the Zeta cell so efficiently. Unfortunately, it just wouldn't work here.)

Table 22-10 shows the results from the time study, with the data segregated into large and small rotors.

The data for the time study was then used to create the line balance chart, Fig. 22-7, showing both the large and small stator cycle times compared to the *takt* of 7.09 minutes. This was then converted to a balancing chart to show operation with a line that is balanced with 50 percent large stators and 50 percent small stators, shown in Fig. 22-8.

Evaluation of the Balance Chart

As we evaluate this line balance chart, we can see there is lots of waiting and that the process stations are not well balanced. However, the critical problem is the bottleneck of the CNC lathe. Not only is it a bottleneck but the cycle time exceeds *takt*. This shows clearly why a great deal of overtime is required. It is obvious that to meet customer demand we will need to improve the cycle time at the CNC lathe.

We Reduce the Lathe Cycle Time

A breakdown of the lathe cycle time is shown in Table 22-11.

Kaizen Opportunities in Synchronization

The following were *kaizen* opportunities we could exploit from improvement ideas in both external and internal synchronization.

- It was clear from the balancing chart that the CNC lathe cycle time was the bottleneck, and both models ran at a cycle time exceeding *takt*. Thus, we needed to reduce the average cycle time. After some analysis, we decided to prepare a

Process Step	Large Stator Work Time (Min)	Small Stator Work Time (Min)	Present Staffing	Ave. C/T Based on Present Staffing Large/Small (Min)
Weld stator	14	11	2	7.0/5.5
Coil insert.	105	65–90	16*	6.5/5.0
Lace and press	36	26	9	4.0/2.9
Hipot	3.5	3.5	1	3.5/3.5
Preheat	7.5	3.75	1	7.5/3.75
Varnish	6.7	3.3		6.7/3.3
Bake	8.75	4.38		8.75/4.38
Polish	3.5	3.5		4.5/3.5
CNC lathe	9.5	8.5	2*	9.5/8.5
Final assy	13	12	4***	6.5/6.0
Final insp.	6	6	1	6.0/6.0
Packaging	2.5	2.5	1	2.5/2.5

*Four cells of four people each.
**Lathe personnel operate in parallel.
***Two cells of two people each.

TABLE 22-10   Time Study

universal holding jig for small motors and one for large motors. This would reduce the number of changeovers to as few as five per day. Also, using Single Minute Exchange of Dies (SMED), quick changeover technology, we were able to reduce the changeover time to 11 minutes.

- Shut down one winder—this was no longer required—and go to three operators for the wire coiling cell.

- Model-mix level the production work. The complexity in leveling was in the coil insertion cells. The work to insert coils on a large stator averaged 105 minutes. The work for an average small stator averaged 80 minutes. To level the model mix, we would need two cells of four operators each on the large stators, which would average a motor every 26 minutes/cell (105 min/4 stations per cell). We would also have two cells of three operators each on the small stators and average a motor every 27 minutes/cell (80 min/3 stations per cell). Overall, we would have a cycle time across all four cells of 6.6 minutes. This was below *takt*, yet close enough to avoid overproduction.

- Other opportunities might be possible in staffing the insertion cells and the lacing island.

- Eliminate the polishing step since it was no longer needed.

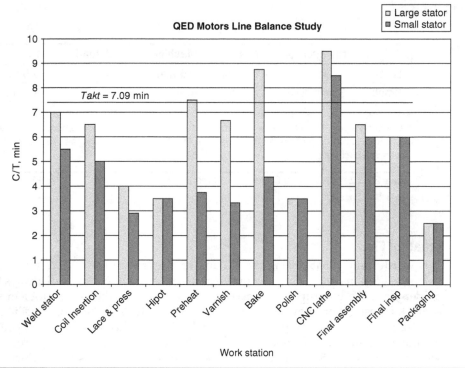

FIGURE 22-7   Line balance chart.

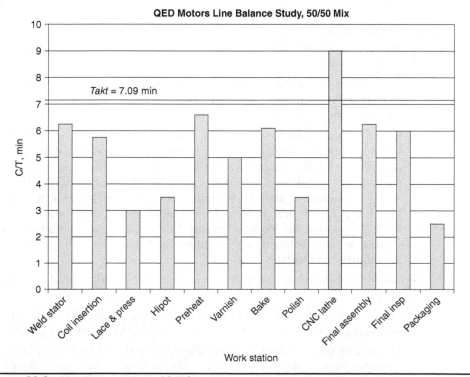

FIGURE 22-8   Line balance chart, 50/50 mix.

	Changeover*	Load	Machine	Unload	Total
Large	3 min	0.5 min	5.5 min	0.5 min	9.5 min
Small	3 min	0.5 min	4.5 min	0.5 min	8.5 min

*Actual changeover (CO) time was 45 minutes, averaging about 15 units per changeover. A unique holding jig was required for 8 of the 10 models. An average of eight COs occurred each day.

TABLE 22-11   CNC Lathe Cycle-Time Elements

Creating Flow

It is worth noting that during the making of the present state VSM, all production stopped at coil insertion, and the flow went to zero. Mother Nature will always find the hidden flaw. We interrupted the making of the VSM and did some problem solving. A quick investigation showed there was no wire made with the proper coiling. Yet all 40 transport trees were 100 percent full. On the floor, we had over 4 days of coiled wire but none of it matched the demand at the cells. This, of course, was the direct result of sending planning information to multiple points on the value stream which was not (and can seldom be) synchronized to the line's needs. In addition to the planning issues, creating flow was achieved by balancing the flow, managing the inventories of both finished goods and WIP, as well as reducing transportation steps and distances.

Kaizen Opportunities Related to Flow

The following were *kaizen* opportunities we could exploit that were related to the process flow improvements.

- The process of updating orders currently had a 2-day delay at the home office. This was for accounting purposes only. Consequently, we would have the customer copy us on the electronic updates to avoid the 2-day delay.

- Lace and press used nine operators, and at a cycle time of less than 4 minutes indicated there was a lot of waiting. The press operation was about 2.5 minutes, regardless of stator size so we decided to transfer the press operation to the subsequent operation of Hipot for better balance and flow. The Hipot/Press station would now have a 6-minute cycle time, consisting of 3.5 minutes of Hipot and 2.5 minutes to press. Lacing alone would now be 33.5 minutes for large stators and 23.5 for small stators. Consequently we could staff the lacing operation with only five operators and still have an average cycle time of 5.7 minutes, comfortably below *takt*.

- Make the flow at lace and press visual with floor marking and in/out marked locations

- Since many tests and inspections that had been done at final test, were no longer necessary, we could eliminate them and combine final test with packing.

- No motors were allowed to "break the flow" and could only be placed in WIP locations at workstations and supermarkets. Rework was sent directly to and isolated to the rework station.

Establishing the Pull-Demand System

Kaizen Opportunities Related to Pull Systems

The following were *kaizen* opportunities we could exploit that were related to the implementation of pull systems.

- Send daily shipping information from local planning to the storehouse, only A key element of lean planning is that for any value stream you try to schedule at one location only and then use flow, pull signals and FIFO lanes to make the entire process flow seamlessly. We could do that here now.
- The planning function will make minor changes via the *kanban* system and *kanban* replenishment to the *heijunka* board will occur twice per shift.
- Production planning will be by *heijunka* board at the insertion cells only.
- Install a *kanban* system for wire supply using the wire trees as *kanban* carts.
- Establish a storehouse of stators at the Insertion cell so they can withdraw when needed.
- Establish FIFO lanes from insertion all the way to packing. Calculate the maximum inventories and document this on the FSVSM.
- Calculate and establish the finished goods inventory.

Making a Spaghetti Diagram

The spaghetti diagrams, shown in Figs. 22-9 and 22-10, display several process opportunities to reduce transportation and make better use of the floor space. The spaghetti diagrams used in this project are detailed further in App. D of Chap. 14.

Kaizen Opportunities Related to the Spaghetti Diagram

The following *kaizen* opportunities were found while studying the spaghetti diagram.

- Move FG Inventory near the aisle on the storehouse side of the process area.
- Move the Hipot operation next to the press.
- Move the CNC lathe to the corner near the curing oven.

Document *Kaizen* Activities on a Future State Value Stream Map

In this case, we documented the major *kaizen* activities on a future state value stream map (FSVSM). See Fig. 22-11.

Additional *Kaizen* Activities

In addition to the *kaizen* opportunities mentioned earlier, we took the following measures:

- We ran off the huge volume of WIP, almost 700 units which took special actions to accomplish this task. First, we temporarily shut down operation of all stations before the CNC lathe and reassigned those who were trained to fill in for all workstations from the CNC lathe through packing. Next, we sent the extra people to the rework station or assigned them to a *kaizen* project and began rescheduling to cover 24-hour operation, including lunch and breaks, for all stations from CNC lathe through packing.

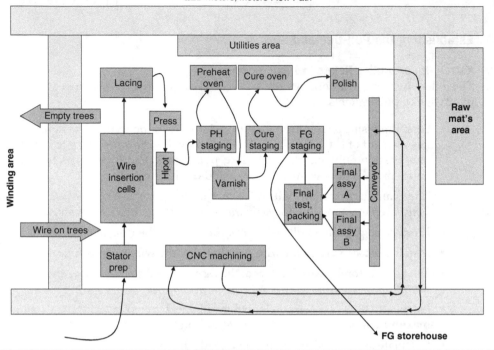

Figure 22-9 Spaghetti diagram: QED motors, before.

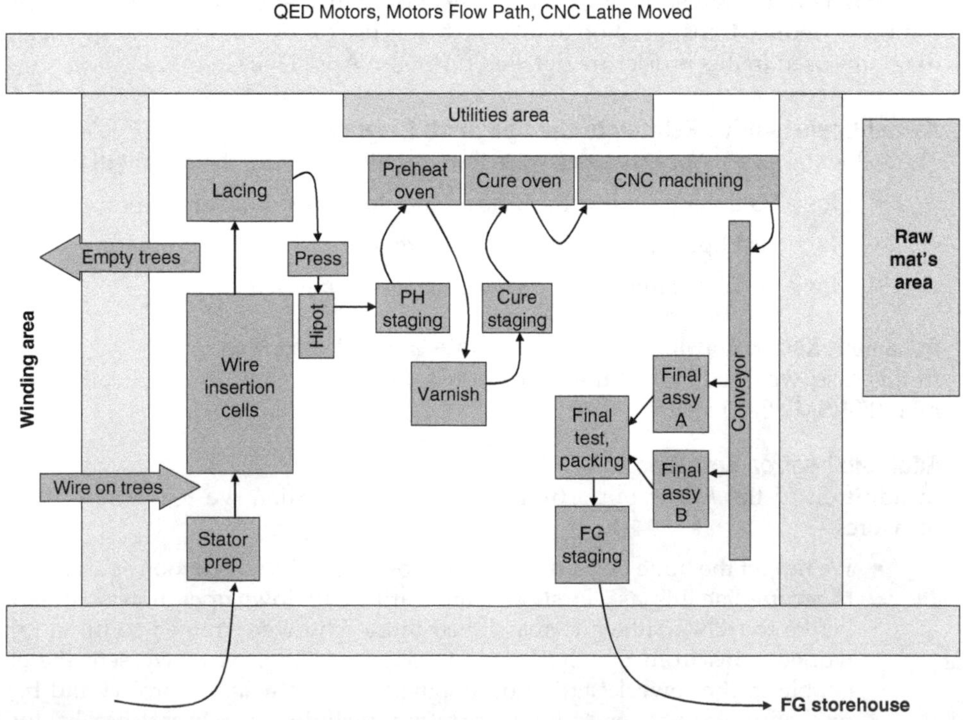

Figure 22-10 Spaghetti diagram: QED motors, after.

- We reduced the cycle time at the CNC lathe to achieve 160 units per day by adjusting work procedures here. To do this we transferred work from the CNC operator to the helper, who was lightly loaded, so as to decrease workload and the cycle times of the CNC operator. This reduced the line bottleneck by a half minute.

- We performed SMED analysis on the CNC changeover process to reduce changeover time and further increase production at the lathe. We began immediately producing the universal holding jigs. Next we implemented other SMED activities; changeover time was reduced to 11 minutes. These steps took 3 days and when complete we were producing 220 units per day at the CNC lathe.

- Final assembly could no longer keep pace, so we increased the number of changeovers at the CNC lathe. This allowed us to decrease the batch size and better level the model mix. At this point, we had all stations running, at least at half rate.

- Next, we cut the conveyor from CNC lathe to final assembly in half and marked with visuals for red/yellow/green for production control.

- We leveled the production at the insertion cell using a *heijunka* box with *kanban* and modified the staffing for the small stators to three operators per cell for better balance.

- After 2 weeks, it was possible to combine final inspection with packing and eliminate another position.

- Final inspection and packing were made visual using floor markings and color coding.

- In addition, over 70 other minor process improvements were executed, mostly by the supervisors, line leaders, and operators in the next 4 weeks.

- Even though capacity was up, we continued to run 7 days per week with two normal 11-hour shifts. This was needed to establish inventories of cycle, buffer, and safety stocks for all 10 models.

- It took over 3 weeks to build up enough inventories so we could take the CNC lathe out of service and move it. We did that over one weekend and the lathe started up on Monday without incident.

- The plant ran 7 days per week for 2 more weeks, and then reverted to 5 days per week operation.

Standardize and Sustain

Since QED motors had been working on Lean for 5 years, they knew all the tools for standardization. Unfortunately, due to the chaotic nature and the poor leadership, very little standardization had actually occurred. We immediately revitalized these efforts and prepared new job breakdown sheets, retrained operators, implemented visuals at every opportunity and reinstituted Leader Standard Work. They already had an active but ineffective management audit system in place, you could see the report-outs on the wall, they called it kamishibai. Unfortunately, it too was just basically wallpaper with no actions taken beyond recording the deficiencies. This too was revitalized.

QED Motors Future State

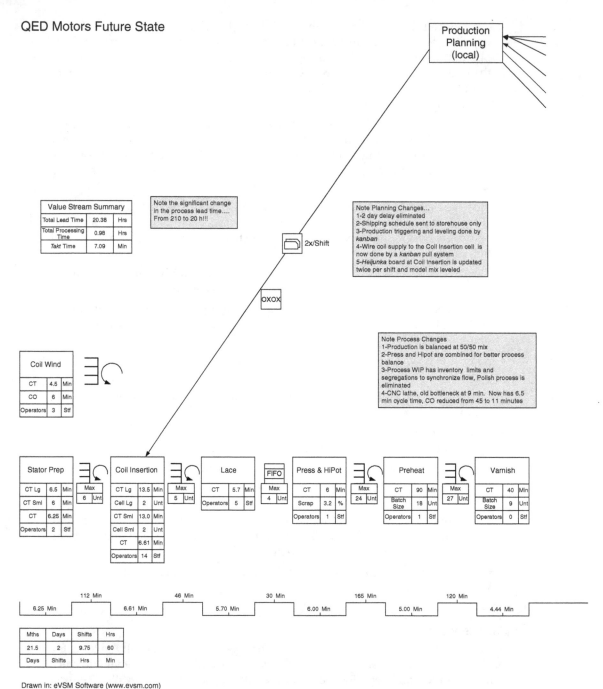

FIGURE 22-11   Future state value stream map: QED motors.

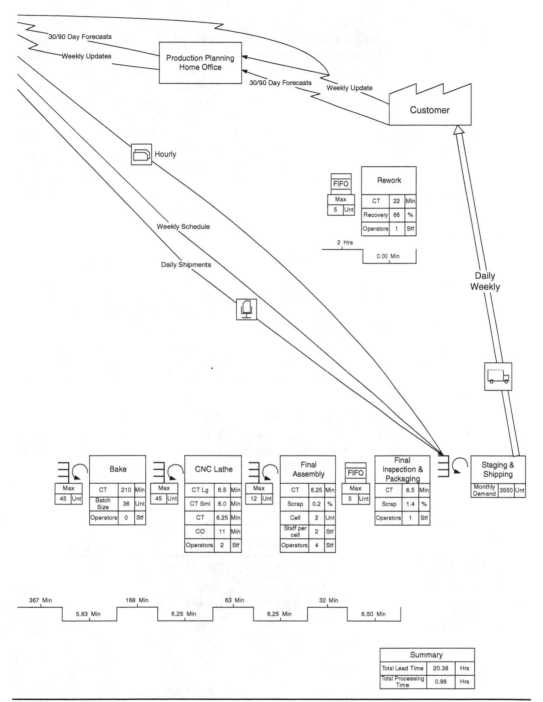

30/90 Day Forecasts

Weekly Updates

Production Planning
Home Office

30/90 Day Forecasts   Weekly Update

Customer

Hourly

FIFO

Max
5 Unt

Rework

CT	22	Min
Recovery	66	%
Operators	1	Stf

2 Hrs

0.00 Min

Weekly Schedule

Daily Shipments

Daily
Weekly

Bake

| Max | 45 | Unt |

CT	210	Min
Batch Size	36	Unt
Operators	0	Stf

CNC Lathe

| Max | 45 | Unt |

CT Lg	6.5	Min
CT Sml	6.0	Min
CT	6.25	Min
CO	11	Min
Operators	2	Stf

Final
Assembly

| Max | 12 | Unt |

CT	6.25	Min
Scrap	0.2	%
Cell	2	Unt
Staff per cell	2	Stf
Operators	4	Stf

FIFO

Max
5 Unt

Final
Inspection &
Packaging

| Max | 5 | Unt |

CT	6.5	Min
Scrap	1.4	%
Operators	1	Stf

Staging &
Shipping

| Monthly Demand | 3550 | Unt |

367 Min   168 Min   63 Min   32 Min

5.83 Min   6.25 Min   6.25 Min   6.50 Min

Summary		
Total Lead Time	20.38	Hrs
Total Processing Time	0.98	Hrs

Figure 22-11   (Continued)

The Results

Our activities had allowed the plant to improve dramatically. Specifically we could now meet demand in a 5-day workweek. We were asked to reduce the production lead time by 50 percent and we actually achieved 89 percent reduction using a make-to-stock, pull, production system operating at *takt*, as requested. We were able to reduce defects by almost 60 percent and improve on-time delivery to over 95 percent, actually achieving 99 percent. In short, by focusing on waste reduction, with an aggressive attack on the seven wastes we met or exceeded all our goals and even achieved two significant gains that were not requested. These included major savings in production labor and space reductions. The labor savings alone showed a 44 percent reduction in direct labor to produce a motor. All of this was achieved by an aggressive waste reduction effort. Except for the minor cost to produce the universal jigs, no capital was expended. This was a massive success and the results are summarized in Table 22-12.

Some Comments on the Results

- The gains were huge and early, all achieved in less than 2 months.
- Of particular note were the quality improvements. Luis was very worried because we had no specific line items to work on any of the defect modes. As it turned out, by reducing inventory and implementing standard work we accomplished quality improvements.
- The facility still needed to work on a number of projects. First, and most importantly, although we cut the defects by 60 percent, line rejects were almost 6 percent. This is not acceptable by today's standards, thus they still need to work on this.
- They were now able to shut down the California plant and did so.
- One weakness in this effort was that we did not implement any kind of a *jidoka* concept. We were not asked to but I was disappointed we were not able to weave it into the project.

	Initial Situation	After First Pass of Lean	Improvement
Production rate	125 U/day	165 U/day	+32%
On-time delivery	76%	99%	+23%
Lead time	92 h (7.5 d)	16 h (0.8 d)	−89%
Line rejects	14.4%	5.8%	−59.7%
Operators per shift*	50	43	−17.5%
Labor consumed (Mh/motor)	9.33	5.21	−44%

*In addition to the staffing on the FSVSM, there were three operators in rework, four materials handlers, and three in material prep. None of this was changed as yet.

TABLE 22-12   QED Motors Results Following the Lean Applications

What Is Next for QED Motors?

As is always the case, these improvements brought about a new present state, which will need to be fully evaluated, including applying the Five Strategies to Becoming Lean, starting with the *takt* equation using a spaghetti diagram, as well as a future state value stream map with *kaizen* activities noted ending in the "standardize and sustain" step so the system can be stressed once again.

The company's future *kaizen* activities will clearly depend upon their specific goals, but we made a starter list based on what we had learned to date. Possible measures included:

- Implement a reasonable *jidoka* concept.

- A new floor layout to take advantage of the space savings achieved.

- Review the replenishment time to see if cycle stocks could be further reduced. Also review production and demand variations to see if buffer and safety stocks can be further reduced.

- Since the major impediment to lead-time reductions was the batch-operated preheat, varnish, and bake operation, they needed to consider converting this to a continuous flow operation. This would shorten lead time significantly, but would require new capital equipment.

- The majority of the defects found at Hipot were due to the manual coil insertion operation. Automatic winders should be explored, which of course would require capital but would substantially reduce lead time and quality dropout. We did some preliminary economic analyses and it appeared to have a rate of return on invested capital exceeding 80 percent, based on labor, rework, and waste savings.

- QED Motors had made significant progress in a very short period of time and they can be proud of their accomplishments. Now they must focus on their future improvement ideas and continue on the journey to becoming Lean. It is a never-ending cycle of assessment, analysis, and improvements through the continuous reduction of waste on the way to their goal of making their facility a better money-making machine; a more secure workplace for all; and the supplier of choice to their customers.

CHAPTER 23

The Precursors to Lean— Not Handled Well

Aim of this chapter … to explain the story of the ABC Widgets Co. that is an example of a company that *wanted to look Lean more than they wanted to be Lean.* Improper mental models, poorly trained management, misplaced motivation, internal conflicts, along with too few on-site resources created a large resistance to becoming Lean. In spite of these deficiencies, we assisted them in making large improvements. However, they are an example of "How *Not* to Implement Lean," in that after 4 years into their lean initiative, basic and large foundational problems still persisted. They are a story of the "Precursors to Lean (see Chap. 24) … not handled well."

Background to the ABC Widgets Story

We were called in by Miguel, the reliability manager. Our task was defined as problem-solving assistance, not lean implementation. The situation was this: A mature value stream, one of four within the plant, previously had nearly 100 percent on-time delivery for 2 years, but was now having serious problems: On-time delivery had dropped to 74 percent and remained there for over 5 months, despite repeated efforts to improve the situation.

Also, 5 months earlier, the daily production demand had increased from 30 units per day to 36 units per day, and the customer, in keeping with their own lean initiative, had begun daily pickups. Previously, customer demand was 150 units per week and would be picked up in a single weekly shipment.

The reliability manager was sure one of the major contributors to the on-time problem was a welding machine with low availability, but he felt powerless to correct it. So, he sought our assistance.

In addition, 2 months earlier, the plant received a visit from the home office, and a lean expert had assisted them in problem-solving activities. After 1 week of analysis and meetings, the plant was given a long and detailed list of 22 projects they were to work on, but after nearly 60 days they had not been able to improve the delivery situation.

Our Challenge

The problem was stated as: "We have low on-time delivery. It is 74 percent and needs to be greater than 99 percent." Thus, we were asked to:

- Survey the situation
- Make recommendations
- Sell the recommendations consistent with their lean initiative
- Assist in the implementation

Our constraints were many, but for the sake of brevity, I will distill them to the critical few:

- There was really no capital available to solve this problem.
- We could not increase inventories. (Miguel noted that they were presently holding 30 units of safety stock to handle the internal supply variations. This volume was calculated to be three sigma of inventory, but they were still missing deliveries.)

Some More Relevant Information

Of course, a little more information is necessary to appreciate this particular situation. This plant was a Maquiladora and it had only a skeleton of technical personnel on-site, with little or no design responsibilities. Like many Maquiladoras, which are the Mexican half of the Twin Plant concept, taking advantage of low-cost Mexican labor, their task was to meet the production schedule at minimum costs. Purchasing was done centrally, while planning was local. Four years earlier, the ABC Widgets Co. had embarked on a lean initiative, but most lean-specific skills were centrally located at the home office. Both the metrics of performance and the actual performance results of the plant were much the same now as they were before the implementation of their lean initiative. They had made great strides in one area: inventories. Both raw materials and finished goods inventories had been dramatically reduced. The entire plant took great pride in this because they were among the leanest plants in the corporation by this measure. Raw materials had been improved from 12 to 35 turns and finished goods for this line improved from 24 to nearly 70 turns. No information was available about WIP (work in process).

> **P**oint of Clarity When discussing variation, the standard deviation has no meaning unless the data show stability.

Another issue was influential in this problem. The reliability manager, Miguel, was the son-in-law of the general manager of their North American manufacturing division. Miguel had graduated with honors, 3 years earlier from the Tec de Monterey and was hired 18 months before we arrived to assist in the lean initiative. Since his arrival, machinery reliability had improved, and availability associated with machinery reliability had risen over 7 percent; these gains were clearly seen in Overall Equipment Effectiveness (OEE) on several value streams. However, "water cooler rumors" of nepotism had arisen and several managers questioned his capabilities. To make things more problematic, Miguel had apparently created a serious rift with the production personnel when, earlier in the year, he had initiated autonomous maintenance as part of their TPM efforts. It did not go well, and that effort was aborted after just a few months due to the friction it caused. So obvious tension remained between the Reliability and Production departments.

We Analyze the Data

Our efforts began with a look at the data associated with this line for the last 30 normal production days. This is shown in Table 23-1.

Even though these data had huge variation, we were assured that these production data were typical. In this 6-week period:

- Six daily shipments had been missed
- Production was well below the 36-unit target
- The production rate had a huge day-to-day variation
- Weekend work (both Saturday and Sunday) was required to replenish inventory

> **P**oint of Clarity   Prior to implementing a lean transformation, processes must show statistical stability. This process does not!

We Pareto-ize the Days of Low Production, We Find Two Problems

We reviewed production data to find the reasons for low-line production. In this 30-day period, there were 9 days where actual production was 10 or more units below demand. (Each tray had 25 production items, with four trays per box, so a box contained 100 items. This was their production jargon: boxes, which they called "production units" or just "units" for short.) In all nine cases, either the automatic welding machine failed or the sensor, a high-cost component, had a stock out, or both occurred. At any rate, their large problems were not quality problems, but availability problems caused by these two items.

We Investigate Further

With this information in hand, we wanted to have another discussion with Miguel to gain more insight. First, the welding machine availability was number one on our list of items to correct. Unfortunately, the home office facilitator had characterized the problem as inadequate maintenance: specifically, the need to train a replacement for Jorge, the welding technician, who had retired 4 months earlier. The plant was not allowed to replace Jorge because they had been asked to reduce manpower. Miguel told us the problem had nothing to do with Jorge, it was a capacity problem with the welder. It would simply overheat at the new rate and the electrode and holder would fail, requiring a shutdown to replace the parts and several hours to complete the setup, which included alignment and testing. Work was underway to implement Single Minute Exchange of Dies (SMED), quick changeover technology on the machine startup, but a practical solution was months away. However, Miguel did not believe that starting up faster was the issue. We did some more investigating and found out he was right. The problem was poor reliability.

As for the sensor stock outs, this sensor was a high-cost component comprising 44 percent of the total raw material cost of the product. The home office had implemented strict inventory guidelines and aggressively reduced the inventory levels of all components, including the sensor. After the ramp-up, the supplier could meet demand, but only by working overtime. Since these issues were managed by the central purchasing group, Miguel thought it would be futile to attempt to increase inventory levels, although that would surely solve the problem of stock outs.

> **"L**et's not work hard to get good at something which should not be done at all. **"**
> J. Keating

	Base Case					
	Daily Prod	Dev. from Goal[1]	Avg. for the Week[2]	SD for the Week[3]	Total Weekly Prod'n	Weekend Run[4]
1	35	−1				
2	19	−17				
3	38	2				
4	8	−28				
5	36	0	27.20	13.14	136	44
6	22	−14				
7	36	0				
8	18	−18				
9	35	−1				
10	29	−7	28.00	7.91	140	40
11	38	2				
12	12	−24				
13	32	−4				
14	28	−8				
15	27	−9	27.40	9.63	137	43
16	29	−7				
17	39	3				
18	40	4				
19	35	−1				
20	31	−5	34.80	4.82	174	6
21	37	1				
22	0	−36				
23	14	−22				
24	29	−7				
25	37	1	23.40	16.10	117	63
26	40	4				
27	18	−18				
28	17	−19				
29	27	−9				
30	40	4	28.40	11.28	142	38
		Avg.	28.20	10.66		

[1]Deviation from the planning goal
[2]Average daily production for the 5-day week
[3]The daily variation measured as the standard deviation (SD)
[4]The production over the weekend, needed to make the weekly demand

TABLE 23-1   Production Data, Base Case

We also inquired about the line capacity in relation to the recent increase from 30 to 36 units per day. The nameplate bottleneck on the line would limit the line to 40 units in a 24-hour work day. Line OEE would need to be at 90 percent, and these levels had not been previously achieved.

We asked for the background on the recent increase to 36 units per day. We were told the customer had demanded a 15 percent price reduction over 2 years and would then give the plant additional demand. According to Miguel, the line was very profitable and management jumped at the chance with very little review, not even an update of the Process Failure Mode Effects Analysis (PFMEA). Management was very confident the line would perform. Prior to the ramp-up, the customer returns were less than 500 ppm, on-time delivery was practically 100 percent, and the line name plate capacity was adequate. It appeared to be a win-win situation....that is until they could not meet demand. And that problem cropped up almost immediately. In addition, with all the weekend overtime and expediting, profits had eroded to practically zero. Miguel was convinced they were not on the correct path to return to profitability.

We Investigate More and Fix the Welding Machine, the First Problem

With this background, we did some digging and discovered the following. The welding machine had numerous previous problems, but those problems were masked by excess capacity in the line and the inventory situation, specifically the weekly pickups. Earlier, they could have an off-day or two and still be able to make up volume with the excess capacity and even a little overtime. Also, it was only recently that they achieved the low inventory levels. The higher inventory levels also helped mask the problem of the welder's poor reliability. We asked the technician in charge of solving this problem to bring in the manufacturer's rep for the welder. The welder's rep readily confirmed that the machine was probably capable of 30 units per day, but would surely overheat at 36 units per day. He suggested a minor upgrade to the cooling system; and for $1500, he would expect greater than 99 percent reliability at our needed cycle times. The equipment was ordered and installed. Table 23-2 shows the next 25 days of production.

In this 5-week period, significant progress had been made. The welder did not fail once. In addition:

- Only one shipment was missed
- Production had improved by over 3 units per day
- The production variation shrank from 10.7 to 7.6 units
- Weekend work was significantly reduced

The Second Problem, Sensor Stock Outs Is Addressed

The second problem, the sensor stock outs, proved to be a little more slippery. The supplier was contacted and he was doing all he could but was severely stretched. His production facility was at capacity and earlier he had informed the buyer, who did not pass on this information. They had plans to increase capacity, but any future increases were 6 months away. It seemed the only viable

Point of Clarity The only purpose of inventory is to protect sales, not production, but sales. So where should the inventory be located? Inventory is a waste, so minimize the inventory cost, be it in raw materials or finished goods.

	Daily Prod	Dev. from Goal	Avg. for the Week	SD for the Week	Total Weekly Prod'n	Weekend Run
			Base Case—Welding Machine Now Reliable			
1	36	0				
2	32	−4				
3	39	3				
4	21	−15				
5	37	1	33.00	7.18	165	15
6	20	−16				
7	36	0				
8	35	−1				
9	33	−3				
10	17	−19	28.20	8.98	141	39
11	40	4				
12	36	0				
13	33	−3				
14	14	−22				
15	30	−6	30.60	9.99	153	27
16	33	−3				
17	36	0				
18	37	1				
19	16	−20				
20	34	−2	31.20	8.64	156	24
21	39	3				
22	33	−3				
23	34	−2				
24	35	−1				
25	36	0	36.00	2.30	177	3
		Avg.	31.68	7.62		

TABLE 23-2   Production after Welding Machine Upgrade

short-term solution was to increase inventory levels to account for the supplier's variation. No one wanted to hear that.

Nonetheless, we made an XL model from our data, which showed that although sensor inventory costs would need to increase by 25 percent, finished goods safety stock inventory could be reduced by 60 percent. It turned out that for each $1.00 of sensor inventory that was added, $8.00 of finished goods inventory could be reduced. This was a clear winner in terms of total money tied up in inventory.

Point of Clarity The system optimum is not always the sum of the local optima!

The computer model showed a new financial optimum. This, coupled with our prior success, got us an audience with the general manager (GM) of the North American manufacturing division. We met the GM during his monthly visit and convinced him that although sensor inventory costs would go up, we could reduce finished products inventory by a greater dollar volume so that the overall effect would be a significant savings in money tied up as inventory. For finished goods, we now had a safety and buffer stock inventory of 12 units, which was less than 1 day of production. We got the go-ahead and the supplier was able to pull some strings and use capacity from another plant in order to supply some additional product to create the inventory. After the new sensor inventories had been established, sensor stock outs stopped. The production now looked like that shown in Table 23-3.

In this 6-week period, additional progress had been made, specifically:

- Zero shipments had been missed
- Production had improved by an additional 4 units per day
- The production variation shrank from 7.6 to 2.9 units
- Weekend work was reduced to practically zero

Summary of Results

The summary of results is shown in Table 23-4.

Demand could now be met, overtime practically ceased, there was no need to airship product, labor cost was reduced, and the product was once again profitable. As a result, almost everyone was happy.

How Did the Management Team from ABC Widgets Handle the Cultural Change Leading Indicators?

For just a moment, think back to the major problems they had prior to the ramp-up. They had supply and process capability problems of which they were unaware. They had inadequate machine and line availability. It was only the excess line capacity and excess inventory that masked these problems. The problems were there from the beginning! They simply failed to "see" them and take them on.

A lean plant will see these as opportunities and attack them to reduce space needs, manpower needs, lead-time issues, and inventories, to name just a few. All of this could have been done earlier, but it was not. It wasn't even recognized.

So, let's look at the actions of the management of ABC Widgets in relation to the implementation of any cultural change transformation. How did plant management handle the five issues?

Did They Have the Necessary Leadership?

The leadership was clearly deficient in several areas. First, when they ramped up, they had nothing that resembled a plan. Recall that they did scarcely nothing except hope they could make it work. Next, once they got in trouble, they did not have much of a plan for that either. They counted on the home office for help that did not materialize. Third, when it became obvious they could not solve the problem themselves, management resisted outside involvement. Finally, as you hear the entire story, a grand deficiency

	Daily Prod	Dev. from Goal	Avg. for the Week	SD for the Week	Total Weekly Prod'n	Weekend Run
Base Case—Welding Machine Reliable, Sensor Supply Resolved						
1	35	−1				
2	36	0				
3	38	2				
4	36	0				
5	40	4	37.00	2.00	185	−5
6	34	−2				
7	36	0				
8	38	2				
9	35	−1				
10	32	−4	35.00	2.24	175	5
11	38	2				
12	39	3				
13	36	0				
14	28	−8				
15	37	1	35.60	4.39	178	2
16	31	−5				
17	40	4				
18	36	0				
19	35	−1				
20	34	−2	35.20	3.27	176	4
21	37	1				
22	35	−1				
23	36	0				
24	32	−4				
25	40	4	36.00	2.92	180	0
26	38	2				
27	35	−1				
28	37	1				
29	30	−6				
30	35	−1	35.00	3.08	175	5
		Avg.	35.63	2.89		

TABLE 23-3   Production after Welding Machine and Sensor Countermeasures

	Prod Rate	Std. Dev. of Rate	Missed Shipments	Weekend Production
Base case	28.20	10.66	26%	196
Fix welding machine	31.68	7.62	4%	108
Eliminate stock outs	35.63	2.89	0%	11

TABLE 23-4   Summary Results

in leadership existed at the corporate level. Lean cannot be managed from afar. Lean leadership and the problem solvers—need to be on-site. There is no substitute for this, becoming Lean is a local activity.

Their Motivation

Did they have the motivation to make it work? At one level, the answer to this was "yes," since they were actively trying to solve the problem—that is, the problem of low production, once they got into it. But at the larger level, their motivation was grossly lacking. Motivation is measured by actions. And just what were their actions to make the plant Lean? Superficial and clearly inadequate. Several years into the initiative there were basic, foundational problems that should have been found and resolved during the first wave of activities. Yes, problems need to be both uncovered and resolved. They did not find them because they did not look, nor did they listen to those who knew about the problems. Just for practice, based on these data, do a commitment evaluation on their management (see Chap. 24).

Their Problem Solving

Did they have the necessary problem solvers in place? Clearly not. Regarding this rather straightforward production problem, they could not solve the problems themselves. For example, the "help" from the home office was not helpful, and ultimately, to resolve the issues, they needed outside assistance from us. Finally, their problem-solving methodology was not PDCA (Plan, Do, Check, Act), regrettably they were the poster child for PDHM (Plan, Do, Hope, and Move on; see Chap. 9).

How about Their Engagement?

This was clearly deficient. It was clear the management neither knew what to do nor did they know how to do it, for once they got in trouble as they were obviously resource deficient. This was a classic case of a plant that wanted to look engaged, but was not. In addition from a responsibility standpoint, they exhibited several of the five Big Responsibility Runarounds (see Chap. 2, Lean Killers).

Did They Have a Learning/Teaching Environment?

The answer is a resounding no! There were no signs of either. Theirs was a closed culture (See Chap. 2, Lean Killer No. 9) that did not want to make sound data-based decisions. They had only a superficial problem-solving approach that left many issues unresolved; and they were not staffed in any way to get the problems solved even if they had the awareness to find them. Their sole driving force was short-term economics, using strong central control mechanisms in an attempt to manage the business.

The Real Message

What was the message of the ABC Widgets story? The lessons are multiple but can best be described by a call we received about 3 months after our work with ABC Widgets. The caller was Miguel. We had spoken with him several times since our work together. He would occasionally call for advice or sometimes just to chat. This call was particularly revealing in that he told us he was taking a new position with a competitor of ABC Widgets. He was going to be in charge of their lean transformation. He was changing companies because, as he put it:

> "I was getting so frustrated with our efforts. I could easily see the opportunities but we were not staffed nor were we structured to capture them. I recall some advice you had given me earlier when you said, 'You cannot have a just in time (JIT) materials system without a JIT support system, including JIT maintenance and JIT problem solving.' I began to see the problems caused by our centralized approach, especially the problem-solving issues. When you were at the plant, then and only then did we have sufficient problem-solving support and we made great headway. Once you left, progress ceased. It became obvious how we had completely circumvented lean principles when we increased the line capacity. There were no process analyses, no FMEAs. Only a superficial economic analysis and the hope that things would go well. It was at this point that I decided our efforts were more at *looking* Lean than *being* Lean. I discussed all these issues with our management and there was no interest in changing these areas, so I decided to move on."

Other issues of note here, which are very common in many plants and are always destructive, are the issues of favoritism and politics. These have no place in dispassionate problem solving and lean analyses. Lean decisions need to be data driven and fact based. Here, the relationship of the reliability manager, Miguel, and the general manager, Juan Pablo, his father-in-law, created some problems. We found Miguel to be bright and unbiased. His decisions were fact based and good. Unfortunately for him, many others were biased against him since they thought he got his current job because he was the "son-in-law," rather than because he was a good engineer and manager. It was this bias that was largely the reason the implementation of autonomous maintenance had failed. We found that many decisions here were not fact based. The entire plant was short on problem solvers, and existing personnel were neither trained in, nor accountable for, problem solving. Yet there was no shortage of problems. This is not a good combination!

> **P**oint of Clarity   A JIT manufacturing system needs JIT problem solving and JIT decision making. Hence, central controls are incompatible with Lean!

Chapter Summary

The problem solving worked and we helped the facility fix their problems. We increased capacity, improved reliability, reduced labor costs, and reduced inventory as we let them return to excellent on-time delivery. However, this was not the most revealing part of this story, which is a sad lesson we must all learn.

Earlier, I had characterized this as a "not-so-Lean-but-we-want-to-appear-Lean" plant. I hope it is obvious now what I meant. Superficially, the plant had a very lean appearance. 5S was mature and cells were set up neatly with visual information centers.

Centralized materials handling was in place and, quite frankly, from a visual standpoint the plant had the appearance of being Lean. However, as can now be seen, the plant was anything but Lean. When we helped them solve these problems, it uncovered issues that should not have been lingering in a mature line after 4 years of lean efforts. Go back and check the Six Roll Out Errors in Chap. 3 and the 10 Lean Killers in Chap. 2, and you will see they hit squarely on most of them.

Still, the absolute number one killer issue was that they actually *thought* they were Lean.

Another serious issue should be remembered as well. The entire plant, especially management, refused to "see" many of the problems. During our work there, it was obvious that many of the issues had earlier been "seen" by a number of staff, but there was a marked reluctance to bring them up, as if by ignoring the problem, it would politely stay in the closet. Unfortunately, these problems have a way of surfacing and resurfacing. In this case, the problems found the light of day when an opportunity arose to increase capacity and make money.

> **"In** the choice between changing one's mind and proving there's no reason to do so, most people get busy on the proof. "
>
> **John Kenneth Galbraith**
>
> **"O**pportunity is missed by most people because it is dressed in overalls and looks like work. "
>
> **Thomas Edison**

Like ABC Widgets management, we would all be well advised to listen to the wisdom of those from the past. It is the wisdom quoted earlier in the preface.

The lesson of ABC Widgets is that there are no shortcuts to becoming Lean. We must:

- Address the cultural change leading indicators
- Take care of the foundational issues of quality control
- Implement quantity control measures

This must be done to make our facility a better money-making machine, a more secure workplace for all and the supplier of choice to our customers.

Yet Another Initiative???

Many of you will notice that to describe the efforts of ABC Widgets, I used the term "lean initiative" rather than the more commonly used term of "lean transformation." Here is an example of an "initiative" rather than a "transformation." A lot was "initiated" much less was "transformed", and it is pretty clear it was done "to the people" not "for the people". ABC Widget, like the Bravo line in Chap. 21, have what I sarcastically refer to as a **"Seems to be Lean"** system.

CHAPTER 24

House of Lean, Systemwide Assessments and a Cool Experiment

Aim of this chapter ... to introduce you to the House of Lean, explain the systemwide assessments you will need to use, and introduce you to a cool experiment you can use to explain various aspects of variation and its deleterious effects.

A House of Lean

Figure 24-1 shows a "House of Lean," which was developed for use with a specific client based on their individual needs. This is a good tool that gives a graphic description of your lean transformation to all employees. It is a way to capture the entire program in a pictorial format. I have found that it often makes the overall effort more easily visualized and hence understood. Most of the techniques listed are common but some unique items are also listed. For example, process simplification was a key component for this company. They needed to make process changes to eliminate unneeded steps and also to change some processes that had poor availability due to using outdated technology. These items needed to be taken care of first, or they would have created unnecessary future work. Your *sensei* should be able to identify these issues and help you with the order of activities.

The House of Lean is a metaphor that is designed to show how the various topics discussed in Lean Manufacturing fit together and interact. As with all metaphors, they have limits. Nevertheless, it is a good tool to use in teaching:

- Objectives
- Strategy, tactics, and skills
- Foundational elements

Which ones comprise *your* Lean Manufacturing system?

When we discuss strategies, we speak at the conceptual level, while tactics are normally small group activities that are required to achieve the strategies, and skills are the individual behaviors that must be executed to accomplish the tactics.

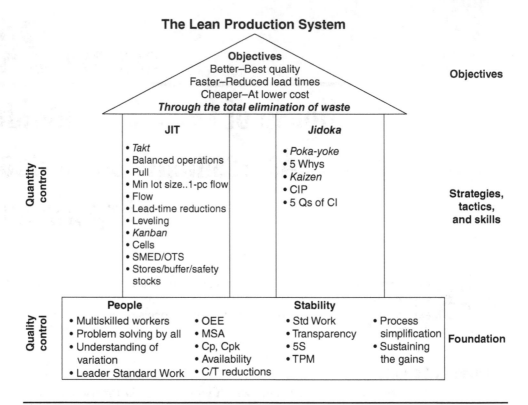

The Lean Production System

FIGURE 24-1   House of Lean.

For example, it is clear from Ohno's writings that the two key strategies of the Toyota Production System are just in time (JIT) and *jidoka*. One of the tactics of JIT is *kanban*. To execute *kanban*, we need a variety of skills, such as the skill of making the *kanban*, sizing the *kanban* volumes, planning the circulation of the *kanban*, and so on.

My advice is to thoroughly evaluate your needs and create your own House of Lean to explain them.

Assessment Tools

Based on our experiences, which cover over 40 years of implementation efforts, I have catalogued a number of problems we have encountered in assisting various companies in their lean transformation efforts. I have documented the key issues in three attachments:

1. The Five Tests of Management Commitment to Lean Manufacturing

2. The Five Precursors to Implementing a Lean Transformation

3. Process Maturity

My hope is that facilities wishing to implement a lean transformation will read, digest, and be guided by these documents. To this end, I hope our experiences will be helpful to others.

The Five Tests of Management Commitment to Lean Manufacturing

Commitment Questions	Yes	No
1-Are you continually and actively studying about, and working at, making your facility leaner and, hence, more flexible, more responsive, and more competitive? (We must be intellectually engaged. All must continue to learn. **No spectators allowed!**)		
2-Are you always willing to listen to critiques of your facility and then understand and change those areas in your facility that are not Lean? (We must be intellectually open.)		
3-Do you always honestly and accurately assess your responsiveness and competitiveness on a global basis? (We must be intellectually honest.)		
4-Are you totally engaged in the lean transition with your time, presence, management attention and support (We must be engaged at the behavioral level; we must be doing it. We must be on the floor, observing and talking to people, and imagining how to do it better. Lean implementation is not a spectator sport.)		
• Time		
• Presence		
• Management attention		
• Support (including manpower, capital, and emotional support)		
5-Are you always willing to ask, answer, and act on the question: How can I make this facility more flexible, more responsive, and more competitive? (We must be inquisitive, and be willing to listen to all personnel, including peers, superiors, and subordinates alike, no matter how painful it may be, and then be willing and able to make the needed changes.)		
A "Yes" to all five questions means you have passed the commitment tests. Any "No" answer means there is an opportunity for management improvement.		

The Five Precursors to Implementing a Lean Transformation

There are Five major Precursors to Implementing a Lean Transformation. These precursors are foundational issues that must be in place to make a lean transformation successful. If they are not in place at the kickoff of the lean transformation, it can still proceed; however, these precursors must be recognized and they must be built into the lean transformation plan in the proper sequence (See Table 24-1):

1. High levels of stability and quality in both the product and the processes
2. Excellent machine availability
3. Talented problem solvers, with a deep understanding of variation
4. Mature continuous improvement philosophy
5. Strong proven techniques to standardize

Needed Trait	0	1	2	3	4	5
1 - High levels of stability and quality in both the product and the processes	Does not meet Level 1 criteria	"Critical" product characteristics identified; the stability and capability are known for these; many unknown losses	All key product and process characteristics are identified; most are stable; some meet minimum Cpk	All key product and process characteristics are identified; are stable and meet minimum Cpk	A shift is in place to eliminate the need to monitor processes and build in process robustness; the need for SPC and numerical techniques is reduced	Significantly reduced need to monitor the process; robustness is built into the process
2 - Excellent machine and line availability	Does not meet Level 1 criteria	Mach availability is unknown; stock outs unknown; most maintenance is reactionary; line losses are known; >8%, not stratified	Avail. known; total losses <8%; stock outs <1%; no formal plan to improve; planned and predictive maintenance done	Total losses <3%; active plan to improve; first 4 elements of TPM in full use	Losses <2%; Level 3 plus the 5th pillar of TPM; early management and MP design	Losses <1%; Level 4 plus actively improving design of production equipment with engineering and suppliers
3 - Talented problem solvers, with a deep understanding of variation	Does not meet Level 1 criteria	Firefighting only; done by supers and engrs. only; little training; few understand variation effects	Mostly firefighting, engrs., super do PS; trained in SPC, Five Whys, 6 Sigma, etc.	Some firefighting; all do PS; training given to all; all given training in variation reduction	Little or no firefighting; all are involved; all are trained in PS and variation reduction	No firefighting; total involvement including customers and suppliers; joint training; lean implemented through suppliers and customers

4 - Mature continuous improvement philosophy	No policy exists	Policy, but no methods to implement	Policy and methods in place; done by engineers, supers, and mgrs.	Level 3, but all are involved; CI is recorded and posted	Everyone is involved at the facility; use policy deployment to get all involved in CI planning	All are involved, including customers and suppliers
5 - Strong proven techniques to standardize	No policy; problems repeat	Policy in place; many problems repeat quickly; no clearly defined methods	Policy and methods are documented; problems repeat but after a significant time span	Policy and methods are documented; some problems repeat	Problems do not repeat	Everyone is involved, including customers and suppliers

TABLE 24-1 The Five Precursors to Lean—An Evaluation Matrix

Process Maturity

Frequently, I receive requests from manufacturing firms to assist them in a lean transformation. Typically, they have done some background research on Lean Manufacturing and want guidance in making their operations Lean. Unfortunately, the most common thing I find is that they do not have the foundational work elements in place to begin a lean transformation. More importantly, they do not recognize this and think that, independent of the current state of the process, a lean transformation can be implemented and techniques such as *kanban* can then be readily applied. Their common perception is that they only need to complete some training and they will be on the road to a lean enterprise. Nothing could be further from the truth. First, a number of precursors must be taken care of. For example, they must have a reasonably stable materials supply, and they must have good machine availability, to name a few. Most importantly, they must have in place a process that already produces at high-quality levels; processes that exhibit both process stability as well as adequate capability. This item is of particular importance to the success of many lean techniques.

It is to those groups who wish to undertake a lean transformation that this document is written. In this format, a specific order is implied. In practice, I have found that those firms that deviate greatly from this basic format spend much more time and much more of their resources in reaching the goal of becoming Lean. Those processes that produce to this lean standard I refer to as "mature."

Process maturity differs from process "goodness." Process or product "goodness" usually means the product meets the needs of the customer. In this regard, "goodness" is usually measured by the process capability indices of Cp and Cpk for a few key product and process characteristics. If these two indices meet minimum standards, for example, they are greater than 1.33, then the customer is satisfied, declares the process to be "good enough" and the supplier is allowed to proceed with greater independence. If, however, the indices fall short of the standard, then the supplier is required to do extra processing, which is usually some form or containment or extra inspection coupled with a specific action plan to achieve the desired levels of Cp and Cpk.

Process maturity goes beyond the measure of "good enough" so that a product is not only good, but is produced with a minimum amount of waste. These processes have other characteristics, such as minimum inventories and short production lead times, to name a few. These processes are now widely referred to as Lean, and in this document, the process that produces a lean product is a mature process.

This treatment addresses levels of process maturity for a typical manufacturing process and does not address some topics, which are outside the scope of this treatment. Other aspects of the process not addressed herein, might include ergonomics, environmental issues, and safety, to name but a few.

A process that is mature has five characteristics, which are:

1. Documentation
2. Flow, that is, a specific process routing
3. Quality understanding and performance
4. Inventory understanding and control
5. Leanness by all 20 measures; exhibit advanced levels of continuous improvements, with lean goals driving the process

In short, process Levels 1 through 3 work to achieve a process that has met all *quality* standards of the customer. Levels 4 through 5 address the issues of *quantity* control.

Generally, the development of a process should follow a natural pattern as outlined here and in this order. Although this is not true for all work stations or even all complex process flows, it is a good general guideline and will serve you well in describing the level of process maturity. The levels are labeled as Level 1 through Level 5 and are characterized by the following:

1. Level 1
   A. A Level 1 process has a set of documentation that allows a firm to design and build a product, as well as a means to assess quality and delivery capabilities. Documentation usually includes:
      i. A complete up-to-date drawing of the part or assembly, including any part or subassembly requirements. All construction goals are clearly understood, with product and process critical characteristics defined and agreed upon
      ii. A test specification, plus all operational definitions to determine if a product is accepted or rejected
      iii. Packaging specifications
      iv. A plot plan
      v. A flow chart
      vi. A PFMEA
      vii. A Part Number Control Plan
      viii. Appropriate work instructions, including all instruction for rework
      ix. Demand rates or projections good enough for production scheduling
   B. Nearly all the 20 lean techniques (listed below) should be designed into the process. For example, cells, multiskilled workers, *kanban, takt* time, leveling, and standard work, to name a few, should be built into the design.
   C. Since a Level 1 process is basically a preproduction condition, successful attainment would include such items as completion of PPAP and run-at-rate to meet customer requirements.

2. Level 2
   A. A Level 2 process has all the characteristics of a Level 1 process, plus the production process flows (has an actual process routing) in full accordance with the Level 1 documentation. Specifically:
      i. Raw materials enter the process only as noted; products leave only as documented.
      ii. No other intermediate assemblies, and so on, enter or leave the process unless documented on the flow.
      iii. All work in process flows just as in the flow chart. All good product follows the process flow diagram. All scrap and potential reworked product are properly segregated from the normal process flow. All scrapped and reworked products follow the prescribed process flow for scrap and rework.
      iv. All critical processes or process steps have completed and acceptable gauge studies.

       v. All process steps are statistically stable.

     vi. Work instructions are in place and adequately describe the work. Work instructions are scrupulously followed.

   vii. All elements of transparency are present so the worker can confirm that he has "stable flow at takt", or in his case, "stable flow at process cycle time"

  B. Again, with a Level 2 process, many lean techniques have been incorporated into the design but the real focus of this step it to institute continuous flow of the product through the process, and to make the process statistically stable. Other lean techniques that are focused on usually include:

   viii. Flow

    ix. *Jidoka*

    x. 5S

   xi. *Takt* time

   xii. Usually only minor *kanban* implementation, such as carts, and so on

  xiii. Balanced operations

  xiv. Transparency

   xv. Standard work

  xvi. *Kaizen*

  xvii. Five Whys

  C. Key process and product variables monitored include production rate, stratified defects, machine availability, first time yield, and OEE.

3. Level 3

  A. A Level 3 process has all the characteristics of a Level 2 process, plus the quality levels are understood and are acceptable to the needs of the customer— in other words, all process steps have adequate Cpks. A continuous quality improvement effort is in place. Specifically:

    i. A modified Part Number Control Plan that addresses the quality issues found.

    ii. Specific measures of internal quality, with long-term graphs available on the shop floor.

   iii. Specific measures of external quality, with long-term graphs available on the shop floor.

   iv. A documented plan of continuous improvement for quality, for both internal and external quality measures.

    v. Continuous improvement quality goals in place, for both internal and external quality measures.

  B. Lean techniques that can now mature at this phase, include:

   vi. More *kanbans*

   vii. Minimum lot sizes

  viii. SMED/OTS

   ix. Store/buffer/safety stocks

    x. *Poka-yoke*

  C. Quality measures are fully mature at this point and used as process improvement drivers.

4. Level 4

  A. A Level 4 process has all the characteristics of a Level 3 process, plus all inventory levels are controlled and minimized.

B. Lean techniques at this level include the following:
   i.   *Kanban* is further improved. *Kanbans* control all in-plant materials flow.
   ii.   SMED/OTS become critical for further lot size reductions.
   iii. Standard inventory is fully reviewed.
   iv.   Autonomation is more fully developed.
   v.   JIT must be fully embraced with a "JIT support system," including the culture change to:
         a. JIT material supply and product production
         b. JIT problem solving
         c. JIT maintenance
   vi.   At Level 4, frequently major improvements are made in the flow, and a cell redesign is often beneficial, with a critical look at staffing, flow, rebalancing, and even the basic layout.
   vii. Also at Level 4, it is common to prepare a meaningful future state value stream analysis map to guide future projects.
C. Key process measures include the previously mentioned ones, especially manufacturing lead time plus inventory turns. Also, it is common to address both value-added (VA) steps as a percentage of total process steps, and VA time as a percentage of manufacturing lead time as process improvement focusing tools.

5. Level 5

   A. A Level 5 process has all the characteristics of a Level 4 process, plus all lean techniques are in full maturity.

Efforts to improve VA steps and VA time are in place and the supply value stream becomes a key focal point, whereby suppliers and customers are included in improving the overall process. A dynamic future state value stream map is a fully functional tool that is guiding process improvements. This will, by necessity, take you outside the bounds of the plan

An Experiment in Variation, Dependent Events, and Inventory

Background

This is a simple experiment that can be done at your desk or in a small group. All that is needed are some dice and a simple form (shown later in this chapter). The experiment is a factory simulation using pull production while **studying the effects of variation and dependent events** on factory performance.

This is a phenomenon that is understood by only a few.

In short, when we have variation, as we do in any process, coupled with dependent events, as we do in a multistep process, then the process will not produce to the average rate of the processing steps, unless we have inventory between the dependent steps. Plus, as the variation is increased, the inventory levels must increase to maintain production. In addition, as the number of sequential steps are increased, the inventory increases by an exponential factor. It is for these reasons that most *factories cannot produce at the nameplate average rate of the equipment, unless inventory levels are extremely large or variation is reduced to very low levels.*

Read that sentence again.

It gives tremendous insight into why, when projects or production schedules are planned, halfway through the project, overtime is needed, and in the end more overtime is needed, and yet we still often need to pay to expedite the shipment.

This experiment explains this phenomenon and more.

Variation and dependent events are everywhere in a factory. Take a simple cell, for example. Let's say we have a six-station cell and all work stations have 60 seconds of work, which is also *takt*. Also, there is one piece at the workstation and there is no inventory between stations—true one-piece-flow. When station 1 finishes a piece, so do stations 2 thru 6, and in unison, all six pieces of in-process work are simultaneously pulled to the next work station every 60 seconds—the perfect synchronization of process flow: the Ideal State.

But, for the moment, let's imagine that the cycle time for station 4, although it averages 60 seconds, varies from 50 to 70 seconds. When station 4 performs at 50 seconds, it finishes its process, and then station 4, has a 10-second wait time before its product is pulled by station 5. Ten seconds of waiting time elapse, which is a waste for station 4, *but this is not a production rate problem*, the cell will still produce to *takt*. It is just that the operator at station 4 will sit around a while. On the other hand, when station 4 takes 70 seconds to produce its work, that subassembly is held up and station 5 is starved for work for 10 seconds. This delay passes through all the work stations of the cell in a wave and that piece is produced on a 70-second cycle time.

So let's recap... If the station that varies—in this case, station 4—operates faster than *takt*, station 4 must wait for the subsequent station to pull the production; however, when station 4 just happened to operate slower than *takt*, station 4 would slow down the whole cell on that cycle and there is no recovery with a resultant loss of production rate.

So even though the station may have a 60-second cycle time *on average*, any time the cycle time is above average, the production rate drops. This concept is known as the effect of variation and dependent events. (The dependency is that the "next step" depends on the "prior step" for supply.)

So the solution is, guess what …? You got it! Add some inventory. We will need to add inventory both before and after station 4, the one with the variation. We need the inventory in front of station 4 so when it produces faster than *takt*, say at 50 seconds, there is raw material available to keep it producing. We also need the inventory after station 4, so when it is operating slower than *takt*, say 70 seconds, there is raw material to supply station 5. Then, station 4 can have the variation *and* maintain production at *takt* on average.

This effect of variation and dependent events is not well understood and is a problem, always. Two solutions can be employed: Either totally remove the variation, or totally remove the dependency. To totally remove the variation is an impossibility. Recall that the definition of variation is, "the inevitable differences in the outputs of a system." Since it is inevitable, total removal is an impossibility. Okay, so let's totally remove the dependency. This means tons of inventory, the exact thing we are trying to eliminate in a lean solution.

So, guess what? The solution is to find the happy medium and it is best done by first reducing the variation to a minimum so inventory can then be reduced accordingly.

That's enough background for now; do the experiment and see firsthand—under controlled conditions—exactly how this phenomenon plays out.

Team No.	No. of People	No. of Dice	Multiplier
1	1	1	6
2	2	2	3
3	2	3	2
4	3	6	1

TABLE 24-2   Team Distribution of Dice

The Experiment

To do the factory simulation experiment, get 12 ordinary dice and some students; four teams would be ideal. If you only have enough for two or three teams, do that. The experiment it totally flexible. We will use dice to get random numbers, and we will vary the number of dice for each team to modify the amount of variation for each team. Distribute the dice as shown in Table 24-2. A little math will show that the average values, independent of the number of dice used will approximate 21. For example, the average for any dice is 3.5, and in each case the multiplier times the number of dice is six, so each team will average 21. Hence, the rate of the production process simulations for all four teams will average 21, but their factory simulations will perform differently because ... well, let's do the experiment.

Creating the Data

1. Roll the dice (or die) and count the total spots.
2. Multiply the total by the multiplier for your team.
3. Enter this number in the oval on the plant simulation spreadsheet, starting with Cycle 1, station 1.
4. Repeat steps 1 thru 3 for Cycle 1, station 2, and so on, through Cycle 1, station 8 ...
5. Do this for 20 cycles at least, which means several copies of the spreadsheet will be needed.
6. To simulate the process, fill in the rectangles, described in this chapter in the section entitled, "Processing the Data."
7. Calculate the production.
8. Sum the totals, as in the summary data table shown in Table 24-2.

Processing the Data

1. Each horizontal row of information is a cycle for this process, and each cycle of production goes through eight processing steps.

The oval (see Figs. 24-2 and 24-3) represents the instantaneous capacity of that work station based on station capacity alone. Since each dice has the potential of numbers 1 through 6, for Team 1 since its multiplier is six, it can be seen that the instantaneous capacity for each station is a value from 6 to 36. And in the high variability case of 1 dice, the only possible values are 6, 12, 18, 24, 30, and 36.

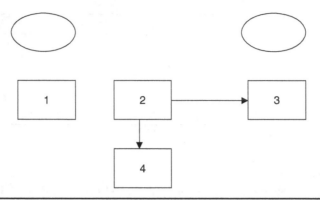

Figure 24-2   The data format for the experiment.

2. The rectangle directly below the oval, Rectangle 1, is the amount of material (think of them as kits) that is available for the next station in the cycle. We assume the warehouse has infinite material, so there is no material constraint at the first station. The following then becomes the two possible constraints on the system:

3. Instantaneous Station capacity given by the dice data

   a. Material availability, which is the kits processed and the kits remaining in WIP (the volume in Rectangle 2)

   b. Next to Rectangle 1 is Rectangle 2. It represents the total available material for the subsequent processing station.

4. Rectangle 3 is then the actual production for that station based on the system constraint. Rectangle 4 then represents the amount of material left by that station in that cycle when the instantaneous station capacity (the oval) is the constraint. This amount of material, the WIP, is then available for the subsequent cycle.

   a. To make this material available for the subsequent cycle, write this value in Rectangle 4 and add it to the value in Rectangle 1 in the subsequent cycle.

   b. Place this value in Rectangle 2. This now represents the total amount of material for this station during the next cycle of production. It is the sum of the material not used in the previous cycle, plus the amount produced by the prior station, in that cycle.

5. At the end of the cycle are three rectangles and a triangle. The first two rectangles are the net production from that cycle. They are the same number. The last rectangle is the accumulated production from that cycle, plus all previous cycles.

6. The triangle is the total WIP that was accumulated in that cycle. It is the sum of all the Rectangle 4s in that cycle.

7. A simple check of your logic and math is to add up the cumulative production, plus the WIP for that cycle. This total should be equal to all the materials withdrawn (station 1 accumulated totals), up to and including that cycle.

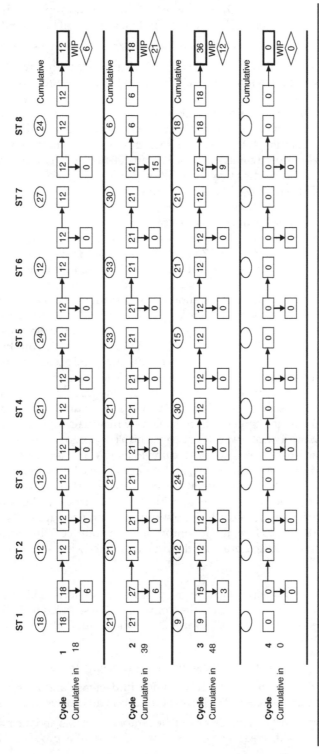

FIGURE 24-3 Experiment spreadsheet.

8. A simpler approach is to write me at www.qc-ep.com and I will send the spreadsheet on a CD. Shipping and handling will be the only charges.

9. For example, I have attached the first three cycles of a spreadsheet.

   a. Review that in Cycle 1, station 1, 18 units are withdrawn from the storehouse. Hence, station 1 does its work and they are processed and transferred to station 2.

   b. However, station 2's pull capacity is only 12, so six units remain in WIP for the next cycle, 12 units are passed to station 3, and so on.

   c. At station 4, the instantaneous capacity is 21 units; however, there are only enough raw materials for 12 units, hence only 12 are processed and passed on. It goes this way thru station 8.

   d. Consequently, 12 units are produced, six remain in WIP, which checks since we started with 18.

   e. In Cycle 2, station 1 has the capacity to produce 21, so 21 units are withdrawn from the storehouse, but six units were left in WIP from the first cycle, so now station 2 has 27 units of material at its disposal. Station 2 has enough raw materials to produce 27 units, but its limit is the instantaneous station capacity of 21, so 21 units are pulled, processed, and sent to station 3. The six units not pulled by station 2 now remain for the next cycle.

   f. In Cycle 2, since the capacity of stations 3 thru 7 all exceed the raw materials availability, they all process and pass on 21 units until we get to station 8.

   g. At station 8, we have enough material for 21 units, but the machine capacity is only six; hence, only six units are produced, leaving 15 units in WIP available for station 8 for the next cycle.

   h. Six units are produced and added to the 12, so we have accumulated 18 units of finished goods. Our WIP totals 21. The sum of finished goods plus WIP is 39, which agrees with the total raw materials input of 39. It checks.

   i. Now look at Cycle 3. Station 1 which only produces nine units, but since it had an inventory of six from the previous cycle, it was able to produce 12 units. Had there been no inventory, the rate would have been limited to nine.

   j. In Cycle 3, the capacity and materials availability limit the production at stations 2, 3, 4, 5, 6, and 7 to 12 units. However, because of the inventory left over from Cycle 2, now station 8 can produce to its capacity and make 18 units. Here, the inventory has "uncoupled" the process and allowed it to produce more than the previous processing step.

The Results

We have done this experiment in many training sessions, and the results are always very similar. I have included results from one of the sessions in Table 24-3.

As expected, there is some variation in the results. Nevertheless, we will always find the following:

- Variation exists in the instantaneous capacity of the work stations.

- As the variation is reduced, WIP will decrease and both the total production and percentage of RM Turned into Sales will always increase.

- As the variation increases, WIP inventory increases and the process steps become uncoupled, creating what we call "island like production." Only then will

Team	No. of Dice	Total Production Yield, Units	Total WIP, Units	% RM Turned into Sales	Avg. Prod Rate (Dice Average)	Std. Dev. of Capacity (Dice Std. Dev.)
1	1	246	168	59.4%	20.59	10.76
2	2	294	162	64.5%	20.79	6.69
3	3	328	98	77.0%	20.29	5.64
4	6	350	36	90.7%	21.18	4.01

TABLE 24-3   Typical Results from the Experiment

production increase, but at the expense of huge inventory accumulations which reduce flow dramatically and consequently increase lead time significantly.

- This all occurs because we have variation and dependent events! This is a demonstration that:

  - If your system has variation, the system will need inventory to sustain production.

The more variation the system has, the more inventory the system will need to sustain with any given level of production.

Glossary

5 Whys A problem-solving method which employs the technique to continue to ask "Why?" to explore the cause and effect relationship, in an attempt to find the root cause. It requires a great deal of experience and a strong knowledge base to do well.

5S A tool in the lean toolbox which focuses on workplace readiness. The 5 S's stand for Sort, Set to order, Shine, Standardize, and Sustain.

Andon A warning device, normally a light, to signal an abnormality; it is a part of the system of transparency.

Autonomation Ohno's word for *Jidoka*. It literally means automation with a human touch designed to supply 100 percent inspection, sort and then initiate problem solving.

Availability The concept that production facilities are capable of producing when they are scheduled to be producing, or when needed.

Balancing Synchronizing operations, generally making sure that each step has the same process cycle times.

Balancing chart Normally a bar chart which shows process steps on the X-axis and cycle times on the Y-axis.

Bottleneck Any activity or process step which limits production.

BRP Business Resource Planning, another version of MRPII.

Buffer stocks A type of inventory which is held to account for external demand variations.

Buffers An excess resource in the production system, usually designed to compensate for some type of variation so demand can be met. There are three types of buffers: capacity, time, and inventory.

Catchball A process used in *Hoshin–Kanri* planning where the boss decides what to do, the subordinate decides how to do it and an interchange is effected to make sure the goals and the means to achieve the goals are focused and aligned.

Cells A manufacturing equipment layout where people and machines are in close proximity to reduce transportation and WIP inventories. Cells are designed to achieve one-piece flow.

Cells, U & C shaped Cells are normally laid out in a U or C shape so the incoming raw materials and outgoing products are near one another. It aids in materials and information handling.

Changeover   Converting a machine, or process to make a different model or different product.

Changeover time   The time it takes from the last good part prior to the changeover to the first good part after the changeover.

CIP   Continuous Improvement Process.

Coaching Question   An open-ended question, asked by a *sensei*, supervisor, or other teacher specifically structured to make the student think in a specific way to guide him in learning.

Compartmentalization   An attempt to separate some aspect of a system from the rest of the system for convenience and simplification. Usually this is an abuse of system logic and done for self-serving reasons.

Constraint   Another word for bottleneck, see bottleneck.

Continuous Improvement Process   A series of sequential steps to forever analyze a product or a process and continue to increase the value-added portion.

Control chart   A statistical tool invented by Walter Shewhart to evaluate the statistical properties of a process. Control charts will allow you to characterize both the variation in your process and if you are producing to the target specification or not.

Correlation and regression   A technique used to study the relationship of cause and effect and the impact that variation has on this relationship.

Cp, Cpk   Process capability indices; quantified measures of "process goodness".

Culture   The combined thoughts, actions, beliefs, artifacts, and language of a group of people. It is "How we do things around here."

Current state, VSM   A current state value stream map, sometimes called an Information and Materials Flow Diagram, see also PSVSM.

Customer   Your client; they are usually defined by four characteristics; they are courted to consume your product; they pay for your product or service; they pick up and use your product or service; and if they are dissatisfied with your product or service, they can cause you immediate discomfort, that is they can complain and get action. Your external customer is the entity which pays you, however, the customer is also the next step in the process and the needs of the internal customers must be met, just as the needs of the external customers must be met.

Defects   Things gone wrong with your products; quality characteristics which are not met.

Deming   W. Edwards Deming, the great statistician and quality guru, creator of *Deming's 14 Obligations of Management* and author of *Out of Crisis and The New Economics*.

DMAIC   The problem-solving methodology used the Six Sigma. It means, define, measure, analyze, improve, and control. Its parallel in lean is PDCA.

DOE   Designs of experiments, an advanced statistical tool used for in-process understanding and optimization.

Downtime   Time that a process or machine is not running.

Effectiveness   The ability to achieve a goal.

Efficiency   Achieving a goal using minimum resources.

ERP   Enterprise resource planning, another version of MRPII.

Excess processing   One of the seven wastes. Performing work on a product beyond what the customer considers value, beyond what they are willing to pay for.

FIFO   Acronym for first in, first out.

FIFO lanes   Processing lanes of goods where FIFO materials handling is practiced.

Finished goods inventory   The completed production of a product held to assure supply to a customer in a make-to-stock supply system. It has three components: stores inventory, buffer stocks, and safety stocks.

First-piece lead time   The time it takes to produce the first piece of a batch. A key factor in quality responsiveness. The objective is to reduce this to a minimum.

First-time yield   A common measure of internal quality at a factory. It is the concept that the first time a part is touched, all activities are done correctly.

Flow   The concept that once started a product continues to move with value-added work being performed, during the entire manufacturing process.

Flow line   A linear arrangement of processing close-coupled equipment, as distinguished from a U cell, for example.

FMEA   Acronym for Failure Mode Effects Analysis, a quality tool designed to sort out quality problems, preproduction, and implement countermeasures.

Foundational element   Any of the basic elements, mostly issues related to quality, that need to be in place to create the foundation for quantity control. See the House of Lean in Chap. 24.

FSVSM   Future state value stream map, a map showing the information and materials flow for a product depicting some hoped for, future state.

Future state, VSM   Value stream map with future conditions designed into it, see Present State VSM.

GAAP   Generally Accepted Accounting Practices, an acronym that defines itself.

Gaming   In the process of evaluations, whether we are speaking of budgets, plant performance metrics, or individual performances, gaming is the intentional distortion of, or avoidance of data to make the situation you are interested in to appear to be more attractive to others than they actually are. It is the manifestation of the phrase that "figures don't lie, but liars can figure."

Hawthorne effect   The positive effect that is achieved in improved performance when attention is paid to people.

Heijunka   Japanese word for leveling, specifically leveling production which means stabilizing the rate in a narrow band; no large ups or downs in rate. A *heijunka* board is a planning board used to level production and become part of the visual system to evaluate production status.

Hoshin–Kanri **planning**   A strategic planning method developed in Japan, it means policy deployment and is one of the very few tangible top management tools in the lean toolbox.

Ideal State   The end result of a total elimination of waste in a system.

Inattentional blindness   It is a psychological lack of attention and is not associated with any vision defects or deficits and is defined as the event in which an individual fails to recognize something that is in plain sight. It is natural and largely unavoidable problem caused as we focus on one aspect in our vision, and as we concentrate on that, we tend to block out other observations. It is a case of not being able to see some of the trees that are there … because of the other trees we happen to be focusing upon.

Inventory   One of the seven wastes. It includes finished goods which have not been picked up by the customer and all the materials in the system which you intend to convert to finished goods, including raw materials and WIP. All inventories are waste, although some is necessary considering the present conditions.

Inventory turns   A measure of the rate at which inventory turns over each year. Twelve inventory turns mean that you have 30 days of inventory on hand.

Ishikawa   A Japanese quality guru who wrote extensively about quality, the creator of the *Ishikawa* diagram or sometimes called a Fishbone or cause–effect analysis.

Island production   A type of production where the work stations are setup far apart, typically with lots of inventory in front of and behind the island. Generally an inferior method of production to the cellular manufacturing.

Jidoka   One pillar of the TPS. *Jidoka* is a method to prevent bad material from advancing in the production system and to find system weaknesses and fix them.

JIT   Just in time, the other pillar of the TPS. The concept is to avoid waste by supplying exactly the right quantity of materials to exactly the right location at exactly the right time. It is quantity control.

JUSE   Japanese Union of Scientists and Engineers.

Just in time   See JIT.

Kaikaku   Radical change, it is super fast, super large *kaizen*.

Kaizen   The concept of making continual product and process improvements, usually small and typically done by the entire workforce.

Kanban   *Kanban* means card; it is the method the JIT pillar uses to minimize inventory and follow pull-demand system rules to reduce wastes.

Lead time   The elapsed time it takes, from start to end, to produce a product through whatever process you are speaking of, most commonly it is used as production lead time or order lead time. However, there can be a cycle lead time, a production lean time, or even a value stream lead time. It is good to add a noun to the front of lead time so you can specifically describe the lead time you are speaking of.

Lead time, first piece   The elapsed time it takes for one piece to completely flow through the production process.

Lean   Short form of Lean Manufacturing System. The generic name given to the Toyota Production System. Through use and misuse, it has come to have many applications, some good, some bad, but the popular concept is to be able to make more product, using less resources. This addresses many of the technical aspect of Lean. However, to get the full definition of Lean, refer to Chap. 2.

Leanspeak The unique language used in Lean Manufacturing using common words like waste, pull, and flow, in a different sense than the typical definition, plus the use of unique lean terms such as "catchball" and "autonomation."

Learn-do-reflect A teaching mantra, used between student and teacher, whose purpose is to remind people to reflect after all activities to enhance comprehension. More formal and more structured methods include the Japanese practice of *hansei* or the U.S. Army uses After-Action Reports.

Leveling, model mix Avoiding the batch production of models of a given product.

Leveling, production Avoiding the unnecessary changes in production rates.

Make-to-order A production system with no finished goods inventory. Production does not start until a specific order is received and they are shipped directly. A system with no finished goods inventory except those materials awaiting pickup.

Make-to-stock A production system with finished goods inventory and all production is sent to a storehouse for holding prior to shipment.

Mass production systems Production systems designed to produce in large volumes using large batch philosophy in an attempt to be cost effective, which it seldom is. Characterized by push production systems using large batches, long production runs, large inventories, and island production. It is characterized by inflexible, nonresponsive systems with long lead times and a very low percentage of value-added work; as compared to Lean.

MassProd The abbreviation for Mass Production Systems.

Metric The measurements used to evaluate plant performance. For example, OEE, On-time delivery, and lead time are all metrics.

Minimum lot size An attempt to shrink lot sizes to reduce inventory, specifically WIP inventory, and to improve flow and reduce lead time for the production lot AND the first piece.

Movement One of the seven wastes, movement of people.

MRPII Manufacturing Resource Planning Two is a planning program designed to integrate business needs down to the production floor and among other things, create a meaningful production plan. In short, it is too slow for production floor planning and is uniformly overused in a lean facility but has other necessary functions. For the purposes of Lean, it can be defined as an inadequate planning tool for hourly, daily, and sometimes even weekly production plans.

MSA Measurement System Analysis, a statistical method to determine the usefulness of the measurement system for both products and processes. Its chief benefit is the ability to find and classify variation in the measurement system.

NVA Non-value-added, the opposite of VA.

OEE Overall Equipment Effectiveness. A means to numerically describe production effectiveness, the ability to produce good product. Within OEE are characterized the three key production losses: quality, availability, and cycle-time losses.

Ohno   *Taiichi Ohno*, long-time Chief Engineer for Toyota and accepted architect of the TPS.

One-piece flow   This concept starts at the customer, whereby the customer purchases a single piece and the manufacturing system should replenish only that piece. Hence the Lean system strives to make just one piece at a time; this is true one-piece flow.

Optimum, local   An optimum condition for some local situation.

Optimum, system   An optimum condition for the overall system, and it must not be subordinated to any local optima.

OTED   One-Touch Exchange of Dies, see OTS.

OTS   One-Touch Setup, see OTED.

Overproduction   The largest of the seven wastes, it includes all excess production and production made too soon.

Pacemaker step   The step of the process which determines the process rate and the process model mix. It is the step where scheduling will send production orders.

Paradigm   Your mental image of a concept, often developed unconsciously but paradigms often shape how you act.

PDCA   Plan-Do-Check-Act. This is the iterative process improvement cycle which is inherent within the *kaizen* improvement process.

PFEP   A Lean acronym, Plan for every part, used in *Kanban* design, for example.

PFMEA   Process Failure Mode Effects Analysis, a structured process to determine, before final design, which aspects of a process need additional controls so the production process will be more safe, stable and have a higher yield.

Pitch   A time interval equal to *takt* time multiplied by pack out size, normally the minimum quantity released from the pacesetter and the practical extent to leveling, considering the current packaging.

Poka-yoke   Error proofing. For example, most cars have thousands of *poka-yokes*. While filling your gas tank, there could be several such, for example, a device to connect your gas cap to the car so you do not lose it; an automatic shutoff on the gas pump; a ratcheting device to prevent over tightening of the gas cap; and a warning light on the door to warn you if it is not closed properly.

PPAP   Production Part Approval Process, a method developed by the Automotive Industry Action Group, to standardize the process of obtaining customer approval of a product prior to mass production.

Process   A sequential series of steps which are designed to produce a product or a service.

Process cycle time   The time it takes to complete the work in a process or a process step. Generally lead time is the term used instead of cycle time when we speak of the entire production process.

Product family   A group of products which have the same basic complement of parts and are produced using the same basic production process.

PSVSM   Present state value stream map, a map showing the information and materials flow for a product using present state condition.

Pull   The lean production supply concept; production should only occur when the customer removes a product, the opposite of a push system.

Push   Production is determined by schedules, resource rates, and goals which are generally designed to create an optimum condition at the production source, but it ignores the system optimum. Production will continue, regardless of usage, until the planning system is tweaked to modify the release of jobs.

QFD   Quality Function Deployment; a technique to connect customer needs to process parameters.

Replenishment   To restock. However, in Lean the replenishment concept is JIT.

Root cause   The term used in problem solving. It is the primary source of the variation which you are attempting to remove using problem-solving tools. The generally accepted definition is "that source of variation, which reasonable people would accept, that if removed would permanently eliminate the undesirable effects."

Safety stock   A type of stock, the volume of which is statistically determined. It is designed to take care of internal variations in a make-to-stock-system.

SAP   Another version of MRPII.

Sensei   A teacher, literally one who has gone before, hence the concept of wise and experienced.

Shingo   *Shigeo Shingo*, one of the architects of the TPS. Credited with much of the technology of SMED and *Poka-yokes*, wrote extensively on these two topics.

Skills   Individual behaviors necessary to execute work.

SKU   Another term for a unique part number.

SMED   Single Minute Exchange of Dies, the quick changeover methodology, largely developed by *Shingo*, and absolutely necessary in most plants to avoid large batch production.

SPC   Statistical Process Control, a series of technical tools often equated to *Ishikawa*'s Seven Tool, but more and more equated to just control charting. See control chart.

Standard inventory   The inventory designed to be at any given work station, documented on the Work Instructions and Standard Work also.

Standard work   Not standardized operations, Standard Work is a document written for the manager and the engineer, not the line worker. It contains three elements: the work sequence, the standard inventory, and the cycle time. It is part of the system of visual management, transparency system.

Statistical stability   A technical term, developed by Walter Shewhart and operationally it means the process is in statistical control when placed on a control chart. In lay terms, it means the system is predictable.

Stores inventory   The inventory of finished goods which is built up between customer pickups.

Strategies   Concepts of how you intend to attack a problem or situation; supported by tactics, which are in turn supported by skills; usually expressed in the form of a plan.

Supermarket   A controlled volume of inventory to be replenished by the upstream process, also called stores.

Supplier Those entities which provide resources, usually, raw materials to a process. We have external suppliers and our own employees are internal suppliers.

SWCT Standard Work Combination Table.

Synchronized production The concept that all process steps take the same cycle time. So in theory, in a cell, all parts are completed simultaneously and consequently are moved to the next step simultaneously. A concept to be achieved, rather than a reality.

Synchronized supply The concept of supplying the product to your customer not only in the volume he desires on the delivery date he desires but also producing it at the rate he consumes it even if he has periodic pickups. This concept provides maximum flexibility and responsiveness. This is the manifestation of the concept of leveling and a key batch destruction strategy.

Tactics Small groups of people acting together to comply with the strategy.

Takt German word for rhythm. In Lean Manufacturing the formula is, the available work time divided by the customer demand, over a time interval such as a month, week, or day. It is the "normalized" rate of supply to the customer. It is normalized to your production schedule.

TPM Total Productive Maintenance, (not Preventive but Productive); a methodology to eliminate the six maintenance losses.

TPS Toyota Production System.

Transparency A concept for management which allows you to "see" what is happening in production without using computers, charts tables, or graphs. See Visual management.

Transportation One of the seven wastes, movement of inventory, WIP and finished goods, including all activities necessary to achieve the transportation including packaging.

TWO DIME A mnemonic for the seven wastes: transportation, waiting, overproduction, defective parts, inventory, movement, and excess processing.

Uptime The time that a process or a machine is running.

VA Value-added. In leanspeak, it refers to something the customer is willing to pay for.

Value What the customer is willing to pay for.

Value-added work Those work steps which add value to the product; processing which augments the form, fit, or function of a product.

Value stream The process flow which applies value to the raw materials. The value stream culminates in a product for the customer.

Value stream mapping A technique to graphically describe the value stream so a system review of lead time and value-added time can be made. A key tool in the battle of waste reduction.

Visual management The placing of tools, materials, and information in plain view using simple tools so the status of the process or product can be understood at a glance. Transparency.

VSM Value stream mapping.

Waiting   One of the seven wastes, it is the waiting of people for any reason, including waiting for information, parts, or machines.

Waste   Things the customers is not willing to pay for; the focus of the TPS; "the absolute elimination of waste," T. Ohno.

WIP   Work in process. All materials in the production process once they are withdrawn from the storehouse until they are stored as finished goods. One of the three basic forms of inventory which are raw materials, WIP, and finished goods.

Work element stack   A graphic tool to show how work elements combine so line balancing can be achieved.

Yokoten   The concept of sharing process improvement ideas with others and applying these in other applications beyond the original concept.

Index